# ELEMENTARY NUMBER THEORY

# ELEMENTARY NUMBER THEORY

## A PROBLEM ORIENTED APPROACH

JOE ROBERTS

THE MIT PRESS

CAMBRIDGE, MASSACHUSETTS &

LONDON, ENGLAND

Second printing 1978

This book was printed and bound
in the United States of America.

Library of Congress catalog card number:
76-46738

ISBN 0-262-68-028-0

To Marian, Mark, Katy, & Ruth.

## Table of Contents

# Preface

This book is designed so that it may be used in several ways: it can be used for self study, as a guide for tutorially directed work, or as a supplementary text or source of problems for an ordinary first or second course in number theory. The aim of the book is similar to that of *Aufgaben und Lehrsätze aus der Analysis* by Pólya and Szegö.

A considerable part of the work consists of sets of problems culminating in well known theorems. In this way much of the material of an elementary course in number theory is covered. Moreover, many theorems not often met in such elementary courses, but which require little or no greater sophistication, are included.

A large part of the book may be read by a student with little or no college mathematics. In the earlier parts of the book such a student would only infrequently find it necessary to skip a problem because of its dependence on some special mathematics not in his background.

Later in the book, especially in the last half of XIII and in XVI, XXI, XXIII, XXIV the reader will need a fairly good working knowledge of limiting processes as met in elementary and advanced calculus. Some chapters, such as VI, XV, and XIX, are quite technical though not advanced so far as the mathematical techniques used are concerned. Most chapters are independent of one another and even a mathematical beginner should find it relatively easy to dip and choose at random. Nevertheless, each chapter is written with the thought that most readers will wish to work it through in detail.

The solution section (pp 15-356 s) is designed to serve two functions: the first is to complete the problem section in a way so as to make of the two sections together a self contained exposition of the topics discussed; the second is to offer to the student wishing to work on his own an opportunity to (sparingly) use it for hints and ideas. This section should be well thumbed rather than well read. After saying this it should be added that many of the problems are of considerable difficulty and a reader unable to make any headway with a problem should not feel guilty about turning to the solutions for help.

Appended to the text is a rather extended list of references, most of which have some direct bearing on at least one problem. It must, however, be emphasized that the list is not intended to be complete and contains only those references familar to the author and felt to be particularly relevant to the material presented. Further references on virtually every topic may be found in the extraordinarily useful compendium LeVeque [1974]. Symbols such as VII, VII 22, VII R appearing at the end of a reference indicate, respectively, the reference is a general one for much or all of Chapter VII, is relevant to problem 22 of Chapter VII, or is mentioned in the remarks for Chapter VII.

Finally, a word concerning the format and style of the book is in order. It has long been the author's opinion that the format of a mathematics book is of greater importance than is generally recognized. Consequently, when the opportunity arose to have the manuscript hand calligraphed it was decided to proceed with this even though it was necessary to begin before the entire manuscript was completed. This has led to some stylistic disadvantages in the final text. However, though their occurrence is regretable, they do not seem to be a serious deterrent to the general aims of the presentation.

Though the author can make no claim to have written a book on a par with that by Polya and Szegö, mentioned above, that work has consistently been considered as a model for excellence. It has been an inspiration from the beginning.

Great thanks are due to Gregory Maskarinec for undertaking the arduous task of calligraphing the manuscript from a hand written manuscript of quite different appearance. Throughout, our working relation has been excellent and left nothing to be desired. Thanks also are due to my many students who, over the years, have worked through various versions of parts of this material and to helpful colleagues for their criticisms.

All comments from readers designed to help in the improvement of the work will be gratefully received.

Joe Roberts

Portland, Oregon 1975

# Special Symbols and where 1st used or defined.

Other notations used in the text.

For integers:

$a \mid b$ means there is an integer $c$ such that $b = a \cdot c$;

$a \equiv b \pmod{m}$ means $m \mid a - b$;

$n! = 1 \cdot 2 \cdot 3 \cdot \cdots \cdot n$; $0! = 1$;

$\binom{a}{b} = \frac{a!}{(a-b)!\,b!}$, $0 \le b \le a$;

$\mathbb{Z}$ is the set of positive integers;

gcd stands for "greatest common divisor";

$[u, v]$ is the least common multiple of $u$ and $v$;

LHS (RHS) left (right) hand side.

ELEMENTARY NUMBER THEORY

# I The Game of Euclid & the Euclidean Algorithm

Consider the sequence of sets (duplicate elements are permitted):

$$\{78,35\}\to\{43,35\}\to\{8,35\}\to\{8,11\}\to\{8,3\}\to\{2,3\}\to\{2,1\}\to\{0,1\}.$$

Each set in the sequence may be obtained from the preceding one by subtracting some positive integral multiple of one of its elements from the other. When a set $\{a,b\}$ of non-negative integers arises in this way from another such set $\{m,n\}$ we say it is a *derived set* of $\{m,n\}$. A sequence of sets, like the above, in which each set is a derived set of the preceding set and in which the last set contains a zero will be called a *derived sequence*.

If $\{a,b\}$ is a derived set of $\{m,n\}$ with least value for $a+b$ we call it a *minimal derived set* of $\{m,n\}$. In the above sequence $\{43,35\}$

is not a minimal derived set of $\{78, 35\}$ while $\{2, 3\}$ is a minimal derived set of $\{8, 3\}$. The passage from any set to a derived set is called a *move* and a move to a set one element of which is 0 is called a *winning move*.

Throughout, all integers are to be non-negative and $m \le n$. Further, $\tau = \frac{1 + \sqrt{5}}{2}$.

1. Noting that $\{m, n\} = \{n, m\}$ for all $m, n$ we see that:

*i*) $\{m, n\}$ has $t$ derived sets, where $t$ is the largest positive integer for which $tm \le n$ is true;

*ii*) $\{m, n\}$ has exactly one minimal derived set, which is $\{m, n - tm\}$, where $t$ is as in (*i*);

*iii*) if $\{a, b\}$ is a derived set of $\{m, n\}$ then the greatest common divisor of $a$ and $b$ is equal to the greatest common divisor of $m$ and $n$; in symbols, $(a, b) = (m, n)$;

iv) every derived sequence starting with $\{m,n\}$ ends with $\{0,(m,n)\}$.

2. If two players, say A and B, start with $\{m,n\}$ and alternately make the moves of a derived sequence, A moving first and each desiring to make the winning move of the sequence then we call the play resulting "the game of Euclid". The following assertions are true of this game:

i) if at any stage of the game a set occurs in which one element is a positive integral multiple of the other then the player next to move can win by moving to the minimal derived set;

ii) it is *not* always to a player's advantage to move to a minimal derived set;

iii) if there is a winning strategy for A then at each play he must select one or the other of:

the minimal derived set, or, the derived set whose only derived set is the minimal derived set ;

*iv*) when $1 < \frac{a}{m} < \tau$ there is a unique move from $\{a, m\}$ and that is to a set $\{r, m\}$ where

$$\frac{m}{r} > \tau .$$

3. *i*) The player moving first in the game of Euclid, starting from $\{m, n\}$, $0 < m < n$, can force a win for himself if and only if $\frac{n}{m} > \tau$ ;

*ii*) when a game starts with $\{m, n\}$ then player A may force a win if $\frac{n}{m} = 1$ or $\frac{n}{m} > \tau$ while if neither of these is true player B may force a win .

4. An efficient method of computing the greatest common divisor (hereafter denoted gcd) of two positive integers $a$ and $b$ is to compute a derived sequence beginning with $\{a, b\}$ and in which each other element of the sequence is the minimal

derived set of the preceding one. Thus if $a > b$ and $a = qb + r$, $0 \leq r < b$, where $q$ and $r$ are integers, the first move would be $\{a, b\} \to \{b, r\}$. Putting $a = r_0$, $b = r_1$, $q = q_0$, $r = r_2$, etc. one finds the gcd of $a$ and $b$ is $r_n$ when

$$
\begin{array}{ll}
r_0 = q_0 r_1 + r_2 & 0 < r_2 < r_1 \\
r_1 = q_1 r_2 + r_3 & 0 < r_3 < r_2 \\
r_2 = q_2 r_3 + r_4 & 0 < r_4 < r_3 \\
\vdots & \\
r_{n-2} = q_{n-2} r_{n-1} + r_n & 0 < r_n < r_{n-1} \\
r_{n-1} = q_{n-1} r_n + 0 &
\end{array}
$$

This process is called the *Euclidean Algorithm*. In the process the gcd of the starting numbers is the last non-zero "remainder".

5. Using the Euclidean algorithm it is not hard to see that given any positive integers $a$ and $b$ there exist positive integers $x$ and $y$ for which

$$(a, b) = ax - by .$$

We call expressions like $ax - by$ or $ax + by$ linear combinations of $a$ and $b$.

6. It is interesting to ask how efficient the Euclidean algorithm is for the determination of the gcd of two numbers. Information about this question is given in a theorem due to Lamé. To prove the theorem we will make use of the Fibonacci sequence $u_0, u_1, u_2, \cdots$ defined by:

$$u_0 = u_1 = 1,$$

$$u_{n+2} = u_{n+1} + u_n \text{ for } n > 0.$$

i) For $n \geq 1$, $u_{5n+1} > 10^n$, so $u_{5n+1}$ has at least $n+1$ base 10 digits;

ii) if $n$ steps are used in the Euclidean algorithm determining the gcd of $r_0$ and $r_1$, $r_0 > r_1 > 0$, using $r_1$ as the first divisor, then

$$r_1 \geq u_n;$$

iii) (Lamé [1844]) the number of divisions needed by the Euclidean algorithm in finding the gcd of two numbers does not exceed five times the number of base 10 digits in the smaller of the two numbers;

iv) the maximum number of divisions allowed by (iii) is actually used in computing the gcd's of (8, 13), (89, 144), (987, 1597) by the Euclidean algorithm; note that all numbers involved are in the Fibonacci sequence;

v) if the Euclidean algorithm in computing the gcd of $a$ and $b$, $a > b$, $b$ having $t$ base 10 digits, takes $5t$ steps then the number of base 10 digits of $u_{5t}$ is $\leq t$;

vi) $\left| \frac{u_{n+1}}{u_n} - \tau \right| < \frac{1}{u_n^2}$;

vii) $u_{n+5} > 10 u_n$ for $n \geq 4$;

viii) for $t \geq 4$, $u_{5t} > 10^t$ and, therefore, $u_{5t}$ has more than $t$ base 10 digits;

*ix*) the Euclidean algorithm when applied to two numbers the smaller of which has at least 4 base 10 digits never takes as many divisions as allowed by Lamé's theorem ; i.e. Lamé's theorem is not "best possible" when applied to numbers the smaller of which is $\geq 10^3$ .

*Remarks.*

The game of Euclid is due to Cole & Davie [1969] and has been further analysed by Spitznagel [1973]. The theorem of Lamé was first proved by him in 1844. The result in #6(*ix*) is far from the best known result of this kind. The interested reader might consult Dubisch [1949], Dixon [1971], Brown [1967], or Plankensteiner [1970] for further information and references.

## II The Golden Mean

The point C divides

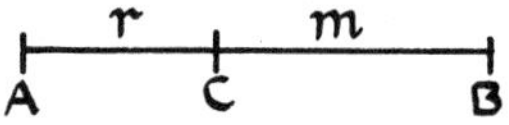


a line segment AB into "extreme and mean ratio" (Euclid, Book IV, Definition 3) when

$$\frac{m}{r} = \frac{m+r}{m}.$$

Such a division of a line segment is sometimes called a *golden section* or a *golden cut*. The ratio $\frac{m}{r}$ for such a division is called the *golden mean* or the *golden ratio*. A rectangle whose sides are in this ratio is a *golden rectangle*. In the following we again use $\tau$ for the irrational number $\frac{1+\sqrt{5}}{2}$ and use $\tau'$ for its "conjugate" $\frac{1-\sqrt{5}}{2}$.

1. If $0 < r < m$ and $\frac{m}{r} = \frac{m+r}{m}$ then $\frac{m}{r} = \tau$.

2. $\tau^2 = 1 + \tau$, $\tau^{-1} = \tau - 1$, and $\frac{1}{\tau} = -\tau'$.

3. If $r$ and $m$ are positive numbers with $\frac{m}{r} \neq \frac{m+r}{m}$ then $\tau$ lies strictly between $\frac{m}{r}$ and $\frac{m+r}{m}$. Further, no other number shares this property with $\tau$, even if $r$ and $m$ are constrained to be integers.

4. Consider the sequence

$$\frac{m}{r},\ \frac{m+r}{m},\ \frac{2m+r}{m+r},\ \frac{3m+2r}{2m+r},\ \frac{5m+3r}{3m+2r},\ \ldots$$

where each term has a numerator which is the sum of the previous numerator and denominator and has a denominator which is the previous numerator.

i) if $c = \min\{r, m\}$, i.e. $c = r$ if $r \leq m$ and $c = m$ if $m < r$, then the numerator of the $n^{th}$ term is $\geq nc$ and, therefore, when $r$ and $m$ are positive both numerator and denominator increase without bound;

ii) given 3 consecutive terms of the sequence,

say $\frac{a}{b}, \frac{c}{d}, \frac{e}{f}$, it is true that

$$ad - bc = -(cf - de);$$

iii) if $\alpha = |m^2 - mr - r^2|$ then the sequence of moduli of the successive differences in the given sequence is

$$\frac{\alpha}{mr}, \frac{\alpha}{m(m+r)}, \frac{\alpha}{(m+r)(2m+r)}, \cdots ;$$

iv) the sequence converges to $\tau$.

5. Consider the sequence of #4 in the special case $m = r = 1$:

$$\frac{1}{1}, \frac{2}{1}, \frac{3}{2}, \frac{5}{3}, \frac{8}{5}, \frac{13}{8}, \cdots$$

The sequence $1, 1, 2, 3, 5, 8, \cdots$ of denominators is the Fibonacci sequence and is denoted by $u_0, u_1, u_2, \cdots$ (see I #6).

i) $u_{n+2} = u_{n+1} + u_n$ for $n \geq 0$;

ii) $(u_{n+1}, u_n) = 1$ for $n \geq 0$;

iii) $u_n^2 - u_{n-1}u_{n+1} = (-1)^n$ for $n \geq 1$;

iv) $u_n \geq n$;

v) $\left| \frac{u_{n+1}}{u_n} - \tau \right| < \frac{1}{u_n^2}$ ;

vi) $\frac{u_{n+1}}{u_n} \to \tau$ as $n \to \infty$ .

6. i-a) $\tau^{n+1} = u_{n-1} + u_n\tau$ for $n \geq 1$ ;

b) $(-1)^n \tau^{-(n+1)} = u_n\tau^{-1} - u_{n-1}$ for $n \geq 1$ ;

ii) ( Binet 1843 )

$$u_n = \frac{1}{\sqrt{5}}\{\tau^{n+1} - \tau'^{n+1}\} \text{ for } n \geq 0 .$$

7. Consider the triangle inscribed in a rectangle as shown.

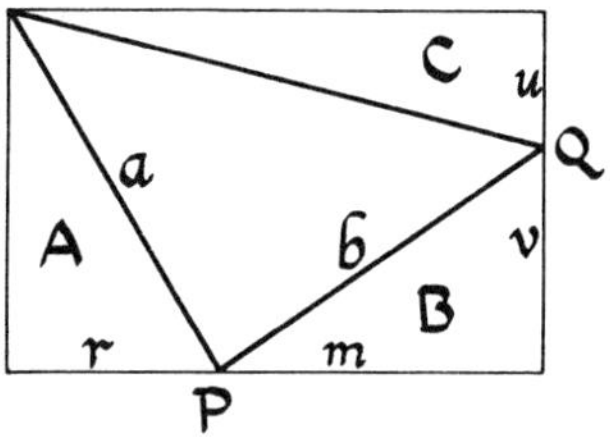


i) if the triangles A, B, C are equal in area then P and Q cut their respective sides in the golden ratio;

ii) if, in addition, a = b then the large rectangle is golden.

8. The diagonal of a regular pentagon with side 1 is $\tau$ .

9. The lengths of the segments of the dark zigzag line in the "star-pentagram" are as indicated. Further, the process may be continued indefinitely both in the inward and outward directions. The diagonal of the large pentagon is of length $\tau^3$.

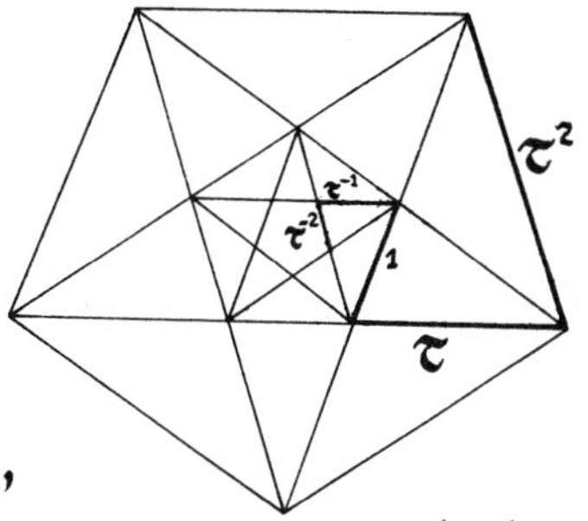


10. One may cut a square into four pieces, as indicated, in such a way that the four pieces may be reassembled into a non-square rectangle.

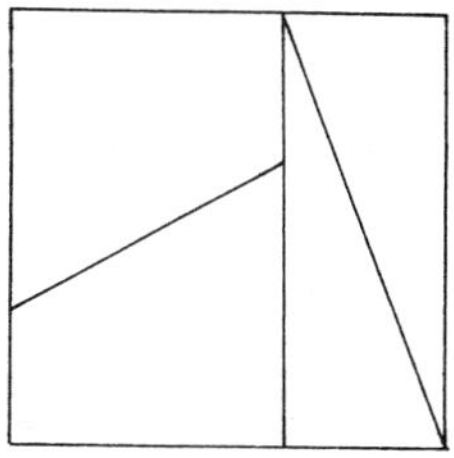

11. (Schlegel) Attempting to carry out the decomposition of #10 with dimensions as shown at the right leads to a surprising result when one constructs a model.

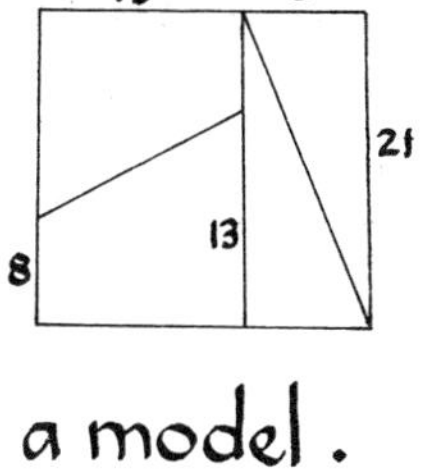

12. i) $\tau = 1 + \frac{1}{\tau} = 1 + \frac{1}{1+\frac{1}{\tau}} = 1 + \frac{1}{1+\frac{1}{1+\frac{1}{\tau}}} = \cdots,$

and the "pieces" of the limiting "continued fraction" $1 + \cfrac{1}{1 + \cfrac{1}{1 + \cfrac{1}{1+\cdots}}}$ are

$$1,\ 1+\frac{1}{1} = 2,\ 1 + \frac{1}{1+\frac{1}{1}} = \frac{3}{2},\ 1 + \frac{1}{1+\frac{1}{1+\frac{1}{1}}} = \frac{5}{3}, \cdots$$

with the general one being $\frac{u_{n+1}}{u_n}$;

ii) it is entirely reasonable to write

$$\tau = 1 + \cfrac{1}{1 + \cfrac{1}{1 + \cfrac{1}{1+\cdots}}}$$

13. For $m \geq 1$, $n \geq 1$:

i) $u_{m+n} = u_{m-1}u_{n-1} + u_m u_n$;

ii) $u_{n-1}$ divides $u_{nm-1}$;

iii) $(u_{n-1}, u_{m-1}) = u_{(n,m)-1}$.

14. Using matrix multiplication one has

$$\begin{pmatrix} 1 & 1 \\ 1 & 0 \end{pmatrix}^n = \begin{pmatrix} u_n & u_{n-1} \\ u_{n-1} & u_{n-2} \end{pmatrix} \text{ for } n \geq 2.$$

15. Let $A_n$ be the set of all those subsets of $\{1, 2, \cdots, n\}$ containing no pair of consecutive integers. Further, let $q(n)$ be the cardinality of $A_n$ and let $f(n, k)$ be the number of elements in $A_n$ having exactly $k$ elements. Then

*i)* $q(n) = q(n-1) + q(n-2)$ for $n > 2$ ;

*ii)* $q(n) = u_{n+1}$ for $n \geq 1$ ;

*iii)* the number of strings of $k$ 1's and $n-k$ 0's in which no two 1's are consecutive is just $f(n, k)$ ;

*iv)* the number of ways of placing $k$ 1's into $n-k+1$ boxes so that no box has more than 1 element is exactly $f(n, k)$ ;

*v)* $f(n, k) = \binom{n-k+1}{k}$ when $2k \leq n+1$ and is 0 otherwise ;

*vi)* setting $\binom{s}{t} = 0$ when $s < t$ we have

$$u_n = \sum_{k=0}^{n-1} \binom{n-k}{k}, \text{ for } n \geq 1.$$

vii) the sums in the indicated slant rows of Pascal's triangle are consecutive terms of the Fibonacci sequence.

| | | | | | | | |
|---|---|---|---|---|---|---|---|
| 1 | | | | | | | |
| 1 | 1 | | | | | | |
| 1 | 2 | 1 | | | | | |
| 1 | 3 | 3 | 1 | | | | |
| 1 | 4 | 6 | 4 | 1 | | | |
| 1 | 5 | 10 | 10 | 5 | 1 | | |
| 1 | 6 | 15 | 20 | 15 | 6 | 1 | |
| 1 | 7 | 21 | 25 | 25 | 21 | 7 | 1 |

16. Let $\mathcal{U}$ be the power series

$$1 + x + 2x^2 + 3x^3 + 5x^4 + 8x^5 + 13x^6 + \cdots$$

Then:

i) $\mathcal{U}$ converges to $\frac{1}{1-x-x^2}$ for $|x| < \tau^{-1}$;

ii) $$\frac{1}{1-x-x^2} = \frac{1}{r-s}\left\{\frac{r}{1-rx} - \frac{s}{1-sx}\right\},$$

where $r + s = 1 = -rs$ ;

iii) from (ii) one sees

$$u_n = \frac{1}{\sqrt{5}}\left\{\tau^{n+1} - \tau'^{\,n+1}\right\} ;$$

(compare with #6 (ii)) ;

iv) $$\frac{10000}{9899} = 1.0102030508132134559\cdots ;$$

v) $1 + 2 + 3 + \cdots + u_n = u_{n+2} - 2$ .

Remarks.

1. The Fibonacci sequence seems to have originated in connection with the famous rabbit problem posed by Leonardo of Pisa (Fibonacci) in 1202. One phrasing of the problem is as follows.

> One places a pair of rabbits in a confined area. How many pairs of rabbits can be produced in a year if every month each pair begets a new pair which from the second month itself becomes productive?

It will be noted that the sequence of numbers obtained for the numbers of pairs of rabbits at the ends of consecutive months is just the Fibonacci sequence.

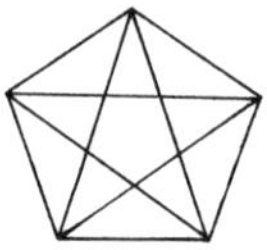

2. The Pythagoreans were so taken by the properties of the star-pentagram (see e.g. #9) that they used it as a symbol of recognition and brotherhood. In his book *Science Awakening* [1954 p. 101] van der Waerden tells a charming story of this.

3. The "paradoxical" decomposition of #11 seems to go back to Schlegel [1868]. See also Coxeter [1953].

4. The frequent occurrence, in a wide variety of settings, of the golden mean and the Fibonacci numbers has lead in recent years to a new mathematics journal, *The Fibonacci Quarterly*. Besides this journal the interested reader

might consult any of the following for further information: Coxeter [1953], Gardner [1959], Pacioli [1509, reprint 1956], Huntley [1970], Archibald [1918], Thompson [1952].

## III Prime Factorizations & Primes

A positive integer $n$ which satisfies an equation $n = ab$, where $a$ and $b$ are integers larger than 1, is called a *composite integer*. If $n$ is neither composite nor equal to 1 it is called *prime*.

1. Every integer larger than 1 has at least one prime divisor. (This prime divisor may be the number itself.)

2. Each integer larger than 1 is either prime or a product of two or more primes; i.e. each integer larger than 1 has a prime factorization.

3. If a prime number divides a product of two integers then that prime must divide one or the other of the two integers.

4. The prime factorization of an integer larger than 1 is unique except for the order in which the factors occur.

5. Given any integer $n$ there is a prime factor of $1 + n!$ exceeding $n$. Therefore there are infinitely many primes.

6. The last conclusion of #5 also follows from observing that every prime factor of $1 + p_1 \cdots p_k$, where each of $p_1, \cdots, p_k$ is prime, differs from each of $p_1, \cdots, p_k$. (This proof was given by Euclid in his Elements.)

7. Given a positive integer $k \geq 2$ there exists a string of $k$ consecutive composite integers.

8. If $p_1, \cdots, p_k$ is any finite collection of primes the number $4p_1 \cdots p_k - 1$ contains a prime factor

of the form $4k+3$ and this prime differs from each of $p_1, \cdots, p_k$. Therefore, there are infinitely many $4k+3$ primes.

9. Let $F_n = 2^{2^n} + 1$ for $n = 0, 1, 2, \cdots$. These numbers are called *Fermat numbers*.

i) The base 10 unit's digit of $F_n$, $n \geq 2$, is 7;

ii) if $2^m + 1$ is prime then $m$ is a power of 2; i.e. $2^m + 1$ is a Fermat number whenever it is a prime;

iii) though Fermat thought all $F_n$ to be prime this is not the case since, as Euler first observed, 641 is a prime divisor of $F_5$;

iv-a) $\prod_{0 \leq n < m} F_n = F_m - 2$;

b) $(F_n, F_m) = 1$ for $n \neq m$;

v) (iv-b) implies the infinitude of the number of primes.

10. No integral polynomial has only prime values for all sufficiently large integers. (By "integral polynomial" we mean a polynomial with integer coefficients.)

11. (Luthar) Write $x_n = p_1 + \cdots + p_n$, $n \geq 1$, where $p_j$ is the $j^{th}$ prime number.

i) $p_{n+1} > 2n+1$ for $n \geq 4$;

ii) $x_n > n^2$ for $n \geq 1$;

iii) $p_{n+1} \leq 2(n+k)+1$ implies

a) $k > 0$;

b) $p_{n-j} \leq 2(n+k) - (2j+1)$ for $0 \leq j < n$;

c) $x_n < (n+k)^2$;

iv) $(n+k)^2 \leq x_n < (n+k+1)^2$ implies $p_{n+1} > 2(n+k)+1$;

v) for $n \geq 1$ there is a square strictly between $x_n$ and $x_{n+1}$.

12. (Grimm) For each $k$, $2 \le k \le n$, put

$$q_k = \begin{cases} k \text{ if } \frac{n}{2} < k \le n \text{ and } k \text{ is prime;} \\ \text{any prime factor of } \frac{n!}{k} + 1 \text{ otherwise.} \end{cases}$$

i) $q_k \mid n! + k$;

ii) $q_k \mid n! + j$, $2 \le j \le n$, implies $j = k$;

iii) $q_2, \cdots, q_n$ are pairwise distinct;

iv) it is possible to select $n-1$ pairwise distinct prime divisors, one from each of

$$n!+2, \cdots, n!+n .$$

Remarks.

1. Results such as the one proved in # 8 are very special cases of a general theorem of Dirichlet to the effect that every arithmetic progression $a, a+b, a+2b, a+3b, \cdots$ for which $(a,b) = 1$ contains infinitely many primes (see XXIV).

2. The only known prime $F_n$ (see #9) are those with $n = 0, 1, 2, 3, 4$. There are 38 values of $n$ for which $F_n$ is known to be composite. These are: 5 through 16, 18, 19, 23, 36, 38, 39, 55, 58, 63, 73, 77, 81, 117, 125, 144, 150, 207, 226, 228, 260, 267, 268, 284, 316, 452, 1945. The number $F_{17}$ has more than 30 000 digits and its character is not known. Until very recently (1971) $F_7$ was known to be composite but its factorization was not known. In 1971 this number was factored by Morrison and Brillhart and it was found that $F_7$ =

340 282 366 920 938 463 463 374 607 431 768 211 457

= (59 649 589 127 497 217)(5 704 689 200 685 129 054 721).

The number of digits in $F_{1945}$ exceeds $10^{582}$ but nevertheless it is known that $5 \cdot 2^{1947} + 1$ is its smallest prime divisor. Further information about Fermat primes may be found in XIX. The interested reader might also consult Sierpinski [1964 a,b].

3. Despite the truth of the result in #10 it has recently been proved, as a consequence of Matijasevich's solution of Hilbert's Tenth Problem, that there do exist integral polynomials whose *positive* range consists precisely of the prime numbers. (See Davis [1973].)

4. Problems #11, 12 are due, respectively, to Luthar [1969], and Grimm [1961, 1969]. For related work to #12 see Just [1972] and Cijsouw, Tijdeman [1972].

# IV Square Brackets

The largest integer not exceeding $x$ is denoted by $[x]$. Thus $[\pi] = 3$, $[\frac{1}{2}] = 0$, $[-\pi] = -4$, etc. This function appears in a number of diverse settings. In this chapter we set forth a number of its properties as well as use it in expressions for various other number theoretic functions. Throughout the chapter we use $m, n, k$ for integers and $\alpha$ and $\beta$ for arbitrary real numbers.

1. $\alpha - 1 < [\alpha] \leq \alpha$ and $[\alpha] \leq \alpha < [\alpha] + 1$.

2. $[\alpha + n] = [\alpha] + n$.

3. $\frac{m+1}{n} \leq \left[\frac{m}{n}\right] + 1$, $n > 0$.

4. $\left[\frac{[\alpha]}{n}\right] = \left[\frac{\alpha}{n}\right]$, $n > 0$.

5. No integer is closer to $\alpha$ than $[\alpha + \frac{1}{2}]$.

6. $-[-\alpha]$ is the smallest integer not less than $\alpha$.

7. $[\alpha]+[\beta] \le [\alpha+\beta] \le [\alpha]+[\beta]+1$.

8. $[\alpha+\beta]+[\alpha]+[\beta] \le [2\alpha]+[2\beta]$.

9. $[\alpha][\beta] \le [\alpha\beta] \le [\alpha][\beta]+[\alpha]+[\beta]$,
for $\alpha > 0$, $\beta > 0$.

10. When $0 < k \le \alpha$, $[\frac{\alpha}{k}]$ is the number of positive integral multiples of $k$ not exceeding $\alpha$.

11. When $\alpha > \beta$, $[\alpha]-[\beta]$ is the number of integers $m$ satisfying $\beta < m \le \alpha$.

12. $[\alpha]+[-\alpha] = \begin{cases} 0 & \text{if } \alpha \text{ is an integer;} \\ -1 & \text{otherwise.} \end{cases}$

13. $[\alpha] - 2\left[\frac{\alpha}{2}\right]$ is either 0 or 1.

14. $\left[\frac{n}{2}\right] - \left[-\frac{n}{2}\right] = n$.

15. $\lim\limits_{n\to\infty} \frac{[n\alpha]}{n} = \alpha$.

16. $\left[\sqrt[k]{\alpha}\,\right] = \left[\sqrt[k]{[\alpha]}\,\right]$ for $\alpha \geq 0$.

17. $[\alpha] + \left[\alpha + \frac{1}{n}\right] + \left[\alpha + \frac{2}{n}\right] + \cdots + \left[\alpha + \frac{n-1}{n}\right] = [n\alpha]$.

18. $\left[\frac{\alpha}{n}\right] + \left[\frac{\alpha+1}{n}\right] + \cdots + \left[\frac{\alpha+n-1}{n}\right] = [\alpha]$.

19. $[m\alpha] + \left[m\alpha + \frac{m}{n}\right] + \cdots + \left[m\alpha + \frac{(n-1)m}{n}\right]$
$= [n\alpha] + \left[n\alpha + \frac{n}{m}\right] + \cdots + \left[n\alpha + \frac{(m-1)n}{m}\right]$.

20. When $n$ and $m$ are of opposite parity
$$\int_0^1 (-1)^{[nx]+[mx]} \binom{n-1}{[nx]}\binom{m-1}{[mx]}\, dx = 0.$$

21. $[\tau^2 n] = [\tau[\tau n] + 1]$ , when $\tau = \frac{1+\sqrt{5}}{2}$.

22. (Skolem) $\left[\sqrt{2}\left[\left(1+\frac{1}{\sqrt{2}}\right)n + \frac{1}{2}\right]\right] = \left[\left(1+\sqrt{2}\right)n\right]$.

23. $u_n = \left[\frac{1}{\sqrt{5}}\left(\frac{1+\sqrt{5}}{2}\right)^{n+1} + \frac{1}{2}\right]$ , where $u_n$ is the $n+1^{st}$ Fibonacci number.

24. If $p$ is a prime number then the highest power of $p$ in $n!$ is $\left[\frac{n}{p}\right] + \left[\frac{n}{p^2}\right] + \left[\frac{n}{p^3}\right] + \cdots$.

25. Problem #24 may be used to show:

i) $\binom{n}{m}$ is an integer ;

ii) $\frac{(n_1 + \cdots + n_k)!}{n_1! \cdots n_k!}$ is an integer ;

iii) (Catalan) $\binom{m+n}{m}$ divides $\binom{2m}{m}\binom{2n}{n}$ .

26. $n$ is a prime if and only if $\sum_{m=1}^{\infty}\left(\left[\frac{n}{m}\right] - \left[\frac{n-1}{m}\right]\right) = 2$.

27. (R. Algee) The number of primes not exceeding $n$ is $\sum_{n=2}^{m}\left[\frac{1}{\sum_{k=2}^{n}\left[\frac{[n/k]}{n/k}\right]}\right]$.

28. $\lim_{m\to\infty}[\cos^2 m!\,\pi x] = \begin{cases} 0 & \text{for } x \text{ irrational;} \\ 1 & \text{for } x \text{ rational.} \end{cases}$

29. If $N$ is the number of solutions of the system $xy \le n$, $0 < x$, $0 < y$, then
$N = \left[\frac{n}{1}\right] + \left[\frac{n}{2}\right] + \cdots + \left[\frac{n}{n}\right] = 2\sum_{k=1}^{[\sqrt{n}]}\left[\frac{n}{k}\right] - [\sqrt{n}]^2$.

30. For $b$ odd there is an integer $q$ such that:

i) $0 \le x - bq < \frac{b}{2}$ if and only if $\left[\frac{2x}{b}\right]$ is even;

ii) $-\frac{b}{2} < x - bq < 0$ if and only if $\left[\frac{2x}{b}\right]$ is odd.

31. i) $\sum_{n=1}^{b-1}\left[\frac{an}{b}\right] = \frac{(a-1)(b-1)}{2} + \frac{d-1}{2}$, where $d = (a, b)$;

ii) (Eisenstein)

$$\sum_{n=1}^{\frac{b-1}{2}}\left[\frac{an}{b}\right] + \sum_{n=1}^{\frac{a-1}{2}}\left[\frac{bn}{a}\right] = \frac{(a-1)(b-1)}{4},$$

when $a$ and $b$ are relatively prime odd positive integers.

32. Let $(a,b)=1$ and suppose $ax_0+by_0=1$. Further, consider the equation

$$(*) \quad ax+by=k.$$

A pair of integers $x, y$ satisfying $(*)$ is called a *solution* of the equation and if, in addition, both $x$ and $y$ are non-negative, we call $x, y$ a *non-negative solution*.

i) If $x, y$ is a solution of $(*)$ then there is an integer $t$ such that

$$x=kx_0+bt,$$
$$y=ky_0-at;$$

ii) the number of non-negative solutions of $(*)$ is given by $N=1+\left[\frac{kx_0}{b}\right]+\left[\frac{ky_0}{a}\right]$;

iii) for $k\geq 0$, $(*)$ has no non-negative solutions precisely when there exist $r, s$, $0\leq r<b$, $0\leq s<a$ such that

$$k=ar+bs-ab;$$

iv) (*) always has a non-negative solution when $k > ab - a - b$ but does not have a non-negative solution when $k = ab - a - b$ ;

v) for exactly $\frac{(a-1)(b-1)}{2}$ positive values of $k$ does (*) fail to have a non-negative solution.

Remarks.

1. There is a wide literature on square brackets. The interested reader might consult the following : Bang [1957]; Beatty [1927] ; Coxeter [1953]; Fraenkel [1969]; Fraenkel, Levitt, Shimshoni [1972]; Graham, Pollack [1970] ; Graham [1973] ; Skolem [1957] ; Watson [1956].

2. The result in #32 (iv) goes back to Frobenius and Schur. Similar results have been sought for the general linear forms

$$a_1x_1 + \cdots + a_kx_k = n$$

but even for $k=3$ the general solution is not known. For an introduction to the literature the reader might consult Brauer, Shockley [1962], Erdös, Graham [1971], Hofmeister [1966], Lewin [1972, 1973], Roberts [1956], Bateman [1958], and Note 14 by Skolem in Netto [1927].

## V Kronecker Theorems

For $x$ a real number we write $(x)$ for the fractional part of $x$; i.e. $(x) = x - [x]$.

1. Let $\alpha$ be irrational and put $P_n = (n\alpha)$. Then:

i) $0 < P_n < 1$, $n = 1, 2, \ldots$;

ii) $P_n \neq P_m$ for $n \neq m$;

iii) given $\epsilon > 0$ there are positive integers $n$ and $r$ such that $|P_n - P_{n+r}| < \epsilon$;

iv) given $\epsilon > 0$ there is an $r$ such that

$$P_r < \epsilon \text{ or } 1 - \epsilon < P_r \;;$$

v) (Kronecker's one dimensional theorem) $\{P_1, P_2, \cdots\}$ is dense in the open unit interval.

2. Define the mapping $f$ of the plane into the unit square by $f(x, y) = ((x), (y))$.

Further, suppose $\alpha, \beta$ irrational and $P_n = f(n\alpha, n\beta)$, $n = 1, 2, \cdots$. Write $PQ$ for the vector from $P$ to $Q$ and $|PQ|$ for the length of this vector.

Then:

i) $P_n \neq P_m$ for $n \neq m$ ;

ii) if $P_m Q = P_n P_{n+r}$ then $f(Q) = P_{m+r}$ ;

iii) if $P_1 Q = m P_1 P_{1+r} + n P_1 P_{1+s}$ then

$$f(Q) = P_{1+mr+ns} ;$$

iv) if $P_1 P_{1+r}$ and $P_1 P_{1+s}$ are not parallel and $L$ is the greatest of their lengths then every point of the unit square is within $L$ of some point of the form $P_{1+mr+ns}$, where $m$ and $n$ are non-negative integers.

3. Let $P_1, P_2, \cdots$ be as in #2 and suppose that the only triple of integers $r, s, t$ for which $r\alpha + s\beta + t = 0$ is $0, 0, 0$ ;

i.e. $\alpha, \beta, 1$ are rationally independent.

Then:

i) $\alpha$ and $\beta$ are irrational;

ii) given $\epsilon > 0$, there are integers $n$ and $r$ such that $|P_n P_{n+r}| < \epsilon$;

iii) for $0 < \epsilon < \min\{(\alpha), 1-(\alpha), (\beta), 1-(\beta)\}$ and $n, r$ as in (ii) the vector $P_1 P_{1+r}$ equals $P_n P_{n+r}$;

iv) for $\epsilon$ as in (iii) there are infinitely many positive integers $r$ such that $|P_1 P_{1+r}| < \epsilon$;

v) it is not possible that infinitely many of the vectors $P_1 P_{1+r}$ appearing in (iv) be parallel;

vi) (Kronecker's two dimensional theorem) $\{P_1, P_2, \cdots\}$ is dense in the unit square.

Remark.

For expositions of the theorems in this chapter see Niven [1963] and Hardy and Wright [1962].

## VI Beatty, Skolem Theorems

Let $S(\alpha)$ be the sequence $[\alpha], [2\alpha], [3\alpha], \cdots$ and let $A(\alpha)$ be the set of distinct elements of $S(\alpha)$. (In the following we use $\mathbb{Z}$ for the set of positive integers and, as before, $\tau$ for $\frac{1+\sqrt{5}}{2}$.) Then (for $\alpha, \beta$ positive):

1. $A(\alpha)$ is precisely the set of non-negative integers when $0 < \alpha < 1$.

2. $A(\alpha) \cap A(\beta) = \phi$ implies $\alpha > 1$ and $\beta > 1$.

3. $A(\alpha) \subseteq A(\frac{\alpha}{m})$.

4. $A(1+\sqrt{2}) \subsetneqq A(\sqrt{2})$.

5. $A(\alpha) \cap A(\beta)$ is an infinite set when both $\alpha$ and $\beta$ are rational.

6. (Beatty) If $\alpha$ is positive and irrational and $\frac{1}{\alpha}+\frac{1}{\beta}=1$ then every positive integer is in exactly one of $S(\alpha)$, $S(\beta)$ and these sequences have no duplicate terms.

7. $A(\sqrt{2}) \cap A(2+\sqrt{2}) = \phi$ and

$$A(\sqrt{2}) \cup A(2+\sqrt{2}) = \mathbb{Z} .$$

8. $A(\tau) \cap A(\tau^2) = \phi$ and

$$A(\tau) \cup A(\tau^2) = \mathbb{Z} .$$

9. (Skolem) The three sequences $(n \geq 1)$

$$\{[\tau[\tau n]]\},\quad \{[\tau[\tau^2 n]]\},\quad \{[\tau^2 n]\}$$

are mutually disjoint and their union is $\mathbb{Z}$.

10. If $A_0 = A(\tau)$ and $A_{m+1} = \{[\tau^2 n] \mid n \in A_m\}$ for $m \geq 0$, then the $A_j$ are disjoint in pairs and their union is $\mathbb{Z}$.

11. If $\alpha$ is positive and irrational and $\frac{1}{\alpha}+\frac{1}{\beta}=1$ and if $A_0=A(\alpha)$, $A_{m+1}=\{[\beta n] \mid n \in A_m\}$ for $m \geq 0$, then the $A_j$ are disjoint in pairs and their union is $\mathbb{Z}$.

12. If $A(\alpha) \cap A(\beta)$ is finite and $A(\alpha) \cup A(\beta)=\mathbb{Z}$ then $\frac{1}{\alpha}+\frac{1}{\beta}=1$.

13. $A(\alpha) \cap A(\beta)$ non-empty and finite is incompatible with $A(\alpha) \cup A(\beta)=\mathbb{Z}$.

14. (Bang) A neccessary and sufficient condition for $S(\alpha)$, $S(\beta)$ to be complementary (i.e. $A(\alpha) \cap A(\beta)=\phi$, $A(\alpha) \cup A(\beta)=\mathbb{Z}$) is that $\alpha$ and $\beta$ be positive irrational numbers such that $\frac{1}{\alpha}+\frac{1}{\beta}=1$.

15. (Uspensky) An interesting result along the lines of #6 and #9 is the following theorem proved by Uspensky in 1927.

> There do not exist 3 or more numbers $\alpha_1, \cdots, \alpha_n$ such that $S(\alpha_1), \cdots, S(\alpha_n)$ are non-empty disjoint sequences which taken together contain each positive integer precisely once.

We prove this following Graham [1963]. In fact we shall assume: $n \geq 3$, $\alpha_1 < \cdots < \alpha_n$ and $S(\alpha_1), \cdots, S(\alpha_n)$ are non-empty disjoint sets exhausting the integers without duplication and shall show that this leads to a contradiction. Throughout, $m$ is the least positive integer not in $S(\alpha_1)$.

i) $\alpha_1 = 1 + \delta$ where $0 < \delta < 1$ ;

ii) $S(\alpha_1)$ does not miss any pair of consecutive integers;

iii) $(m-1)\delta < 1 \leq m\delta$ ;

iv) $m$ is the first element of $S(\alpha_2)$ and

$$\alpha_2 = m + \epsilon, \quad 0 \leq \epsilon < 1 ;$$

v) if $x$ is a positive integer not in $S(\alpha_1)$ the next positive integer not in $S(\alpha_1)$ is

$$\text{either } x + m \text{ or } x + m + 1 ;$$

vi) the next element after $[n\alpha_2]$ in $S(\alpha_2)$ is either $[n\alpha_2] + m$ or $[n\alpha_2] + m + 1$ ;

vii) the $k^{th}$ positive integer missing from $S(\alpha_1)$ is the $k^{th}$ element in $S(\alpha_2)$ ;

viii) the assumption is false.

16. Let $\alpha$ and $\beta$ be positive irrational numbers and suppose $a, b, c$ are integers such that $\frac{a}{\alpha} + \frac{b}{\beta} = c \neq 1$, $a > 0$, $(a, b, c) = 1$. Further, let $d = (a, b)$ and denote the shaded rectangle in the diagram by $S$.

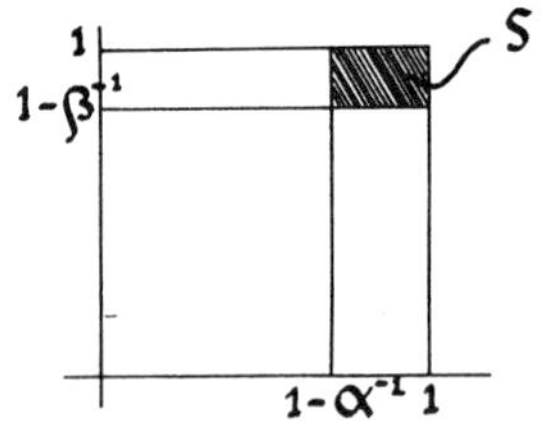

i) if $b<0$ and $c=0$ then $A(\alpha)\cap A(\beta)\neq\phi$ ;

ii) if $b<0$ then $ax+by=a+b$ passes through $S$;

iii) if $b>0$ and $c>1$ then $ax+by=a+b-1$ passes through $S$.

17. Let $\alpha,\beta,a,b,c,d,S$ be as in #16. Then:

i) there are integers $u$ and $v$ such that
$$a\left(u+\frac{d}{\alpha}\right)=-b\left(v+\frac{d}{\beta}\right);$$

ii) if $w_n=\frac{nd}{b}\left(u+\frac{d}{\alpha}\right)-\left[\frac{nd}{b}\left(u+\frac{d}{\alpha}\right)\right]$,
$x_n=\frac{b}{d}w_n$, $y_n=-\frac{a}{d}w_n$ then $\{w_1,w_2,\cdots\}$ is dense in $[0,1]$ while the points $(x_n,y_n)$ are dense on the line segment joining $(0,0)$ to
$$\left(\frac{b}{d},-\frac{a}{d}\right);$$

iii) if $c\neq 0$ and $q$ is a fixed integer there are integers $t,s,u_1,v_1$ such that $dt+sc=q$
$$\text{and } au_1+bv_1=dt;$$

iv) if $x_{nm}=x_n+\frac{mb}{d}+u_1+\frac{s}{\alpha}$,
$$y_{nm}=y_n-\frac{ma}{d}+v_1+\frac{s}{\beta} \text{ then}$$
$(x_{nm},y_{nm})$ is always a point on the line $ax+by=q$;

v) the points $(x_{nm}, y_{nm})$ are dense on

$$ax+by=q\ ;$$

vi) if either $b<0, c\neq 0$ or $b>0, c>1$ then there are infinitely many points $(x_{nm}, y_{nm})$ in $S$ ;

vii) if $b<0$ or if $b>0, c>1$ then

$$A(\alpha)\cap A(\beta)\neq\phi\ .$$

18. If $1, \frac{1}{\alpha}, \frac{1}{\beta}$ are rationally independent (i.e. if there does not exist a triple $a, b, c$ of integers not all zero such that

$$a+b\frac{1}{\alpha}+c\frac{1}{\beta}=0\ )$$

then $A(\alpha)\cap A(\beta)\neq\phi$ .

19. (Skolem) If $\alpha$ and $\beta$ are positive irrational numbers then $A(\alpha)\cap A(\beta)=\phi$ if and only if there are positive integers $a$ and $b$ such that $\frac{a}{\alpha}+\frac{b}{\beta}=1$ .

20. (Skolem) This special case of Uspensky's theorem (see #15) was proved by Skolem [1957].

There do not exist positive irrational numbers $\alpha, \beta, \gamma$ such that $A(\alpha)$, $A(\beta)$, $A(\gamma)$ are pairwise disjoint.

21. (Bang) If $\alpha$ and $\beta$ are positive irrational numbers then $A(\alpha) \cap A(\beta) \neq \phi$ if and only if the line segment joining $(\alpha, 0)$ and $(0, \beta)$ passes through a lattice point.

*Remarks.*

The material of this chapter is drawn primarily from Skolem [1957], Bang [1957], and Graham [1963]. The interested reader might also consult Niven [1963], Connell [1959, 1960], Uspensky [1927], Graham [1973] and the references given in the 1st remark at the end of IV.

For #10, 11 see Roberts [1973].

## VII The Game of Wythoff

Consider the sequence of sets (duplicate elements are permitted):

$$\{78,35\} \longrightarrow \{70,35\} \longrightarrow \{70,25\} \longrightarrow \{50,5\}$$
$$\longrightarrow \{5,5\} \longrightarrow \{0,0\}.$$

Each element in the sequence may be obtained from the preceding one by subtracting a positive integer from one or the other of the two elements or by subtracting one and the same positive integer from each of the two elements. When a set $\{a,b\}$ of non-negative integers arises from a set $\{m,n\}$ in one of these three ways we say $\{a,b\}$ is a *derived set* of $\{m,n\}$. A sequence of sets, as above, in which each set is a derived set of the preceding set and which ends with $\{0,0\}$ is called a *derived sequence*. The passage from any set to a derived set is called a move and a move to $\{0,0\}$ is called a winning move. If two players, A and B, start with $\{m,n\}$ and alternately make the moves of a derived sequence, A moving first, and each desiring to make the winning move of the sequence, then we call the play resulting the game of Wythoff. We are interested in knowing the conditions under which the player moving first, A for us, can force a win for himself. Noting that $\{m,n\} = \{n,m\}$ and assuming all integers are non-negative we may prove:

1. If A can leave any of the following pairs to B then, regardless of B's move, A can win:

$\{1,2\}$, $\{3,5\}$, $\{4,7\}$, $\{6,10\}$, $\{8,13\}$, $\{9,15\}$, $\{11,18\}$.

2. If $\{a, b\}$ is a set of distinct non-zero integers not in the list in #1 and if the smaller of $a, b$ is $< 12$ then there is a move taking $\{a, b\}$ into one of the sets listed in #1.

3. In the game of Wythoff starting with $\{m, n\}$, $m \le 18$, $n \le 18$, A can force a win for himself if and only if $\{m, n\}$ does not appear in the list in #1.

4. There exists an infinite sequence of sets, of which the first 7 are those listed in #1, such that A can always force a win for himself if and only if he starts from a set not in the sequence.

5. The sequence given in #4 is just $\{[n\tau], [n\tau^2]\}$, $n \ge 1$, $\tau = \frac{1+\sqrt{5}}{2}$.

Remarks.

The game of Wythoff was first introduced by Wythoff [1907] as a variant of the game of Nim (see Bouton [1902]). It is discussed in Coxeter [1953] and has been generalized by Holladay [1968] and Connell [1959].

# VIII $\tau, \sigma, \varphi$

The number theoretic functions $\tau, \sigma, \varphi$ are defined as follows:

$\tau(n)$ = number of positive integral divisors of $n = \sum_{d|n} 1$ ;

$\sigma(n)$ = sum of the positive integral divisors of $n = \sum_{d|n} d$ ;

$\varphi(n)$ = number of positive integers not exceeding $n$ and relatively prime to $n = \sum_{\substack{a \\ (a,n)=1}} 1$ .

1. If $(a,b)=1$ then:

   i) $\tau(ab) = \tau(a)\tau(b)$ ;

   ii) $\sigma(ab) = \sigma(a)\sigma(b)$ ;

   iii) $\varphi(ab) = \varphi(a)\varphi(b)$ .

2. If $\alpha$ is a non-negative integer and $p$ is a prime then:

i) $\tau(p^{\alpha}) = \alpha + 1$ ;

ii) $\sigma(p^{\alpha}) = \dfrac{p^{\alpha+1} - 1}{p - 1}$ ;

iii) $\varphi(p^{\alpha}) = p^{\alpha}(1 - \frac{1}{p})$ .

3. Let $n = p_1^{\alpha_1} \cdots p_k^{\alpha_k}$. Then:

i) $\tau(n) = (\alpha_1 + 1) \cdots (\alpha_k + 1)$ ;

ii) $\sigma(n) = \prod_{j=1}^{k} \dfrac{p_j^{\alpha_j+1} - 1}{p_j - 1} \left(= \prod_{p|n} \dfrac{p^{\alpha+1} - 1}{p - 1}\right)$ ;

iii) $\varphi(n) = n \prod_{p|n} (1 - \frac{1}{p})$ .

4. If $\tau(n)$ is odd then $n$ is a square .

5. $\prod_{d|n} d = n^{\frac{1}{2}\tau(n)}$ .

6. $\tau(2^n - 1) \geq \tau(n)$ .

7. $\tau(2^n + 1) > \tau^*(n)$, where $\tau^*(n)$ is the number of positive odd divisors of $n$ .

8. $\sum_{d|n} \tau^3(d) = \left(\sum_{d|n} \tau(d)\right)^2$ .

9. For $n > 0$,

$$\tau(1) + \tau(2) + \cdots + \tau(n) = \left[\frac{n}{1}\right] + \left[\frac{n}{2}\right] + \cdots + \left[\frac{n}{n}\right]$$
$$= 2\sum_{d=1}^{[\sqrt{n}]} \left[\frac{n}{d}\right] - [\sqrt{n}]^2 .$$

10. If $\sigma_t(n) = \sum_{d|n} d^t$ then $\sigma_t(n) = \prod_{p|n} \frac{p^{(\alpha+1)t} - 1}{p - 1}$ .

11. If $a > 0$, $b > 1$ then

$$\frac{\sigma(a)}{a} < \frac{\sigma(ab)}{ab} \leq \frac{\sigma(a)\sigma(b)}{ab} .$$

12. If $a > 0$, $b > 0$ then

$$\sigma(a)\sigma(b) = \sum_{d|(a,b)} d\,\sigma\left(\frac{ab}{d^2}\right) .$$

13. $\sigma(1) + \sigma(2) + \cdots + \sigma(n) = \left[\frac{n}{1}\right] + 2\left[\frac{n}{2}\right] + \cdots + n\left[\frac{n}{n}\right]$ .

14. $\varphi(5186) = \varphi(5187) = \varphi(5188) = 2592$ .

15. i) For $n \geq 1$, $\varphi(n^2) = n\varphi(n)$ ;

ii) for $n \geq 2$, $\varphi(n) < n$ ;

iii) for $n \geq 3$, $\varphi(n^2) + \varphi((n+1)^2) < 2n^2$.

16. For $n > 2$ we have

$$\sum_{\substack{(m,n)=1 \\ 1 \leq m \leq n}} m = \frac{1}{2} n \varphi(n) .$$

17. If $\varphi(n) | n$ then $n$ is of one of the forms

$$1, 2^\alpha, 2^\alpha \cdot 3^\beta .$$

18. If $a$ and $b$ are larger than 1 and $c$ is the product of the distinct prime factors of $(a, b)$ then $\varphi(ab) = \varphi(a)\varphi(b)\frac{c}{\varphi(c)}$.

19. $\sum_{d|n} \varphi(d) = n.$

20. $\sum_{d=1}^{n} \varphi(d)\left[\frac{n}{d}\right] = \frac{1}{2} n(n+1).$

21. $\sum_{n=1}^{\infty} \frac{\varphi(n) x^n}{1 - x^n} = \frac{x}{(1-x)^2}$ .

22. The formula for $\varphi(n)$ given in #3 implies that the number of primes is infinite.

23. (Schinzel) Let $N_m$ be the number of solutions of the equation $\varphi(x) = m$. Then the sequence $N_1, N_2, \cdots$ is not bounded, as can be seen from the fact that $\varphi(p_1 \cdots p_k - \frac{1}{p_j} p_1 \cdots p_k)$ is independent of $j$, $1 \le j \le k$. Here $p_1, \cdots, p_k$ are the first $k$ primes in their natural order.

24. Let $n = p_1^{\alpha_1} \cdots p_k^{\alpha_k}$ and write $\varphi(x, n)$ for the number of positive integers not exceeding $x$ and relatively prime to $n$. Then:

i) (Legendre) $\varphi(x, n) =$

$$[x] - \sum_i \left[\frac{x}{p_i}\right] + \sum_{\substack{i,j \\ i<j}} \left[\frac{x}{p_i p_j}\right] - \cdots + (-1)^k \left[\frac{x}{p_1 \cdots p_k}\right];$$

ii) the expression for $\varphi(n)$ given in #3 (iii) is a special case of (i);

iii) $\pi(x)$, the number of primes not exceeding $x$, satisfies :

$$\pi(x) = \pi(\sqrt{x}) - 1 + \varphi(x, p_1 \cdots p_t),$$

where $p_1, \cdots, p_t$ are all the primes not exceeding $\sqrt{x}$ .

25. If $\sigma(n) = 2n$, one calls $n$ a perfect number .

i) 6, 28, 496, 8128 are perfect ;

ii) $2^n - 1$ prime and $n$ prime imply $2^{n-1}(2^n - 1)$ is perfect ;

iii) if $n$ is even and perfect then there is a $k$ for which $n = 2^{k-1}(2^k - 1)$ and each of $k$ and $2^k - 1$ is prime ;

iv) in antiquity it was often stated that every even perfect number ends in 6 or 8 and that no two consecutive even perfect numbers have the same base 10 final digit; the 1st though not the 2nd of these assertions is true ;

v) $\sum_{d|n} \frac{1}{d} = 2$ if and only if $n$ is perfect ;

vi) if $n$ is odd and has no more than 2 distinct prime factors then $n$ is not perfect.

26. Let $H(n)$ be the harmonic mean of the divisors of $n$ ; i.e. $\frac{1}{H(n)} = \frac{1}{\tau(n)} \sum_{d|n} \frac{1}{d}$ . Then :

i) $H(n) = \frac{n\tau(n)}{\sigma(n)}$ and $H$ is multiplicative ;

ii) $H(n) > 1$ for $n > 1$ and $H(n) > 2$ except for $n = 1, 4, 6$ or $n$ prime ;

iii) if $m = 2^{n-1}(2^n - 1)$ is perfect then $H(m) = n$ ;

iv) if $n = 2^{H(n)-1}(2^{H(n)} - 1)$ is even then

$$H(2^{H(n)} - 1) < 2 \quad ;$$

v) (Laborde) if $n$ is even and $n = 2^{H(n)-1}(2^{H(n)} - 1)$ then $n$ is perfect.

27. i) If $f$ is a multiplicative arithmetic function, i.e. if $f(ab) = f(a)f(b)$ for $(a,b) = 1$, then the function $g$ defined by $g(n) = \sum_{d|n} f(d)$ is also multiplicative ;

ii) the multiplicativity of all of the following functions follows from (i):

$$\tau(n),\ \sigma(n),\ \sum_{d\mid n}\tau(d),\ \sigma_t(n),\ \sigma^\circ(n),$$

where $\sigma^\circ(n)$ is the sum of the odd divisors of $n$.

Remarks.

1. The result in #23 will be found in Sierpinski [1964a].

2. As seen in #25 (iii) the even perfect numbers are all of the form $2^{n-1}(2^n-1)$ where $2^n-1$ is prime. Primes of this form are called *Mersenne primes* and will be discussed in XIX. There are only 24 Mersenne primes known and they are for the following values of $n$: 2, 3, 5, 7, 13, 17, 19, 31, 61, 89, 107, 127, 521, 607, 1279, 2203, 2281, 3217, 4253, 4423, 9689, 9941, 11213, 19937.

It is interesting to note that until 1952 the largest known prime was $2^{127}-1$, a number of 39 digits, while the largest known prime today is $2^{19937}-1$, a number of 6012 digits. There are only the 24 even perfect numbers corresponding to these Mersenne primes known and it is not known if odd perfect numbers exist. See McCarthy [1957]. The result of #25(vi) may, however, be considerably improved. Also it has been shown that no odd perfect number $< 10^{50}$ exists (see Hagis [1973]).

3. The main result in #26 is due to Laborde [1955].

## IX Fermat, Wilson, Chevalley

1. Let $p$ be a prime. Then:

   i) $p \mid (m+n)^p - (m^p + n^p)$ ;

   ii) $p \mid m^p - m$ if and only if $p \mid (m+1)^p - (m+1)$ ;

   iii) ( Fermat's "little" theorem)

   $$p \mid m^p - m \text{ for all } m \geq 1 .$$

2. For $p$ a prime,

$$p \mid (m_1 + \cdots + m_k)^p - (m_1{}^p + \cdots + m_k{}^p) ,$$

and Fermat's little theorem is an immediate consequence of this.

3. When $p$ is an odd prime then:

$$p \mid m^p + n^p \text{ implies } p^2 \mid m^p + n^p .$$

4. (Golomb) Let there be given a collection of beads of $n$ different colors from which we wish to make non-one-color necklaces of

exactly $p$, $p \le n$, beads. The number $p$ is to be a prime. Then:

i) there are $n^p - n$ linear $p$ length strings of non-one-color beads;

ii) the number of distinguishable necklaces of the desired type is $\frac{n^p - n}{p}$;

iii) $n^p \equiv n \pmod{p}$ (Fermat's theorem);

iv) $n^p \equiv n \pmod{2p}$ for $p$ odd.

5. Let $n$ be an arbitrary integer larger than 1 and put $N = (n!)^2$. Then:

i) every prime factor of $N+1$ is odd and greater than $n$;

ii) $N+1 \mid N^m + 1$ for $m$ any positive odd integer;

iii) if $p$ is a $4k+3$ prime factor of $N+1$ then $p \mid N^{2k+1}$ and this contradicts Fermat's theorem;

iv) there are infinitely many $4k+1$ primes.

6. Let $p$ be a prime and suppose $(n,p)=1$.

i) if $na \equiv nb \pmod p$ then $a \equiv b \pmod p$;

ii) $n^{p-1}(p-1)! \equiv (p-1)! \pmod p$;

iii) $n^{p-1} \equiv 1 \pmod p$ (Fermat's theorem).

7. Let $a_1, \cdots, a_{\varphi(m)}$ be relatively prime to $m$ and also be incongruent modulo $m$ in pairs. Further, suppose $(n,m)=1$.

i) if $na_i \equiv na_j \pmod m$ then $i=j$;

ii) $n^{\varphi(m)} a_1 \cdots a_{\varphi(m)} \equiv a_1 \cdots a_{\varphi(m)} \pmod m$;

iii) (Euler) $n^{\varphi(m)} \equiv 1 \pmod m$ for $(n,m)=1$;

iv) Fermat's theorem is a special case of (iii).

8. When $a$ is an odd integer

i) $a^2 \equiv 1 \pmod 8$;

ii) $a^{2^{\alpha-2}} \equiv 1 \pmod{2^\alpha}$ for $\alpha > 2$.

9. Define $\chi$ by:

$$\chi(p^\alpha) = \begin{cases} \varphi(p^\alpha) & \text{for } p \text{ an odd prime} \\ & \text{and for } p = 2,\ 0 \le \alpha \le 2; \\ \frac{1}{2}\varphi(p^\alpha) & \text{for } p = 2,\ \alpha > 2. \end{cases}$$

$$\chi(p_1^{\alpha_1} \cdots p_k^{\alpha_k}) = \operatorname{lcm}\{\chi(p_1^{\alpha_1}), \dots, \chi(p_k^{\alpha_k})\}.$$

For $m > 1$, $m$ odd, $(a, m) = 1$ :

i) $a^{\chi(m)} \equiv 1 \pmod m$;

ii) $m$ a prime implies $\varphi(m) \mid m-1$;

iii) $\varphi(m) \mid m-1$ implies $\chi(m) \mid m-1$;

iv) $\chi(m) \mid m-1$ implies $2^{m-1} \equiv 1 \pmod m$;

v) the converses of (iii) and (iv) are false as can be seen by taking $m = 561$ and $m = 341$.

(No example of $\varphi(m) \mid m-1$ for composite $m$ is known.)

10. If $p$ is an odd prime then

$$2^{\frac{p-1}{2}} \equiv 1 \pmod p \text{ or } 2^{\frac{p-1}{2}} \equiv -1 \pmod p.$$

11. i) Suppose $n > 6$, $n = ab$ with $1 < a \le b$. Then:

a) $b < n-3$ ;

b) if $a = b$ then $2a < n-3$ ;

ii) if $n$ is composite and $\ge 6$ then

$\frac{(n-2)!}{n}$ is an even integer.

12. Suppose $(a,m) = 1$ and $p$ is a prime. Then:

i) $ax \equiv 1 \pmod{m}$ has a unique solution modulo $m$ ;

ii) if, in (i), $a \not\equiv \pm 1 \pmod{m}$ then $x \not\equiv \pm 1 \pmod{m}$ and $x \not\equiv a \pmod{m}$ ;

iii) $2 \cdot 3 \cdots (p-2) \equiv 1 \pmod{p}$ ;

iv) (Wilson's theorem) $(p-1)! \equiv -1 \pmod{p}$ ;

v) for $n > 1$, $n$ is a prime if and only if $n \mid (n-1)! + 1$.

13. Wilson's theorem may be used to show that the congruence $x^2 + 1 \equiv 0 \pmod{p}$ is solvable when $p$ is a $4k+1$ prime.

14. Let $p$ be an odd prime and mark $p$ points uniformly spaced on a circle. Let $T$ and $R$ be the sets of all $p$-gons and the regular $p$-gons respectively. Then:

i) the cardinality of $T$ is $\frac{1}{2}(p-1)!$ ;

ii) the cardinality of $R$ is $\frac{1}{2}(p-1)$ ;

iii) the cardinality of $T-R$ is divisible by $p$ ;

iv) $(p-1)! \equiv -1 \pmod{p}$ (Wilson's theorem).

15. (Clement) Let $m, n$ be positive integers.

i) $(m+n-1)! \equiv (-1)^n n!(m-1)! \pmod{m+n}$ ;

ii) $(n!)^2((m-1)!+1)+(n!-1)(n-1)!m \equiv$

$$n!((-1)^n(m+n-1)!+1) \pmod{m+n} ;$$

iii) if $p$ and $p+k$ are odd primes then

$(p,k)=1$, $k$ is even, and

$$(k!)^2((p-1)!+1)+(k!-1)(k-1)!p \equiv 0 \pmod{p(p+k)} ;$$

iv) the converse of (iii) may be false even though $(p,k)=1$ and $k$ is even ;

v) the converse of (iii) is true when $p$ and $p+k$ are prime to $k!$ ;

vi) let $n$ be an odd integer $>1$ ; then

$$4((n-1)!+1)+n \equiv 0 \pmod{n(n+2)}$$

if and only if $n$ and $n+2$ are odd primes.

16. Let $(a,m)=1$ and suppose $s$ is the smallest integer $t$ for which $a^t \equiv 1 \pmod m$. Then if $a^n \equiv 1 \pmod m$, $s \mid n$.

17. i) $a^{m-1} \equiv 1 \pmod m$ for all $a$, $(a,m)=1$, does not imply $m$ is prime, as one can see with

$$m = 561 ;$$

ii) if $a^{m-1} \equiv 1 \pmod m$ for some $a$ such that $a^t \not\equiv 1 \pmod m$ for any $t$, $0<t<m-1$, $t \mid m-1$, then $m$ is prime.

18. $2^p \equiv 1 \pmod{p^2}$ for the primes $p = 1093$ and $p = 3511$.

19. A composite $n$ which divides $2^n - 2$ is called a pseudoprime.

i) 341, 561, and 161 038 are pseudoprimes;

ii) every composite Fermat number $F_n = 2^{2^n} + 1$, $n \geq 0$, is a pseudoprime;

iii) if $n$ is an odd pseudoprime then $2^n - 1$ is a larger one;

iv) (Erdös [1950]) if $n = \frac{2^{2p} - 1}{3}$, where $p$ is a prime $> 3$, then $n$ is a pseudoprime;

v) there are infinitely many odd pseudoprimes.

20. Let $F$ and $G$ be polynomials in $n$ variables with integral coefficients. We say $F$ *is congruent to* $G$ *modulo* $p$, and write $F \equiv G \pmod p$, if respective coefficients in $F$ and $G$ are congruent modulo $p$. We say $F$ *is equivalent to* $G$ *modulo* $p$, and write $F \sim G \pmod p$, if for all integral choices $c_1, \cdots, c_n$ it is true that

$$F(c_1, \cdots, c_n) \equiv G(c_1, \cdots, c_n) \pmod p.$$

we say that $F$ is *reduced mod p* if no variable appears in $F$ to a power larger than $p-1$. Here $p$ will always be a prime.

i-a) $F \equiv G \pmod{p}$ implies $F \sim G \pmod{p}$;

b) the converse of ( a ) is false ;

c) every polynomial $F$ is equivalent mod $p$ to a reduced polynomial $F^*$, where

$$\deg F^* \leq \deg F;$$

d) if $F$ and $G$ are reduced polynomials in one variable then $F \sim G \pmod{p}$ implies

$$F \equiv G \pmod{p};$$

e) if $F$ and $G$ are reduced polynomials in any finite number of variables the implication in (d) is valid ;

ii) let $F$ be a polynomial in $n$ variables and suppose the congruence $F(x_1, \cdots, x_n) \equiv 0 \pmod{p}$ has exactly one solution $(x_1, \cdots, x_n) = (a_1, \cdots, a_n)$ modulo $p$ ( i.e. the components are taken modulo $p$ ).

Define $H$ and $G$ by:

$$H(x_1,\cdots,x_n) = \prod_{i=1}^{n} (1-(x_i - a_i)^{p-1}),$$

$$G(x_1,\cdots,x_n) = 1 - F^{p-1}(x_1,\cdots,x_n).$$

Then:

a-1) $H(a_1,\cdots,a_n) \equiv 1 \pmod p$;

2) if for some $j$, $1 \le j \le n$, $x_j \not\equiv a_j \pmod p$ then $H(x_1,\cdots,x_n) \equiv 0 \pmod p$;

b) $H \sim G \pmod p$;

c) $H \equiv G^* \pmod p$, where $G^*$ is the reduced form of $G$;

d) $\deg H = n(p-1) = \deg G^* \le \deg G = (\deg F)(p-1)$ so $n \le \deg F$;

iii) (Chevalley) if $F$ is a polynomial in $n$ variables with degree smaller than $n$ then the congruence $F(x_1,\cdots,x_n) \equiv 0 \pmod p$ may not have exactly one solution;

iv) if $F$ is a non-constant form in $n$ variables (i.e. if all the terms of $F$ are of the same degree) and if $\deg F < n$ then $F(x_1, \cdots, x_n) \equiv 0 \pmod{p}$ has a non-trivial solution (i.e. a solution with not all $x_j \equiv 0 \pmod{p}$);

v) (Warning) let $F$ be a polynomial in $n$ variables of degree $r$, where $r < n$, and let $p$ be a prime. Suppose $F(x_1, \cdots, x_n) \equiv 0 \pmod{p}$ has exactly $s$ solutions, say $(a_1^{(i)}, \cdots, a_n^{(i)})$, $1 \leq i \leq s$. Then:

a) if $H(x_1, \cdots, x_n) = 1 - F^{p-1}(x_1, \cdots, x_n)$ then the reduced form of $H$, say $H^*$, is

$H^*(x_1, \cdots, x_n) = \sum_{i=1}^{s} \prod_{j=1}^{n} (1 - (x_j - a_j^{(i)})^{p-1})$;

b) the highest degree term in $H^*$ is

$$(-1)^n s\ x_1^{p-1} \cdots x_n^{p-1};$$

c) since $r < n$ and degree $H < r(p-1)$ it must be true that $p \mid s$;

d) if $F$ is a polynomial in $n$ variables with $\deg F < n$ then the number of solutions of $F(x_1, \cdots, x_n) \equiv 0 \pmod{p}$ is divisible by $p$.

vi) By careful examination of (v-a) one may prove, as in (v), the following theorem.

If $F$ is a polynomial in $n$ variables with $\deg F < n$ and $(a_1^{(i)}, \cdots, a_n^{(i)})$, $1 \le i \le s$, are all the solutions of $F(x_1, \cdots, x_n) \equiv 0 \pmod p$ then for each pair $j, k$ $(1 \le j \le n,\ 0 \le k \le p-2)$ the prime $p$ divides the sum $\sum_{i=1}^{s} (a_j^{(i)})^k$.

vii) Let $F_1, \cdots, F_m$ be polynomials in $n$ variables with respective degrees $r_1, \cdots, r_m$ $r_1 + \cdots + r_m < n$. Suppose, further, the system

(*) $F_1(x_1, \cdots, x_n) \equiv 0 \pmod p, \cdots, F_m(x_1, \cdots, x_n) \equiv 0 \pmod p$

has at least one solution. Then:

a) the system has at least two solutions;

b) the number of solutions of (*) is divisible by $p$.

Remarks.

1. The argument in #4 was given by Golomb [1956].

2. In respect to #9(iii), as we observed, no example of a composite m for which $\varphi(m) \mid m-1$ is known. However in 1932 Lehmer showed that such an m would have to be odd, squarefree, and have at least 7 prime factors. The 7 has since been raised to 11. If, in addition, one assumes 3 divides m then m must have at least 212 prime factors. (See Lieuwens [1970])

3. Gauss in his *Disquisitiones Arithmeticae* had the following to say about Wilson's theorem (see #12, 14).

> It was first published by Waring and attributed to Wilson ... But neither of them was able to prove the theorem,

> and Waring confessed that the demonstration was made more difficult because no *notation* can be devised to express a prime number. But in our opinion truths of this kind should be drawn from the ideas involved rather than from notation.

The proof of Wilson's theorem in #14 goes back to the Danish mathematician J. Peterson who proved it in this way in 1872. The English mathematician A. Cayley, apparently independently, gave a similar proof about 10 years later. (See Dickson's *History* V. I pp. 75-6.)

4. The result in #15 (vi) is due to Clement [1949] and that in #15 (iii) to Tkačev and Shinzel (see MR 32 #1159, erratum p. 1754). There has been considerable work on related problems (see LeVeque [1974] V. 1 A50).

5. Primes $p$ with $2^{p-1} \equiv 1 \pmod{p^2}$, see #18, are of interest in connection with Fermat's last theorem (do there exist integers $x, y, z$ with $xyz \neq 0$ and $x^n + y^n = z^n$ for $n > 2$) since in 1907 Wieferich showed that if $p$ is a prime and $x^p + y^p = z^p$, $xyz \neq 0$, then $p$ satisfies this congruence. For more recent information and further references see Brillhart, Tomascia, Weinberger [1971].

6. For further information on pseudo-primes see Beeger [1951], LeVeque [1974 V.1 A18], and Rotkiewicz [1972].

7. Further extensions of the Chevalley-Warning theorems, see #20, may be found in Borevich, Schafarevich [1966].

## X Divisibility Criteria

Let $S_k(n)$ be the base $k$ digit sum of $n$.

1\. i) $3 \mid n - S_{10}(n)$ and, therefore,

$3 \mid n$ if and only if $3 \mid S_{10}(n)$;

ii) $9 \mid n - S_{10}(n)$ and, therefore,

$9 \mid n$ if and only if $9 \mid S_{10}(n)$;

iii) suppose $d \mid k-1$; then $d \mid n - S_k(n)$, and, therefore, $d \mid n$ if and only if $d \mid S_k(n)$.

2\. (Alvis) Let $p$ be a prime larger than 7. Then:

i) $(6, S_7(p)) = 1$;

ii) the smallest $p$ with composite $S_7(p)$ is 4801;

iii) for $p < 100\,000$ the only possible composite value of $S_7(p)$ is 25.

3\. Let $E_k(n)$ ($O_k(n)$) be the sum of the digits of the even (odd) powers of $k$ in the base $k$

expansion of $n$. Then:

i) $11 \mid n - (E_{10}(n) - O_{10}(n))$ and, therefore,
$11 \mid n$ if and only if $11 \mid E_{10}(n) - O_{10}(n)$;

ii) suppose $d \mid k+1$; then $d \mid n - (E_k(n) - O_k(n))$ and, therefore, $d \mid n$ if and only if
$d \mid E_k(n) - O_k(n)$.

4. Given $n$ write $Q(n)$, $R(n)$ for the quotient and remainder obtained when one divides $n$ by 1000. Thus $n = 1000\,Q(n) + R(n)$, $0 \le R(n) < 1000$. Then:

i) $Q(n) = \left[\frac{n}{1000}\right]$, $R(n) = n - 1000\,Q(n)$;

ii) if $c = 7, 11$, or $13$ then $c \mid n$ if and only if
$c \mid Q(n) - R(n)$;

iii) the above leads to a workable divisibility criterion for determining the divisibility of a number exceeding 1000 by 7, 11, or 13.

5. Let $T_k(n) = \frac{n - S_k(n)}{k-1}$, where $S_k(n)$ is as above. Then:

i) $T_k(n)$ is an integer;

ii) if $k$ is prime then $T_k(n)$ is the highest power of $k$ dividing $n!$;

iii) the highest power of 2 in $n!$ is $n - \nu$, where $\nu$ is the number of 1's in the base 2 expansion of $n$;

iv) if $k$ is a prime and $n = a_0 + a_1 k + \cdots + a_s k^s$, $0 \le a_j < k$, then $k \mid \frac{n!}{(-k)^{T_k(n)}} - a_0! \cdots a_s!$.

## XI Squares

1. The following equalities are special cases of a simple algebraic identity.

$$3^2 + 4^2 = 5^2$$

$$10^2 + 11^2 + 12^2 = 13^2 + 14^2$$

$$21^2 + 22^2 + 23^2 + 24^2 = 25^2 + 26^2 + 27^2$$

$$36^2 + 37^2 + 38^2 + 39^2 + 40^2 = 41^2 + 42^2 + 43^2 + 44^2.$$

2. (Sprague)

i) 128 is not a sum of unequal squares;

ii) if $129 \leq n \leq 192$ then $n$ is a sum of unequal squares all $\leq 10^2$;

iii) if $129 \leq n \leq 256$ $(= 1^2 + \cdots + 10^2 - 129)$ then $n$ is a sum of unequal squares all $\leq 10^2$;

iv) if $129 \leq n \leq 256 + 11^2$ then $n$ is a sum of unequal squares all $\leq 11^2$;

v) if $129 \leq n \leq 256 + 11^2 + 12^2$ then $n$ is a sum of unequal squares all $\leq 12^2$;

vi) every integer larger than 128 is a sum of unequal squares.

3. Let $C$ be the unit circle with center at the origin and let $L_\lambda$ be the straight line of slope $\lambda$ passing through $(-1,0)$. Further, let $C'$ be $C$ with the point $(-1,0)$ removed and let $P_\lambda$ be the intersection of $C'$ and $L_\lambda$. Then:

i) as $\lambda$ runs over all rational numbers the point $P_\lambda$ runs in a one to one fashion over all points of $C'$ both coordinates of which are rational; in fact, the correspondence is

$$\lambda \leftrightarrow \left(\frac{1-\lambda^2}{1+\lambda^2}, \frac{2\lambda}{1+\lambda^2}\right);$$

ii) if $x, y, z$ are non-zero integers with gcd unity and if $x^2+y^2=z^2$ then there exist relatively prime integers $u, v$ of opposite parity, such that either

$$x = v^2-u^2, \quad y = 2uv, \quad z = u^2+v^2$$

or the same expressions with $x$ and $y$ interchanged.

4. The sum of 2 odd squares is never a square.

5. (Thue) Suppose $p$ is a prime not dividing $a$ and $A = \{(m,n) \mid 0 \le m < \sqrt{p}, 0 \le n < \sqrt{p}\}$. Then:

i) there are distinct elements of $A$, say $(m,n)$ and $(m',n')$, such that $am + n \equiv am' + n' \pmod{p}$;

ii) there is an element of $A$, say $(x,y)$ such that $xy \ne 0$ and either $ay \equiv x \pmod{p}$ or $ay \equiv -x \pmod{p}$, i.e. there exist $x, y$ such that $0 < x < \sqrt{p}$, $0 < y < \sqrt{p}$, $ay \equiv \pm x \pmod{p}$.

6. (Generalization - Vinogradoff, Scholz - Shoenberg)

i) If $(a,m) = 1$ and $e, f$ are integers larger than 1 satisfying $e \le m < ef$, $f \le m < ef$ then there exist $x, y$ such that $0 < x < e$, $0 < y < f$,

$$ay \equiv \pm x \pmod{m};$$

ii) #5 (ii) is a special case of (i).

7. (Fermat) As we know from IX #13, when $p$ is a $4k+1$ prime, there is an $a$ such that $a^2+1 \equiv 0 \pmod{p}$; selecting such an $a$ and then choosing $x, y$ as in #5(ii) we conclude $x^2+y^2=p$; thus every $4k+1$ prime is the sum of two squares.

8. i) If $p$ is a $4k+3$ prime then $p=x^2+y^2$ is not solvable in integers $x, y$;

ii) if $p$ is an odd prime then $p$ is representable as a sum of two squares if and only if

$$p \equiv 1 \pmod{4}.$$

9. Let $p$ be an odd prime and suppose $(a,b)=1$, $a^2+b^2 \equiv 0 \pmod{p}$. Then:

i) for all $u, v$

$$(au+bv)^2+(av-bu)^2 \equiv 0 \pmod{p};$$

ii) $x^2+1 \equiv 0 \pmod{p}$ is solvable;

iii) all odd divisors of a sum of two relatively prime squares are of the form $4k+1$.

10. i) The result in #9(iii) guarantees the existence of infinitely many $4k+1$ primes;

ii) all prime factors of the Fermat numbers $F_n = 2^{2^n}+1$ are of the form $4k+1$ and from this we may also conclude the existence of infinitely many primes of the form $4k+1$.

11. i) The set of positive integers which are sums of two squares is closed under multiplication as can be seen by multiplying out the left side of the congruence in #9(i) and then factoring;

ii) the formulae of #3(ii) may be obtained from the identity implicit in (i).

12. Let the canonical prime factorization of $n$ be given by $n = 2^{\alpha} p_1^{\alpha_1} \cdots p_s^{\alpha_s} q_1^{\beta_1} \cdots q_t^{\beta_t}$, where the $p_i$ are $4k+1$ primes and the $q_i$ are $4k+3$ primes. Then:

i) if $n$ is representable as the sum of 2 squares then all $\beta_j$, $1 \le j \le t$, are even;

ii) if all $\beta_j$, $1 \le j \le t$, are even each of $2^{\alpha}$, $p_1^{\alpha_1}, \cdots, p_s^{\alpha_s}, q_1^{\beta_1}, \cdots, q_t^{\beta_t}$ is a sum of 2 squares and, therefore, $n$ is a sum of 2 squares;

iii) an integer is the sum of 2 squares if and only if its canonical prime factorization contains no $4k+3$ prime to an odd power.

13. We write $n = \boxed{4}$ if $n$ is representable as a sum of 4 squares. Thus $25 = \boxed{4}$ and $30 = \boxed{4}$ since $25 = 0^2 + 0^2 + 0^2 + 5^2$, $30 = 1^2 + 2^2 + 3^2 + 4^2$. The product of two sums of 4 squares is itself a sum of 4 squares, as can be seen by verifying the identity:

$$(a_1^2+a_2^2+a_3^2+a_4^2)(A_1^2+A_2^2+A_3^2+A_4^2) =$$
$$(a_1A_1+a_2A_2+a_3A_3+a_4A_4)^2+(a_1A_2-a_2A_1-a_3A_4+a_4A_3)^2$$
$$+(a_1A_3+a_2A_4-a_3A_1-a_4A_2)^2+(a_1A_4-a_2A_3+a_3A_2-a_4A_1)^2.$$

14. Let $p$ be an odd prime. Then:

i) if $A=\{n^2 \mid 0\le n\le \frac{p-1}{2}\}$, $B=\{-1-m^2 \mid 0\le m\le \frac{p-1}{2}\}$ then there is an element of $A$ which is congruent modulo $p$ to an element of $B$;

ii) there exists an $s$, $0<s<p$, such that $sp=a_1^2+a_2^2+a_3^2+a_4^2$, for suitable $a_1, a_2, a_3, a_4$;

iii) for $s$ and the $a_j$ in (ii), if $s>1$ there exist $A_1, A_2, A_3, A_4$ such that $a_j \equiv A_j \pmod{s}$, $-\frac{1}{2}s < A_j \le \frac{1}{2}s$, $1\le j\le 4$, and, for suitable $r$, $0<r<s$, $rs = A_1^2+A_2^2+A_3^2+A_4^2$;

iv) for $r$ and $s$ as in (ii) & (iii), $rs^2p = \boxed{4}$, where the summands on the right are all congruent to 0 modulo $s^2$ and, therefore, $rp = \boxed{4}$;

v) $p = \boxed{4}$;

vi) every positive integer may be represented as a sum of 4 squares.

15. A triple of integers $x, y, z$ for which $x^2+y^2=z^2$ is called a *Pythagorean triple*. When the integers have greatest common divisor 1 we call the triple *primitive*. A triangle whose side lengths form such a triple is called a *Pythagorean triangle*. It is clear that all Pythagorean triples are integral multiples of primitive triples. Define the three matrices $U, A, D$ by:

$$U=\begin{pmatrix} 1 & 2 & 2 \\ -2 & -1 & -2 \\ 2 & 2 & 3 \end{pmatrix}, \quad A=\begin{pmatrix} 1 & 2 & 2 \\ 2 & 1 & 2 \\ 2 & 2 & 3 \end{pmatrix}, \quad D=\begin{pmatrix} -1 & -2 & -2 \\ 2 & 1 & 2 \\ 2 & 2 & 3 \end{pmatrix}.$$

Show that $(x', y', z')$ is a primitive Pythagorean triple if and only if $(x', y', z') = (3, 4, 5)\mathcal{A}$, where $\mathcal{A}$ is a finite product of matrices each factor of which is one of $U, A, D$. I.e. show that every Pythagorean triple is in the following array where the lines leading to the right from any triple correspond to applying to that triple either the matrix $U$ (for up), $A$ (for across) or $D$ (for down).

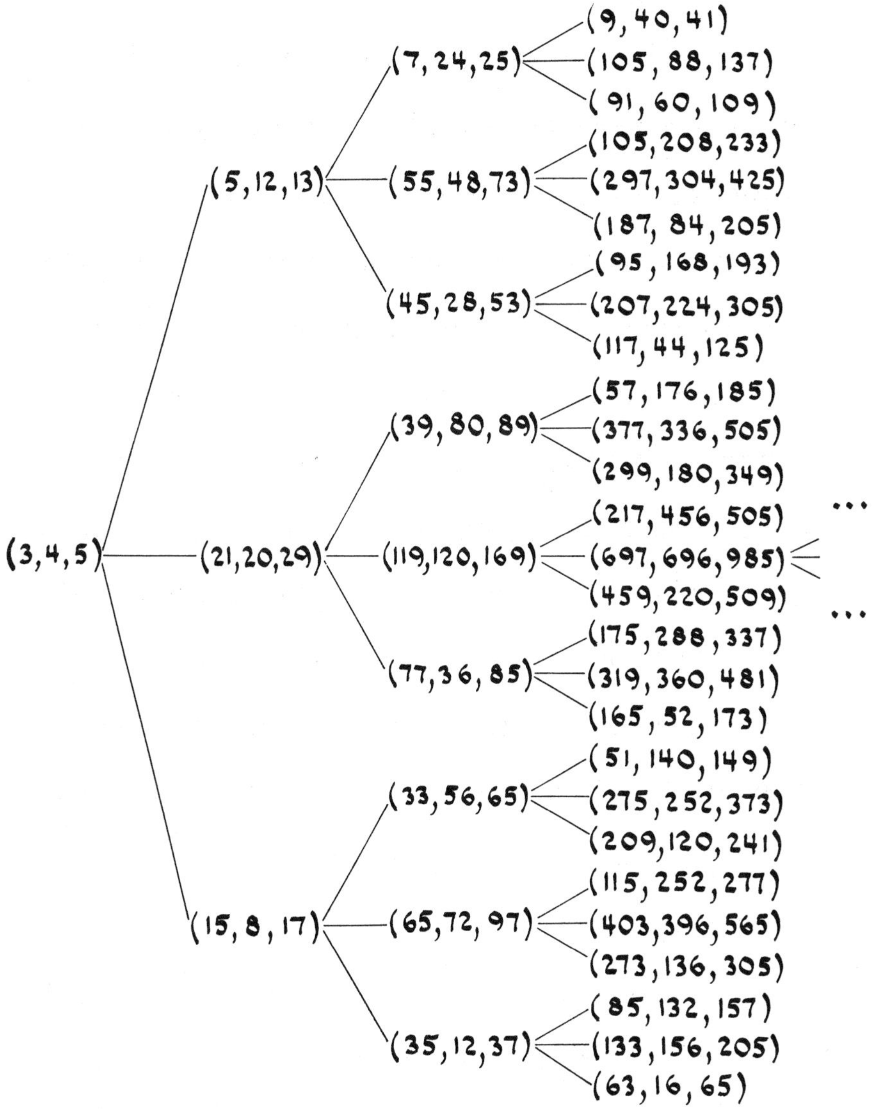
(3,4,5)
(5,12,13)
(21,20,29)
(15,8,17)
(7,24,25)
(55,48,73)
(45,28,53)
(39,80,89)
(119,120,169)
(77,36,85)
(33,56,65)
(65,72,97)
(35,12,37)
(9,40,41)
(105,88,137)
(91,60,109)
(105,208,233)
(297,304,425)
(187,84,205)
(95,168,193)
(207,224,305)
(117,44,125)
(57,176,185)
(377,336,505)
(299,180,349)
(217,456,505)
(697,696,985)
(459,220,509)
(175,288,337)
(319,360,481)
(165,52,173)
(51,140,149)
(275,252,373)
(209,120,241)
(115,252,277)
(403,396,565)
(273,136,305)
(85,132,157)
(133,156,205)
(63,16,65)
...
...

Remarks.

1. Implicit in the solution of #11 is an identity showing that the product of two sums of 2 squares is itself a sum of 2 squares. In #13 there is an identity showing the same thing for the product of two numbers each of which is a sum of 4 squares. There is also an 8 square identity, though, as Hurwitz first proved in 1898, there can be no such identity for values of $n$ other than $n = 1, 2, 4, 8$. (See Curtis [1963].) Dickson [1919] cites Degen as having been the first to give (in 1818) such an 8 square identity. Coxeter [1946] formulates the identity as follows:

"$(a_0^2 + a_1^2 + \cdots + a_7^2)(b_0^2 + b_1^2 + \cdots + b_7^2)$

$= (a_0 b_0 - a_1 b_1 - a_2 b_2 - \cdots - a_7 b_7)^2 +$

$\Sigma(a_0 b_1 + a_1 b_0 + a_2 b_4 + a_3 b_7 - a_4 b_2 + a_5 b_6 - a_6 b_5 - a_7 b_3)^2,$

where the $\Sigma$ implies summation of seven squares given by cyclic permutation of the suffix numbers 1, 2, 3, 4, 5, 6, 7 leaving 0 unchanged."

2. For a number of other interesting results concerning sums of squares see Pall [1933], Taussky [1966, 1970, 1971], Zassenhaus, Eichhorn [1966], and the references at the end of XV.

3. As we showed in #2 (following Sprague [1947-9 (b)]) the largest integer not the sum of unequal squares is 128. (See also Dressler [1972, 1973].) A recent computer proof (see Dressler, Parker [1974]) has been given to show that 12 758 is the largest integer not the sum of unequal cubes. That a similar largest integer exists for $k^{th}$ powers is proved in XII.

4. The result of #6 will be found in Scholz, Shoenberg [1966].

5. The beautiful display of all Pythagorean triples is due to Hall [1970].

## XII Sums of Powers

1. (Tarry) If $b_1^t + \cdots + b_n^t = c_1^t + \cdots + c_n^t$ for all $t$ satisfying $0 \le t \le m$, we write

$$b_1, \cdots, b_n \overset{m}{=} c_1, \cdots, c_n .$$

For example, $1, 4, 6, 7 \overset{2}{=} 2, 3, 5, 8$ since

$$1^0 + 4^0 + 6^0 + 7^0 = 2^0 + 3^0 + 5^0 + 8^0, \quad 1 + 4 + 6 + 7 = 2 + 3 + 5 + 8,$$
$$1^2 + 4^2 + 6^2 + 7^2 = 2^2 + 3^2 + 5^2 + 8^2 .$$

i) If $b_1, \cdots, b_n \overset{m}{=} c_1, \cdots, c_n$ then for all $h$,
$b_1, \cdots, b_n, c_1 + h, \cdots, c_n + h \overset{m+1}{=} c_1, \cdots, c_n, b_1 + h, \cdots, b_n + h$ ;

ii) $b_1, \cdots, b_n \overset{m}{=} c_1, \cdots, c_n$ if and only if for all $x$
$(b_1 + x)^m + \cdots + (b_n + x)^m = (c_1 + x)^m + \cdots + (c_n + x)^m$ ;

iii) for every positive integer $m$ there exists a positive integer $n$ and integers $b_1, \cdots, b_n$, $c_1, \cdots, c_n$ such that $b_1, \cdots, b_n \overset{m}{=} c_1, \cdots, c_n$ .

2. Define a sequence $a_0, a_1, a_2, \cdots$ by:

$$a_n = \begin{cases} 0 & \text{if the base 2 representation of } n \text{ has an even digit sum;} \\ 1 & \text{otherwise.} \end{cases}$$

using #1, starting with $1 \overset{0}{=} 2$, and taking $h$ successively equal to $2, 2^2, 2^3, \ldots$ one obtains

$$1, 4 \overset{1}{=} 2, 3$$

$$1, 4, 6, 7 \overset{2}{=} 2, 3, 5, 8$$

$$1, 4, 6, 7, 10, 11, 13, 16 \overset{3}{=} 2, 3, 5, 8, 9, 12, 14, 15$$

$$1, 4, 6, 7, 10, 11, 13, 16, 18, 19, 21, 24, 25, 28, 30, 31 \overset{4}{=}$$
$$2, 3, 5, 8, 9, 12, 14, 15, 17, 20, 22, 23, 26, 27, 29, 32;$$

and, in general,

$$\sum_{n=1}^{2^{k+1}} (1 - a_{n-1}) n^t = \sum_{n=1}^{2^{k+1}} a_{n-1} n^t, \quad 1 \le t \le k.$$

3. i) With the $a_n$ as in #2 and arbitrary integers $r$ and $s$,

$$\sum_{n=1}^{2^{k+1}} (1 - a_{n-1})(rn+s)^t = \sum_{n=1}^{2^{k+1}} a_{n-1}(rn+s)^t \text{ for } 1 \le t \le k;$$

ii) in (i) we may replace $(rn+s)$ by $P(n)$, where $P$ is any polynomial of degree not exceeding $k$.

4. i) The odd integers from 1 to $2^{k+2}-1$ inclusive may be split into two disjoint equinumerous classes $\{b_1, \cdots, b_{2^k}\}$, $\{b_{2^k+1}, \cdots, b_{2^{k+1}}\}$ so that for all $x$

$$(b_1+x)^k+\cdots+(b_{2^k}+x)^k = (b_{2^k+1}+x)^k+\cdots+(b_{2^{k+1}}+x)^k;$$

ii) for all the $b_j$ of (i) there exist even integers $d_1, \cdots, d_k$ so that no two of the $k \cdot 2^{k+1}$ numbers $b_j + d_i$ are equal, where $1 \le j \le 2^{k+1}$, $1 \le i \le k$;

iii) let $b_j$ and $d_i$ be as in (i) & (ii) and define $L_i$, $R_i$, $1 \le i \le k$, by:

$$L_i = (x+d_i+b_1)^k+\cdots+(x+d_i+b_{2^k})^k$$

$$R_i = (x+d_i+b_{2^k+1})^k+\cdots+(x+d_i+b_{2^{k+1}})^k;$$

then $L_i = R_i$ for each $i$, $1 \le i \le k$, and, further, the $2^k$ products $U_1 \cdots U_k$, where each $U_i$ is either $L_i$ or $R_i$, are equal;

iv) each of the products $U_1 \cdots U_k$ in (iii) is a sum of $k^{th}$ powers of terms of the form $(x+d_1+b_{i_1})\cdots(x+d_k+b_{i_k})$; each $i_j$ satisfies

$$1 \le i_j \le 2^{k+1};$$

further, for $x$ sufficiently large and even, all of these terms are odd and distinct from each other;

v) let $b_j, d_i, x$ be as in (i)-(iv), and put $s = L_1 \cdots L_k$; then $s$ may be written as a sum of odd $k^{th}$ powers in $2^k$ ways, no two of which share a common summand.

5. Let $s$ be a number having $2^k - 1$ completely distinct representations as a sum of odd $k^{th}$ powers, and let these sums of odd $k^{th}$ powers be $S_1, \cdots, S_{2^k-1}$.

i) For each $t$, $0 \le t \le 2^k$, the number $ts$ is a sum of odd $k^{th}$ powers;

ii) if the base $2^k$ representation of the positive integer $m$ is given by

$$m = t_0 + t_1 \cdot 2^k + t_2 \cdot 2^{2k} + \cdots, \quad 0 \le t_i < 2^k,$$

then $ms = t_0 s + t_1 s \cdot 2^k + t_2 s \cdot 2^{2k} + \cdots$ is a sum of $k^{th}$ powers;

iii) In (ii) in the representation of $ms$ as a sum of $k^{th}$ powers no two summands are equal ;

iv) given a positive integer $k$ there is always a positive integer $s$ such that all positive integer multiples of $s$ are sums of unequal $k^{th}$ powers.

6. Setting $S_r = s^k + (s+1)^k + \cdots + (rs+1)^k$, $0 \le r < s$, where $s$ is as in #5(iv), we see that every $S_r$ is a sum of unequal $k^{th}$ powers, and, consequently, since every $s^{2k+1} \ge S_r \equiv r \pmod{s}$ and every integer $\ge s^{2k+1}$ may be written in the form $ms + S_r$, we conclude:

(Sprague) Given a positive integer $k$ there is a positive integer $N$ ($= s^{2k+1}$) for which all larger integers are sums of unequal $k^{th}$ powers; i.e. for each $k$ all sufficiently large integers are representable as a sum of unequal $k^{th}$ powers.

Remarks.

The results in #1-3 go back to Prouhet [1851] and have been generalized considerably in recent years - see Lehmer [1947], Roberts [1964], Wright [1959]. The results in #4-6 are from Sprague [1947-9(b)]. Further information about equal sums of like powers, see #2, may be found in Gloden [1944], Lander, Parkin, Selfridge [1967].

The sum of the products obtained from the product $1\cdot x_0\cdot x_1\cdot \cdots \cdot x_n$ by omitting zero or more disjoint pairs of consecutive factors $x_j x_{j+1}$ from the product is denoted by $E(x_0,\cdots,x_n)$. This quantity, as a function of the $x_j$, is called the *Euler bracket function*.

One sees immediately that:

$E(x_0) = x_0$ ;

$E(x_0,x_1) = x_0x_1+1$ ;

$E(x_0,x_1,x_2) = x_0x_1x_2+x_0+x_2$ ;

$E(x_0,x_1,x_2,x_3) = x_0x_1x_2x_3+x_0x_1+x_0x_3+x_2x_3+1$ ;

$E(x_0,x_1,x_2,x_3,x_4) = x_0x_1x_2x_3x_4+x_0x_1x_2+$

$x_0x_1x_4+x_0x_3x_4+x_2x_3x_4+x_0+x_2+x_4$ .

The number of summands appearing in $E(x_0,\cdots,x_n)$ is denoted by $E_{n+1}$. Thus

$$E_1 = 1 \;;$$

$$E_2 = 2 \;;$$

$$E_3 = 3 \;;$$

$$E_4 = 5 \;;$$

$$E_5 = 8 \;.$$

1. Suppose $n \geq 0$. Then (providing in (*v*), (*vi*), & (*vii*) the presence of an $x_{-1}$ or $x_{n+1}$ in $E(\cdots)$ is interpreted as making that bracket equal to 1):

*i*) $E(x_0, \cdots, x_n) = E(x_n, \cdots, x_0)$ ;

*ii*) for $n \geq 1$,

$$E(x_0, \cdots, x_{n+1}) = x_{n+1} E(x_0, \cdots, x_n) + E(x_0, \cdots, x_{n-1}) \;;$$

*iii*) for $n \geq 2$,

$$E(x_0, \cdots, x_n) E(x_1, \cdots, x_{n-1}) - E(x_0, \cdots, x_{n-1}) E(x_1, \cdots, x_n) = (-1)^{n-1} \;;$$

*iv*) for $n \geq 3$,

$$E(x_0, \cdots, x_n) E(x_1, \cdots, x_{n-2}) - E(x_0, \cdots, x_{n-2}) E(x_1, \cdots, x_n) = (-1)^n x_n \;;$$

v) for $0<s<t<n$,

$$E(x_0,\cdots,x_n)E(x_s,\cdots,x_t)-E(x_0,\cdots,x_t)E(x_s,\cdots,x_n)$$
$$=(-1)^{t-s+1}E(x_0,\cdots,x_{s-2})E(x_{t+2},\cdots,x_n);$$

vi) for $m\geq 0$,

$$E(x_0,\cdots,x_m,x_m,\cdots,x_0)=E^2(x_0,\cdots,x_m)+E^2(x_0,\cdots,x_{m-1});$$

vii) for $m\geq 0$, $E(x_0,\cdots,x_m,x_{m+1},x_m,\cdots,x_0)$

$$=E(x_0,\cdots,x_m)\{E(x_0,\cdots,x_{m-1})+E(x_0,\cdots,x_{m+1})\}.$$

2. i) Putting $E_0=1$ we find

$$E_0=E_1=1,\ E_{n+2}=E_{n+1}+E_n \text{ for } n\geq 0;$$

ii) $E_n=u_n$, where $u_n$ is the $n+1^{st}$ Fibonacci number;

iii) $$u_n=\frac{1}{\sqrt5}\left\{\left(\frac{1+\sqrt5}{2}\right)^{n+1}-\left(\frac{1-\sqrt5}{2}\right)^{n+1}\right\}$$
$$=\frac{1}{2^n}\left\{\binom{n+1}{1}+5\binom{n+1}{3}+5^2\binom{n+1}{5}+\cdots\right\};$$

iv) a) $u_{n-1}u_{n+1}-u_n^2=(-1)^{n-1}$;

b) $u_nu_{n+1}-u_{n-1}u_{n+2}=(-1)^n$;

c) $u_{n+1}u_{t-s+1}-u_{t+1}u_{n-s+1}=(-1)^{t-s+1}u_{s-1}u_{n-t-1}$, for $0<s<t<n$;

d) $u_m^2+u_{m+1}^2=u_{2m+2}$ for $m\geq 0$;

e) $u_{m+1}(u_m + u_{m+2}) = u_{2m+3}$ for $m \geq 0$ ;

f) $u_{n-3}\,u_{n-1} + u_{n-2}(3u_{n-1} + u_{n-2}) = u_n^2$ ;

g) $u_{n-3}^2 + u_{n-2}(3u_{n-3} + u_{n-4}) = u_{n-1}^2$ ;

v) Define the sequence $a_0, a_1, a_2, \cdots$ by :

$a_0 = a$ , $a_1 = b$ , $a_{n+2} = a_n + a_{n+1}$ for $n \geq 2$ .

Then $a_n = u_{n-2}a + u_{n-1}b$ for $n \geq 2$ .

Given an arbitrary infinite sequence $a_0, a_1, a_2, \cdots$ of real numbers such that $a_j \neq 0$ for $j \geq 1$, we define two new infinite sequences

$$p_{-2}, p_{-1}, p_0, p_1, \cdots \text{ \& } q_{-2}, q_{-1}, q_0, q_1, \cdots$$

as follows :

$p_{-2} = 0$ , $p_{-1} = 1$ , $p_m = E(a_0, \cdots, a_m)$ for $m \geq 0$ ;

$q_{-2} = 1$ , $q_{-1} = 0$, $q_0 = 1$ , $q_m = E(a_1, \cdots, a_m)$ for $m \geq 1$ .

Noting that $p_m$ and $q_m$ for $m \leq n$ depend only on the first $n+1$ terms of the $a_j$ sequence we see that $a_0, \cdots, a_n$ determine $p_{-2}, \cdots, p_n$ and $q_{-2}, \cdots, q_n$ .

We write $[a_0, \cdots, a_n]$ for the (finite) continued fraction

$$a_0 + \cfrac{1}{a_1 + \cfrac{1}{a_2 + \cfrac{1}{a_3 + \ddots + \cfrac{1}{a_n}}}},$$

and write $[a_0, a_1, a_2, \cdots]$ for the *infinite continued fraction*

$$a_0 + \cfrac{1}{a_1 + \cfrac{1}{a_2 + \cfrac{1}{a_3 + \cdots}}}.$$

It should be noted that while it is clear that a finite continued fraction always denotes a real number in an obvious way it is not at all clear that an infinite continued fraction denotes a real number in any reasonable way. In certain cases we will find that $\lim_{n\to\infty}[a_0, \cdots, a_n]$ exists and, in those cases, we shall denote the limit by $[a_0, a_1, a_2, \cdots]$ and call this fraction *convergent.*

When an infinite continued fraction does not converge we call it *divergent*. For each finite or infinite sequence of $a_j$ the quantities $\frac{p_m}{q_m}$ are called *convergents* to the corresponding finite or infinite continued fraction. The reason for this terminology will become clear in problem #3 below. Thus if $\alpha = [a_0, \cdots, a_n]$ or $\alpha = \lim_{n\to\infty} [a_0, a_1, a_2, \cdots]$, it being assumed in the second case that the limit exists, we call the $\frac{p_m}{q_m}$ *convergents to* $\alpha$. The $a_j$ themselves are often referred to as the *partial quotients* of the continued fraction. The continued fractions introduced thus far are often referred to as *simple* (or *regular*) *continued fractions*. Until we generalize the notion every reference to a continued fraction should be read as a reference to a simple continued fraction. In the following, unless stated to the contrary, all $a_j$, $j \geq 1$, are to be positive.

3. In this problem in every context in which $\alpha$ and $p_j$ or $q_j$ occur together it is to be presumed that $\frac{p_j}{q_j}$ is an existent convergent to $\alpha$.

i-a) $[a_0, \cdots, a_n] = \frac{p_n}{q_n}$ for $n \geq 0$;

b) $[a_n, \cdots, a_1] = \frac{q_n}{q_{n-1}}$ for $n \geq 1$;

ii) $\left.\begin{array}{l} p_k = a_k p_{k-1} + p_{k-2} \\ q_k = a_k q_{k-1} + q_{k-2} \end{array}\right\}$ for $k \geq 0$;

iii) $[a_0, \cdots, a_{n-1}, a_n + \frac{1}{b}] = \dfrac{b p_n + p_{n-1}}{b q_n + q_{n-1}}$;

iv-a) for $n \geq 0$, $p_n q_{n-1} - p_{n-1} q_n = (-1)^{n-1}$

and $\left| \frac{p_n}{q_n} - \frac{p_{n-1}}{q_{n-1}} \right| = \dfrac{1}{q_n q_{n-1}}$;

b) for $n \geq 0$, $p_n q_{n-2} - p_{n-2} q_n = (-1)^n a_n$

and $\left| \frac{p_n}{q_n} - \frac{p_{n-2}}{q_{n-2}} \right| = \dfrac{|a_n|}{q_n q_{n-2}}$;

v) $\frac{p_0}{q_0} < \frac{p_2}{q_2} < \frac{p_4}{q_4} < \cdots < \frac{p_{2m}}{q_{2m}} < \cdots < \alpha < \cdots$

$< \frac{p_{2n+1}}{q_{2n+1}} < \cdots < \frac{p_5}{q_5} < \frac{p_3}{q_3} < \frac{p_1}{q_1}$;

vi) for integral $a_j$, the sequence $\frac{p_{n-1}}{q_{n-1}}, \frac{p_{n-1}+p_n}{q_{n-1}+q_n}, \frac{p_{n-1}+2p_n}{q_{n-1}+2q_n}, \cdots, \frac{p_{n-1}+a_{n+1}p_n}{q_{n-1}+a_{n+1}q_n} \left(= \frac{p_{n+1}}{q_{n+1}}\right), \alpha, \frac{p_{n+1}+p_n}{q_{n+1}+q_n}, \frac{p_n}{q_n}$ is monotone and, therefore, $a_{n+1}$ is the largest positive integer $t$ for which $\frac{p_{n-1}}{q_{n-1}}$ and $\frac{p_{n-1}+tp_n}{q_{n-1}+tq_n}$ are on the same side of $\alpha$;

vii) $\frac{1}{q_n(q_n+q_{n+1})} < \left|\alpha - \frac{p_n}{q_n}\right| < \frac{1}{q_n q_{n+1}}$ ;

viii) a) $\left|\alpha - \frac{p_n}{q_n}\right| < \left|\alpha - \frac{p_{n-1}}{q_{n-1}}\right|$ ;

b) $|q_n\alpha - p_n| < |q_{n-1}\alpha - p_{n-1}|$ ;

ix) $\lim_{n\to\infty} \frac{p_{2n}}{q_{2n}}$ and $\lim_{n\to\infty} \frac{p_{2n+1}}{q_{2n+1}}$ exist but are equal if and only if $q_n q_{n+1} \to \infty$ as $n \to \infty$ ;

x) $\lim_{n\to\infty} [a_0, \cdots, a_n]$ exists if and only if $q_n q_{n+1} \to \infty$ as $n \to \infty$ ;

xi) $\frac{p_n}{q_n} = a_0 + \frac{1}{q_0 q_1} - \frac{1}{q_1 q_2} + \frac{1}{q_2 q_3} - \cdots + \frac{(-1)^{n-1}}{q_{n-1}q_n}$ , for $n \geq 1$ ;

xii) if all $a_j$ are integers and $a_j > 0$ for $j \geq 1$ then:

a) $p_j$ and $q_j$ are relatively prime integers ;

b) $|p_n|$ and $q_n$ tend to infinity with $n$ ;

c) $q_n \geq 2^{\frac{n-1}{2}}$ for $n \geq 2$ ;

d) $\lim_{n\to\infty} [a_0, \cdots, a_n]$ exists ;

xiii) in the following all $a_j, b_j$, and $c_j$ are integers and, for $j \geq 1$, are positive integers.

a) If $a_j = b_j$, for $0 \leq j \leq n$, $a_n < b_n$, and $\alpha = [a_0, a_1, \cdots]$, $\beta = [b_0, b_1, \cdots]$ then $\alpha < \beta$ when $n$ is even and $\alpha > \beta$ when $n$ is odd ;

b) if $c_j \le a_j \le b_j$ for $j \ge 0$ then, for $\alpha = [a_0, a_1, \cdots]$,
$[c_0, b_1, c_2, b_3, c_4, b_5, \cdots] \le \alpha \le [b_0, c_1, b_2, c_3, b_4, c_5, \cdots]$;

c) if $\alpha = [a_0, a_1, \cdots]$, where every $a_j$ is 1 or 2,
then $\frac{1+\sqrt{3}}{2} \le \alpha \le 1 + \sqrt{3}$.

4. If $\alpha$ is a rational real number the Euclidean algorithm may be used to compute integers $n, a_0, \cdots, a_n$ with $a_j > 0$, $j \ge 1$, for which $\alpha = [a_0, \cdots, a_n, 1] = [a_0, \cdots, a_{n-1}, a_n + 1]$. Further, no continued fraction with integral $a_j$ other than these two is equal to $\alpha$.

5. If $\alpha$ is an arbitrary irrational real number there exist integers $a_0, a_1, \cdots$ with $a_j > 0$, $j \ge 1$, for which $\alpha = \lim_{n \to \infty} [a_0, a_1, \cdots, a_n]$.

Further, this sequence is unique.

Remark. Continued fraction expansions of the form $[a_0, a_1, a_2, \cdots]$ (finite or infinite) in which all $a_j$ are integers and all $a_j$, $j \geq 1$, are positive integers are of particular interest. In the following, whenever one refers to the continued fraction expansion of a real number it is that simple continued fraction expansion in which all $a_j$ are integers and all $a_j$, $j \geq 1$, are positive integers. We shall use the abbreviation scf for "simple continued fraction".

6. i) Using the recurrence relations in #3 (ii) devise a simple scheme for the rapid computation of the convergents of a continued fraction.

ii) Use the scheme of (i) to compute all the convergents to $[2, 2, 1, 3, 1, 1, 4, 3]$. (Note that the last convergent equals this continued fraction.)

iii) Use the Euclidean algorithm to find the scf expansions of $\frac{2227}{9911}$ and $\frac{34453}{10349}$ .

iv) Reduce the fractions in (iii) to lowest terms by using the results of (iii), (i) and #3(xii-a).

v) Compute $a_j$, $0 \le j \le 5$, for the scf expansion of $\pi$ and find the 1st five convergents $\frac{p_j}{q_j}$, $0 \le j \le 4$. Then using $\frac{p_4}{q_4} < \pi < \frac{p_3}{q_3}$ , show $\left|\pi - \frac{p_3}{q_3}\right| < 3 \cdot 10^{-7}$.

vi) Using the results of (v) for $\frac{p_0}{q_0}, \frac{p_1}{q_1}$ compute $\frac{p_2}{q_2}$ by making use of #3(vi).

vii) Find the scf expansion of the Golden mean $\frac{1+\sqrt{5}}{2}$ .

viii) Compute $\alpha$ in more familiar terms if $\alpha = [a, a, 2a, a, 2a, \cdots] = [a, \overline{a, 2a}]$ , $a > 0$.

ix) For $a$ a positive integer, expand $\alpha = \sqrt{a^2 - 2}$ in a simple continued fraction; use the result to compute the scf expansion of $\sqrt{23}$ .

x) $[2, \underbrace{1, \cdots, 1}_{n-3 \text{ 1's}}, 3, \underbrace{1, \cdots, 1}_{n-2 \text{ 1's}}] = \left(\frac{u_n}{u_{n-1}}\right)^2$ .

7. (Seidel's convergence theorem of 1846)

Let $a_0, a_1, a_2, \cdots$ be an infinite sequence of positive real numbers with the possible exception of $a_0$, which may be negative, and let $\frac{p_j}{q_j}$ be the typical convergent of the infinite continued fraction $[a_0, a_1, a_2, \cdots]$.

i) Suppose $\sum_{n=1}^{\infty} a_n$ converges. Then:

a) if $a_k < 1$ either

$$q_k < \frac{q_{k-1}}{1-a_k} \quad \text{or} \quad q_k < \frac{q_{k-2}}{1-a_k} \; ;$$

b) there is a $k_0$ and an $A$ such that

$$q_k < q_s(1-a_{i_1})^{-1} \cdots (1-a_{i_r})^{-1} \le A\Big(\prod_{i=k_0}^{\infty}(1-a_i)^{-1}\Big),$$

when $s \le k_0$ and $k = i_1 > \cdots > i_r > k_0$;

c) $\lim_{n\to\infty}[a_0, \cdots, a_n]$ does not exist.

ii) Suppose $\sum_{n=1}^{\infty} a_n$ diverges. Put $c = \min\{q_0, q_1\}$.

Then: a) $q_k \ge c$ for $k \ge 0$;

b) $q_k \ge q_{k-2} + c a_k$ for $k \ge 1$;

c) $q_k + q_{k-1} > c \sum_{n=1}^{k} a_n$ for $k \ge 3$;

d) $q_k q_{k-1} > \frac{c^2}{2} \sum_{n=1}^{k} a_n$ for $k \ge 3$;

e) $\lim_{n\to\infty}[a_0, \cdots, a_n]$ exists.

iii) (Seidel) $\lim_{n\to\infty}[a_0, \cdots, a_n]$ exists if and only if $\sum_{n=1}^{\infty} a_n$ diverges;

iv) $\lim_{n\to\infty}[a_0, \cdots, a_n]$ always exists when the $a_j$ are all integral, $a_j > 0$ for $j \geq 1$.

8. (Best approximations of 1st kind)

A fraction which is closer to a real number $\alpha$ than every other fraction with equal or smaller denominator is called a *best approximation of first kind to* $\alpha$. If $\frac{a}{b}$ is a best approximation of first kind to $\alpha$ we will write "$\frac{a}{b}$ is a BA1 to $\alpha$".

i) Suppose $a, b, c, d, x, y$ are positive integers. Then

a) if $\frac{a}{b} < \frac{x}{y} < \frac{c}{d}$ then $x \geq \frac{a+c}{bc-ad}$ and $y \geq \frac{b+d}{bc-ad}$;

b) if $\frac{a}{b} < \alpha < \frac{c}{d}$, $bc-ad=1$, then at least one of $\frac{a}{b}, \frac{c}{d}$ is a BA1 to $\alpha$ and if one of $\frac{a}{b}, \frac{c}{d}$ is closer to $\alpha$ that one is a BA1 to $\alpha$.

ii) Every convergent to $\alpha$ is a BA1 to $\alpha$.

iii) Find 5 BA1's to $\pi$.

iv) The quotients of consecutive Fibonacci numbers are convergents and, therefore, BA1's to the Golden Mean $\frac{1+\sqrt{5}}{2}$.

v) If $\frac{p_{j-1}}{q_{j-1}}$ and $\frac{p_{j+1}}{q_{j+1}}$ are convergents to a real number $\alpha$ and if $\frac{a}{b}$, $(a,b)=1$, lies between them then $b > q_j$.

9. (Farey fractions) Let $\Phi_n$ be the sequence of all irreducible fractions in $[0,1]$, whose denominators do not exceed $n$, arranged in order of magnitude. This ordered sequence $\Phi_n$ is called the *Farey sequence of order* $n$.

i) Write out $\Phi_n$ for $1 \leq n \leq 6$.

ii) The union of the $\Phi_n$, as sets rather than as ordered sets, taken over all positive integers $n$, is precisely the set of rational numbers in $[0,1]$.

*iii*) Let $\frac{a}{b}$ be in $\Phi_n$ and suppose $\frac{a}{b} \neq 1$. Then:

a) there is a $y_0$ such that

$$ay_0 \equiv -1 \pmod{b} \text{ and } n-b < y_0 \leq n;$$

b) for $y_0$ as in (*i*) and $x_0 = \frac{ay_0+1}{b}$ the fraction next following $\frac{a}{b}$ in $\Phi_n$ is $\frac{x_0}{y_0}$;

c) it is easy to find the fraction following $\frac{79}{101}$ in each of $\Phi_{101}$ and $\Phi_{200}$;

d) if $\frac{a}{b}$ and $\frac{c}{d}$ are irreducible fractions in $[0,1]$ with $bc-ad=1$ then $\frac{a}{b}$ and $\frac{c}{d}$ are consecutive elements in $\Phi_m$, where $m=\max\{b,d\}$.

*iv*) If $\frac{a}{b} < \frac{c}{d}$ and these are neighboring (i.e. consecutive) fractions in $\Phi_n$ then

a) $b+d > n$ ;

b) $bc-ad=1$ ;

c) $(b,d)=1$ ;

d) $b \neq d$ when $n>1$ .

*v*) If $\alpha$ is between neighboring elements of $\Phi_n$ then at least one of these neighboring elements is a BA1 to $\alpha$.

vi) If $\frac{a}{b}$, $\frac{x}{y}$, $\frac{c}{d}$ are consecutive elements in $\Phi_n$ then:

a) $bc - ad = (a+c, b+d)$ ;

b) $\frac{x}{y} = \frac{a+c}{b+d}$ ;

(The fraction $\frac{a+c}{b+d}$ appearing in (b) is called the *Farey mediant* of the fractions $\frac{a}{b}$ and $\frac{c}{d}$. It will be noted that the Farey mediant of two fractions always lies between them.)

vii) All fractions in $\Phi_{n+1} \setminus \Phi_n$ are Farey mediants of fractions in $\Phi_n$.

viii) Write out $\Phi_7$ using (i) & (vii).

ix) Let $\frac{a}{b}$ and $\frac{c}{d}$ be neighboring fractions in $\Phi_n$ and suppose $\frac{a}{b} < \alpha < \frac{c}{d}$. Then:

a) if an element of $\Phi_{n+1} \setminus \Phi_n$ is a BA1 to $\alpha$ then that element is equal to $\frac{a+c}{b+d}$ ;

b) if $\frac{a+c}{b+d}$ is in $\Phi_{n+1} \setminus \Phi_n$ it is a BA1 to $\alpha$ if and only if it is closer to $\alpha$ than each of $\frac{a}{b}$ and $\frac{c}{d}$.

x) Find all the BA1's to $\pi$ (approximately equal to 3.14159) with denominators not exceeding 125.

10. i) Let $\alpha$ lie between neighboring fractions $\frac{a}{b}$ and $\frac{c}{d}$ in $\Phi_n$, $n>1$. Then at least one of the following inequalities is true.

$$\left|\alpha-\frac{a}{b}\right|<\frac{1}{2b^2}, \quad \left|\alpha-\frac{c}{d}\right|<\frac{1}{2d^2}.$$

ii) If $\alpha$ is a real number and $m$ is a positive integer then there are relatively prime integers $s$ and $t$ such that $\left|\alpha-\frac{s}{t}\right|\le\frac{1}{t(m+1)}$, $0<t\le m$.

11. (H.J.S. Smith proof of Fermat's 2 square theorem)

i) With the sole exception of 1, every rational number has a simple continued fraction expansion with last partial quotient $>1$;

ii) Let $p$ be an odd prime and let $1\le t\le s$, where $s=\left[\frac{p}{2}\right]$. Then there are positive integers $a_0,\cdots,a_n$ such that $a_0>1$, $a_n>1$ and

$$\begin{aligned}\frac{p}{t}&=[a_0,\cdots,a_n]\\&=\frac{E(a_0,\cdots,a_n)}{E(a_1,\cdots,a_n)}\end{aligned};$$

iii) Let $p$ and s be as in (ii). Then there are exactly s finite sequences $a_0, \cdots, a_n$ with $a_0 > 1$, $a_n > 1$ and $p = E(a_0, \cdots, a_n)$. Further, $a_n, \cdots, a_0$ is one of these sequences when $a_0, \cdots, a_n$ is;

iv) when $p \equiv 1 \pmod 4$ then s is even and, since $p = E(p)$, there exist $a_0, \cdots, a_n$, $a_0 > 1$, $a_n > 1$ such that $p = E(a_0, \cdots, a_n)$ and $a_j = a_{n-j}$ for $0 \leq j \leq n$; i.e $a_0, \cdots, a_n$ is palindromic;

v) $p = E(a_0, \cdots, a_m, a_{m+1}, a_m, \cdots, a_0)$ in (iv) is impossible and, therefore, $p = a^2 + b^2$ for suitable a and b, when $p \equiv 1 \pmod 4$;

vi) The above proof is constructive. Use it to represent 13 as a sum of 2 squares.

12. i) If in #10 (ii) we take $\alpha = \frac{a}{b}$, $(a, b) = 1$, $m = [\sqrt{b}]$ then the s and t of that result satisfy:

a) $0 < (at - bs)^2 + t^2 < 2b$;

b) $(at - bs, t) = 1$ when b divides $a^2 + 1$;

c) $0 < t \leq \sqrt{b}$;

*ii*) Every divisor of a number of the form $a^2+1$ is a sum of relatively prime squares ;

*iii*) ( Fermat's 2 square theorem again )

Wilson's theorem guarantees that every prime of the form $4k+1$ divides a number of the form $a^2+1$ and , therefore, by (*ii*) must be a sum of relatively prime squares ;

*iv*) If $(a,c)=1$ and $b \mid a^2+c^2$ then there is an integer $u$ such that $b$ divides $(au)^2+1$ ;

*v*) Every divisor of a sum of relatively prime squares is itself such a sum and if it is respectively odd , even must then be of the form $4k+1$ , $4k+2$ .

13. ( Best approximations of 2nd kind )

If $a$ and $b$ are relatively prime positive integers and $|\alpha b-a|<|\alpha d-c|$ for all pairs $c,d$ of relatively prime positive integers satisfying $d \leq b$ and $\frac{c}{d} \neq \frac{a}{b}$ then $\frac{a}{b}$ is called a *best* approximation of the $2^{nd}$ kind to $\alpha$. We abbreviate this last phrase by writing "$\frac{a}{b}$ is a BA2 to $\alpha$".

*i*) Every BA2 to $\alpha$ is a BA1 to $\alpha$;

*ii*) The converse of (*i*) is false;

*iii*) Let $\frac{a}{b}$ be a BA2 to an irrational number $\alpha$. Clearly either $\frac{a}{b}$ is a convergent to $\alpha$ or it lies in one of the open intervals (see diagram) determined by the convergents to $\alpha$.

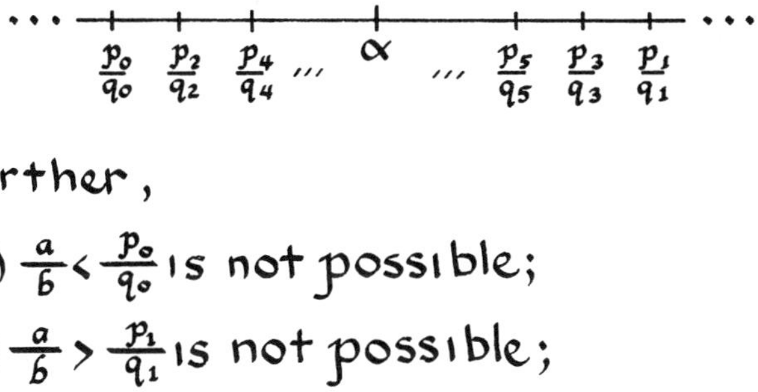


Further,

a) $\frac{a}{b} < \frac{p_0}{q_0}$ is not possible;

b) $\frac{a}{b} > \frac{p_1}{q_1}$ is not possible;

c) if $\frac{a}{b}$ is strictly between $\frac{p_{k-1}}{q_{k-1}}$ and $\frac{p_{k+1}}{q_{k+1}}$ then $b > q_k$, $\left|\alpha - \frac{a}{b}\right| \geq \left|\frac{a}{b} - \frac{p_{k+1}}{q_{k+1}}\right| \geq \frac{1}{b\,q_{k+1}}$,

$$\text{and } \left|\alpha - \frac{p_k}{q_k}\right| < \frac{1}{q_k q_{k+1}};$$

d) the supposition in (c) is unrealizable;

e) every BA2 to $\alpha$ is a convergent to $\alpha$.

w) Let the $\frac{p_j}{q_j}$ below be the convergents to the irrational number $\alpha$. Then:

a) $|\alpha q_n - p_n| < |\alpha q_{n-1} - p_{n-1}|$ ;

b) $q_n|\alpha q_{n-1} - p_{n-1}| + q_{n-1}|\alpha q_n - p_n| = 1$ ;

c) when $\frac{a}{b} \neq \frac{p_{n-1}}{q_{n-1}}$ then

$$b|\alpha q_{n-1} - p_{n-1}| + q_{n-1}|\alpha b - a| \geq 1 \ ;$$

d) when $1 \leq b \leq q_n$ then

$$b|\alpha q_{n-1} - p_{n-1}| + q_{n-1}|\alpha q_n - p_n| \leq 1 \ ;$$

e) for all positive integers $a, b$ with $1 \leq b \leq q_n$, $|\alpha q_n - p_n| \leq |\alpha b - a|$ ;

f) every convergent to $\alpha$ is a BA2 to $\alpha$ ;

(the simplicity of this argument is due to Drobot [1963].)

v) A fraction is a BA2 to an irrational number $\alpha$ if and only if it is a convergent to $\alpha$.

14. At least one, say $\frac{a}{b}$, of each pair of consecutive convergents to $\alpha$ satisfies

$$\left|\alpha - \frac{a}{b}\right| < \frac{1}{2b^2} \ .$$

15. *i*) Let $\alpha$ be a real number and $\frac{a}{b}$ be a reduced fraction satisfying $|\alpha - \frac{a}{b}| < \frac{1}{2b^2}$. Further, let $[a_0, \cdots, a_s]$ be that scf expansion of $\frac{a}{b}$ for which $s$ is even, odd in the respective cases $\frac{a}{b} \le \alpha, \frac{a}{b} > \alpha$. When $\frac{p_{s-1}}{q_{s-1}}, \frac{p_s}{q_s}$ are the last two convergents to $\frac{a}{b}$ and $\alpha' = \frac{\alpha q_s - p_s}{p_{s-1} - \alpha q_{s-1}}$, we have $\alpha' \ge 0$ and

$$\alpha = [a_0, \cdots, a_s + \alpha'] \; ;$$

*ii*) If $\frac{a}{b}$, $\alpha$ and $\alpha'$ are as in (*i*) then $\frac{1}{\alpha'} + \frac{q_{s-1}}{q_s} > 2$ so $0 < \alpha' < 1$ and we conclude $a_0, \cdots, a_s$ are the first $s+1$ partial quotients in the scf expansion of $\alpha$ ;

*iii*) (Dirichlet) If $\frac{a}{b}$ is a fraction such that $|\alpha - \frac{a}{b}| < \frac{1}{2b^2}$ then $\frac{a}{b}$ is a convergent to $\alpha$ ;

*iv*) Every rational number is a convergent for uncountably many real numbers ;

*v*) a positive integer $n$ is a Fibonacci number if and only if $5n^2+4$ or $5n^2-4$ is a square.

16. (Hurwitz' theorem & a result of Markov)

In the following all $\frac{p_j}{q_j}$ are convergents to $\alpha$. Also, we refer to the inequality $\left|\alpha - \frac{p_n}{q_n}\right| \geq \frac{1}{\sqrt{5}\, q_n^2}$ as (*). Finally, we call two real numbers *equivalent* if their scf expansions have the same tails (i.e. if $\alpha = [a_0, a_1, \cdots]$ and $\beta = [b_0, b_1, \cdots]$ then $\alpha$ and $\beta$ are equivalent if for some $s$ and $t$ it is true that $a_{s+j} = b_{t+j}$ for $j \geq 0$).

*i*) a) If (*) is true for $n = s-1$ and $n = s$ then $\frac{1}{\sqrt{5}}\left(\frac{1}{q_{s-1}^2} + \frac{1}{q_s^2}\right) \leq \frac{1}{q_{s-1} q_s}$ ;

b) under the conditions in (a) each of $\frac{q_{s-1}}{q_s}$ and $\frac{q_s}{q_{s-1}}$ is in the open interval

$$\left(\frac{\sqrt{5}-1}{2}, \frac{\sqrt{5}+1}{2}\right) ;$$

c) if (*) is true for $n = s-1$, $n = s$ and $n = s+1$ then $a_{s+1} = \frac{q_{s+1}}{q_s} - \frac{q_{s-1}}{q_s} < 1$ ;

d) at least one of any three convergents to $\alpha$ satisfies $\left|\alpha - \frac{p_n}{q_n}\right| < \frac{1}{\sqrt{5}\, q_n^2}$ ;

e) (Hurwitz) if $\alpha$ is irrational there are infinitely many irreducible rational numbers $\frac{a}{b}$ such that $\left|\alpha - \frac{a}{b}\right| < \frac{1}{\sqrt{5}\, b^2}$ .

ii) Hurwitz' theorem is false if we replace $\sqrt{5}$ by any larger number. In fact, if $0<\beta<1$ then there are only finitely many irreducible rational numbers $\frac{a}{b}$ such that

$$\left|\frac{1+\sqrt{5}}{2} - \frac{a}{b}\right| < \frac{\beta}{\sqrt{5}\, b^2} .$$

iii-a) If $\left|\alpha - \frac{p_n}{q_n}\right| \geq \frac{1}{q_n^2\sqrt{m^2+4}}$ for each of $n=s-1$, $n=s$ and $n=s+1$ then $a_{s+1}<m$ ;

b) if $\alpha$ is irrational either there are infinitely many irreducible fractions $\frac{a}{b}$ satisfying $\left|\alpha - \frac{a}{b}\right| < \frac{1}{b^2\sqrt{m^2+4}}$ or there is an $s_0$ such that $a_s<m$ for all $s>s_0$ ;

c) (Markov) if $\alpha$ is irrational and not equivalent to $\frac{1+\sqrt{5}}{2}$ then there are infinitely many irreducible rational numbers $\frac{a}{b}$ such that $\left|\alpha - \frac{a}{b}\right| < \frac{1}{2\sqrt{2}\, b^2}$ .

(i.e. if one removes all real numbers

equivalent to $\frac{1+\sqrt{5}}{2}$ then the Hurwitz' theorem is no longer best possible and, in fact, the $\sqrt{5}$ of that theorem may be replaced by $2\sqrt{2}$.)

*iv*) Let $[a_0, a_1, \cdots]$ be the scf expansion of an irrational number $\alpha$ and suppose the $\frac{p_n}{q_n}$ are the convergents to $\alpha$. Define

$$\nu(\alpha) = \liminf q_n | q_n\alpha - p_n| .$$

Then:

a) #3 (*vii*) guarantees $\nu(\alpha) \leq 1$ ;

b) Hurwitz' theorem guarantees $\nu(\alpha) \leq \frac{1}{\sqrt{5}}$ ;

c) for $\alpha$ not equivalent to $\frac{1+\sqrt{5}}{2}$, (*iii*-c) guarantees $\nu(\alpha) \leq \frac{1}{\sqrt{8}}$ ;

d) (*iii*-b) guarantees that either $\nu(\alpha) \leq \frac{1}{\sqrt{m^2+4}}$ or for all $a_s$ with s sufficiently large we have $a_s < m$ ;

e) if infinitely many $a_j$ are $\geq 3$ then

$$\nu(\alpha) \leq \frac{1}{\sqrt{13}} ;$$

v) With the same notation as in (iv) and putting $\alpha_n = [a_{n+1}, a_{n+2}, \cdots]$ we have:

a) $\alpha = [a_0, \cdots, a_{n-1}, a_n + \frac{1}{\alpha_n}] = \frac{\alpha_n p_n + p_{n-1}}{\alpha_n q_n + q_{n-1}}$

and, therefore, $q_n|q_n\alpha - p_n| = \frac{1}{\alpha_n + \frac{q_{n-1}}{q_n}}$ ;

b) if $a_j \le 2$ for all but a finite number of $j$ then, for $n$ sufficiently large,

$$\frac{1}{q_n|q_n\alpha - p_n|} = [a_{n+1}, a_{n+2}, \cdots] + [0, a_n, a_{n-1}, \cdots, a_1]$$
$$\le [2,1,2,1,\cdots] + [0,1,2,1,2,\cdots]$$
$$\le 1 + \sqrt{3} + \frac{2}{1+\sqrt{3}} = 2\sqrt{3}$$

and, therefore, $\nu(\alpha) \ge \frac{1}{\sqrt{12}}$ ;

c) $\nu(\alpha)$ cannot lie between $\frac{1}{\sqrt{13}}$ and $\frac{1}{\sqrt{12}}$.

*Remark*. The results in problem #16 are by no means exhaustive of those known. For further details one may consult Cassels [1957] chpt II. The extreme simplicity of the above arguments is due to Forder [1963] and Wright [1964]. For results like those in (v-c) and further references see Cusick [1974].

17. (Periodic scf's)

When $a_j = a_{j+t}$ for $j > s$ we say that $[a_0, a_1, \cdots]$ is *periodic with period* $a_{s+1}, \cdots a_{s+t}$ and write $[a_0, a_1, \cdots] = [a_0, \cdots, a_s, \overline{a_{s+1}, \cdots a_{s+t}}]$. For purely periodic scf's such as $[\overline{a_0, \cdots, a_{t-1}}]$ we conventionally write the above with $s = -1$.

i) If $\alpha$ is a periodic scf then $\alpha$ is a quadratic irrational;

ii) let $\alpha$ be a quadratic irrational which is a zero of the integral polynomial $f(x) = Ax^2 + Bx + C$, where the integers A, B, C have no common factor $> 1$; further, put $\alpha = [a_0, a_1, \cdots]$ and $\alpha_n = [a_{n+1}, a_{n+2}, \cdots]$. Then

a) $\alpha_n$ is a quadratic irrational for all $n \geq 0$;

b) if $\alpha_n$ is a zero of the integral polynomial $A_n x^2 + B_n x + C_n$, $(A_n, B_n, C_n) = 1$, then

$$B_n^2 - 4A_nC_n = B^2 - 4AC;$$

i.e. all $\alpha_n$ have the same discriminant;

c) in (b), $A_n = q_n^2 f\left(\frac{p_n}{q_n}\right)$, $C_n = q_{n-1}^2 f\left(\frac{p_{n-1}}{q_{n-1}}\right)$, and this implies that $A_n C_n < 0$;

d) there are only finitely many distinct triples $A_n, B_n, C_n$;

e) there are positive integers $k$ and $n$ such that $\alpha_{k+n} = \alpha_k$, and, therefore, $\alpha$ is periodic.

*iii*) For P, Q, D integers with D a positive non-square, the number $\alpha = \frac{P+\sqrt{D}}{Q}$ is a quadratic irrational. The *conjugate* $\alpha'$ of $\alpha$ is given by $\alpha' = \frac{P-\sqrt{D}}{Q}$. A quadratic irrational $\alpha$, $\alpha > 1$, satisfying $-1 < \alpha' < 0$ is called *reduced*. Thus $\frac{P+\sqrt{D}}{Q}$ is reduced when

$$-1 < \frac{P-\sqrt{D}}{Q} < 0 < 1 < \frac{P+\sqrt{D}}{Q}.$$

The quantities $\alpha_n$ are defined as in (*ii*).

a) If $\alpha$ has a purely periodic scf expansion then $\alpha$ is reduced;

b) if $\alpha$ is reduced and $\alpha = a_0 + \frac{1}{\alpha_0}$ then $\alpha_0$ is reduced;

c) if $\alpha$ is reduced so also are all $\alpha_n$ ;

d) if $\alpha$ is reduced and has a non-purely periodic expansion

$$\alpha = [a_0, \cdots, a_s, \overline{a_{s+1}, \cdots, a_{s+t}}], \ a_s \neq a_{s+t},$$

then $\alpha_{s-1} - \alpha_{s+t-1} = a_s - a_{s+t}$ and, therefore,

$$\alpha'_{s-1} - \alpha'_{s+t-1} = a_s - a_{s+t} \ ;$$

e) under the hypothesis of (d) the conclusion obtained is impossible. Consequently a reduced quadratic irrational must have a purely periodic scf expansion ;

f) necessary and sufficient conditions that a quadratic irrational $\alpha$ have a purely periodic scf expansion is that it be reduced ; i.e. that

$$\alpha > 1 \text{ and } -1 < \alpha' < 0 .$$

vi) a) Are the scf expansions of $\frac{1+\sqrt{13}}{3}$, $\frac{1-\sqrt{13}}{3}$, $\frac{2+\sqrt{3}}{4}$ purely periodic? ;

b) Compute the scf expansions of

$$\frac{1+\sqrt{13}}{3} \text{ and } -\left(\frac{1-\sqrt{13}}{3}\right)^{-1} ;$$

c) if $\alpha$ is reduced then the period of $-(\alpha')^{-1}$ is the reverse of the period of $\alpha$.

v) Let $Q, D$ be positive integers with $D > Q^2$ and $\alpha = \frac{\sqrt{D}}{Q}$ be irrational. Then the scf expansion of $\alpha$ has the form $[a_0, \overline{a_1, \cdots, a_k, 2a_0}]$.

vi) Let $D$ be a positive integer such that $\sqrt{D}$ is irrational. Then, if $a_0 = [\sqrt{D}]$,

a) $\frac{1}{\sqrt{D} - a_0}$ is reduced;

b) $\sqrt{D} + a_0$ is reduced and its scf expansion period is the reverse of that of $\frac{1}{\sqrt{D} - a_0}$;

c) The scf expansion of $\sqrt{D}$ is of the form

$$\sqrt{D} = [a_0, \overline{a_1, a_2, \cdots, a_2, a_1, 2a_0}].$$

vii) Let $D$ be a positive integer with $\sqrt{D}$ irrational. Then we may write

$$\sqrt{D} = [a_0, \overline{a_1, \cdots, a_{k-1}, 2a_0}] = [a_0, a_1, \cdots],$$

where $k$ is the length of the minimum period.

Further put, as usual, $\alpha_m = [a_{m+1}, a_{m+2}, \cdots]$.
Finally, suppose $x_0^2 - Dy_0^2 = 1$ where $x_0 > 0, y_0 > 0$.
Then

a) $\frac{x_0}{y_0}$ is a convergent to $\sqrt{D}$, say $\frac{x_0}{y_0} = \frac{p_s}{q_s}$;

b) with s as in (a):

A) if $A_s x^2 + B_s x + C_s = 0$ is the integral quadratic equation with $(A_s, B_s, C_s) = 1$ satisfied by $\alpha_s$ then

$A_s = 1$, $B_s^2 = 4(C_s + D)$, $C_s = p_{s-1}^2 - Dq_{s-1}^2$;

B) $\alpha_s = -\frac{1}{2}B_s + \sqrt{D}$;

C) $-\frac{1}{2}B_s = a_0$;

D) $a_j = a_{j+s+1}$ for $j \geq 1$;

E) $s \equiv -1 \pmod{k}$;

c) all positive integral solutions of $x^2 - Dy^2 = 1$ are contained in the sequence $\{(p_{k-1}, q_{k-1}), (p_{2k-1}, q_{2k-1}), (p_{3k-1}, q_{3k-1}), \cdots\}$;

d) $\sqrt{D} = [a_0, a_1, \cdots, a_{tk-1} + \frac{1}{a_0 + \sqrt{D}}]$, and, therefore,

$$p_{tk-1} = a_0 q_{tk-1} + q_{tk-2} \ ,$$
$$Dq_{tk-1} = a_0 p_{tk-1} + p_{tk-2} \ ,$$
$$p_{tk-1}^2 - Dq_{tk-1}^2 = (-1)^{tk} \ ;$$

e) $x^2 - Dy^2 = 1$ has infinitely many positive solutions and the totality of positive solutions consists precisely of the terms in the sequence of (c) with odd subscripts;

f) find the least positive integral solutions to:

A) $x^2 - 22y^2 = 1$ ;

B) $x^2 - 13y^2 = 1$ ;

C) $x^2 - 33y^2 = 1$ ;

g) develop an analogous theory for the equation $x^2 - Dy^2 = -1$.

18. Suppose $a$ and $b$ are positive relatively prime integers with $a > b$ and $\frac{a}{b} = [a_0, \cdots, a_n]$, $a_n \geq 2$.

i) $E_n \leq b$, where $E_n$ is the number of summands in $E(x_1, \cdots, x_n)$ ;

ii) Let $\tau = \frac{1+\sqrt{5}}{2}$ and noting that $\tau^2 = 1+\tau$ deduce $\tau^n < E_{n+1}$;

iii) $n < \dfrac{\log b}{\log \frac{1+\sqrt{5}}{2}}$ ;

iv) if $10^{t-1} \leq b < 10^t$ then $n < 5t$ ;

v) (Lamé) the number of divisions required to find the gcd of two numbers by means of the Euclidean algorithm never exceeds five times the number of base 10 digits of the smaller of the two numbers;

vi) investigate the best possible nature of Lamé's theorem.

19. (The circle diagram)

Corresponding to each irreducible fraction $\frac{a}{b}$ in $[0,1]$ construct the circle, $C(\frac{a}{b})$, of radius $\frac{1}{2b^2}$ and with center at $(\frac{a}{b}, \frac{1}{2b^2})$. Then $C(\frac{a}{b})$ is a circle lying above and tangent to the x-axis at $\frac{a}{b}$.

*i*) $C(\frac{a}{b})$, $C(\frac{c}{d})$ for $\frac{a}{b} \neq \frac{c}{d}$ are either disjoint or tangent and are tangent precisely when $\frac{a}{b}$ and $\frac{c}{d}$ are neighboring fractions in some $\Phi_n$ (see #9);

*ii*) the point of tangency between tangent circles is the rational point $\left(\frac{ab+cd}{b^2+d^2}, \frac{1}{b^2+d^2}\right)$;

*iii*) a vertical line intersecting the $x$-axis at $\alpha$, $0 \leq \alpha \leq 1$, cuts infinitely many of the circles if and only if $\alpha$ is irrational;

*iv*) suppose the vertical line cutting the $x$-axis at the irrational $\alpha$ cuts the curvilinear triangle formed by pairwise tangent circles above $\frac{a}{b}$, $\frac{e}{f}$, $\frac{c}{d}$ where $0 < \frac{a}{b} < \frac{e}{f} < \frac{c}{d} < 1$.

Then

a) $\frac{a}{b} < \alpha < \frac{c}{d}$;

b) $e = a + c$, $f = b + d$;

c) either $0 < d < b < f$ or $0 < b < d < f$;

d) if $0 < b < d < f$ the vertical line above $1-\alpha$ cuts a similar triangle with circles corresponding to some fractions $\frac{a'}{b'}, \frac{e'}{f'}, \frac{c'}{d'}$ with $\frac{a'}{b'} < \frac{e'}{f'} < \frac{c'}{d'}$ and $0 < d' < b' < f'$;

e) rational approximations to $\alpha$ lead to equally good rational approximations to $1-\alpha$ and vice versa.

20. Let $\frac{a}{b} < \frac{e}{f} < \frac{c}{d}$, $0 < d < b < f$, and suppose the circles above these fractions (see #19) are pairwise tangent with A, B, C the x-coordinates of the respective points of tangency of the pairs

$$\frac{a}{b}, \frac{c}{d} \quad ; \quad \frac{a}{b}, \frac{e}{f} \quad ; \quad \frac{e}{f}, \frac{c}{d} \; .$$

Further, suppose $\alpha$ is irrational and that the vertical line above $\alpha$ cuts the curvilinear triangle formed by the three circles.

*i*) Diagrams show clearly the possibility of $A \leq B < C$ as well as of $B \leq A < C$;

*ii*) $A < C$ and $B < C$;

*iii*) *a*) $A < B$ when $\frac{b}{d} > \frac{1+\sqrt{5}}{2}$;

*b*) $A > B$ when $\frac{b}{d} < \frac{1+\sqrt{5}}{2}$;

*iv*) *a*) $A < B$ implies $\left|\alpha - \frac{c}{d}\right| < \frac{1}{\sqrt{5}\,d^2}$;

*b*) $A > B$ implies $\left|\alpha - \frac{e}{f}\right| < \frac{1}{\sqrt{5}\,f^2}$;

*v*) (Hurwitz) for $\alpha$ irrational there are infinitely many irreducible fractions such that $\left|\alpha - \frac{a}{b}\right| < \frac{1}{\sqrt{5}\,b^2}$.

*Remarks*. The geometrical proof of Hurwitz' theorem which is presented in #19, 20 goes back to Ford [1917]. See also Rademacher [1964] for an exposition of the proof. An interesting discussion of these circles, sometimes referred to as Ford circles, will be found in Züllig [1928].

21. (Klein's geometrical interpretation of continued fractions)

Let $\mathcal{L}$ be a line through the origin with irrational slope $\alpha$ and suppose $\alpha = [a_0, a_1, \cdots]$.

*i*) Points $(x, y)$ are below (above) $\mathcal{L}$ precisely when $y < \alpha x$ $(y > \alpha x)$;

*ii*) Consider the points $P_n = (q_n, p_n)$, where $\frac{p_n}{q_n}$ is a convergent to $\alpha$, and show the vector from $P_{n-2}$ to $P_n$ is $a_n$ times the vector from O to $P_{n-1}$;

*iii*) The triangle $OP_{n-1}P_n$ has area $\frac{1}{2}$ and, therefore, contains no lattice points other than its vertices;

*iv*) A thread along $\mathcal{L}$ when pulled to the right or left and constrained to stick at lattice points (other than O) sticks on the lower side of $\mathcal{L}$ precisely on the even convergent points and on the upper side of $\mathcal{L}$ precisely on the odd convergent points.

22. (Some expansions due to Euler & Hurwitz)

i) Let $f_n(x)$ denote the value of $\sum_{s=0}^{\infty} a_{ns} x^{2s}$, where $a_{ns} = \frac{(n+s)!}{s!(2n+2s)!}$, when the series converges.

a) $f_n(x)$ exists for all $x$;

b) $f_n(x) - (4n+2) f_{n+1}(x) = 4x^2 f_{n+2}(x)$

for $n \geq 0$;

c) $\frac{f_0(x)}{f_1(x)} = 2x \frac{e^{2x}+1}{e^{2x}-1}$;

d) $\frac{e^{2x}+1}{e^{2x}-1} = [\frac{1}{x}, \frac{3}{x}, \frac{5}{x}, \frac{7}{x}, \cdots]$ for $x \neq 0$;

ii) Let $\alpha = [a_0, a_1, \cdots]$, where

$$a_0 = 2,$$

$$a_{3n} = a_{3n+1} = 1,$$

$$a_{3n-1} = 2n.$$

Thus, $\alpha = [2,1,2,1,1,4,1,1,6,1,1,8,1,\cdots]$. Further, let the convergents of $\alpha$ be $\frac{p_j}{q_j}$, $j \geq 0$. Then:

a) $\begin{cases} p_{3n+1} = (4n+2)p_{3n-2} + p_{3n-5} & \text{for } n \geq 2; \\ q_{3n+1} = (4n+2)q_{3n-2} + q_{3n-5} & \text{for } n \geq 1; \end{cases}$

b) if $P_n/Q_n$ is the $n^{th}$ convergent to $\frac{e+1}{e-1}$ then $P_n = \frac{1}{2}(p_{3n+1}+q_{3n+1})$, $Q_n = \frac{1}{2}(p_{3n+1}-q_{3n+1})$;

c) $\alpha = e$.

iii) a) $\frac{e^{\sqrt{2}}+1}{e^{\sqrt{2}}-1} = [\sqrt{2}, 3\sqrt{2}, 5\sqrt{2}, 7\sqrt{2}, 9\sqrt{2}, \cdots]$;

b) $\sqrt{2}\left(\frac{e^{\sqrt{2}}+1}{e^{\sqrt{2}}-1}\right) = [2, 3, 10, 7, 18, 11, 26, \cdots]$ is not periodic;

c) $e^{\sqrt{2}}$ is irrational.

(This proof is due to Richard E. Crandall.)

23. (A matricial approach)

I. Let $A = \begin{pmatrix} a & b \\ c & d \end{pmatrix}$ and write $K_1(A) = \frac{a}{c}$, when $c \neq 0$, and $K_2(A) = \frac{b}{d}$, when $d \neq 0$. If $K_1(A_1 \cdots A_n) \to \alpha_1$ we write $K_1(A_1 \cdots) = \alpha_1$. Similarly for $K_2(A_1 \cdots)$. If $K_1(A_1 \cdots) = K_2(A_1 \cdots) = \alpha$ we write $K(A_1 \cdots) = \alpha$. In the following, $\alpha = [a_0, a_1, \cdots]$.

i) if the $\frac{p_j}{q_j}$ are convergents to $\alpha$ then

$$\begin{pmatrix} p_n & p_{n-1} \\ q_n & q_{n-1} \end{pmatrix} = \begin{pmatrix} a_0 & 1 \\ 1 & 0 \end{pmatrix}\begin{pmatrix} a_1 & 1 \\ 1 & 0 \end{pmatrix} \cdots \begin{pmatrix} a_n & 1 \\ 1 & 0 \end{pmatrix} \quad ;$$

ii) $[a_0, a_1, \cdots] = K\left\{\begin{pmatrix} a_0 & 1 \\ 1 & 0 \end{pmatrix}\begin{pmatrix} a_1 & 1 \\ 1 & 0 \end{pmatrix}\cdots\right\}$ ;

iii) if $A = \begin{pmatrix} a & b \\ c & d \end{pmatrix}$, $K_1(A_1 \cdots) = \alpha$, and $c\alpha + d \neq 0$, then $K_1(AA_1A_2\cdots) = \dfrac{a\alpha + b}{c\alpha + d}$ ;

iv) if $kA = \begin{pmatrix} ka & kb \\ kc & kd \end{pmatrix}$ for $k$ a number then $K_1(A_1 \cdots) = \alpha$ implies $K_1((k_1A_1)(k_2A_2)\cdots) = \alpha$, for $\{k_n\}$ an arbitrary numerical sequence;

v) if $K_1(A_1 \cdots) = \alpha$ then $K_1((A_1A_2A_3)(A_4A_5A_6)\cdots(A_{3n-2}A_{3n-1}A_{3n})\cdots) = \alpha$;

vi) let $A_1 \cdots A_n = \begin{pmatrix} p_n & r_n \\ q_n & s_n \end{pmatrix} = P_n$ ; then

$$|K_1(P_n) - K_2(P_n)| = \frac{\prod_{r=1}^{n} |\det A_r|}{|q_n s_n|} ;$$

vii) let $P_n$ be as in (vi), $|\det A_r| = 1$ for all $r$, $K(\lim P_n) = \alpha$, $B = \begin{pmatrix} a & b \\ c & d \end{pmatrix}$ ; then

a) $q_n s_n \to \infty$ as $n \to \infty$ ;

b) $a^2 + c^2 \neq 0$ implies $K_1(P_nB) \to \alpha$ ;

c) $b^2 + d^2 \neq 0$ implies $K_2(P_nB) \to \alpha$ ;

d) $B$ has no zero column implies

$$K(\lim P_nB) = \alpha ;$$

viii) if all $A_r$ have determinant $\pm 1$, B is non-singular, $K(A_1 \cdots) = \alpha$, and $C_r = B^{-1}A_rB$ for all r, then $K(BC_1C_2\cdots) = \alpha$;

ix) let $ad - bc = \pm 1$, $0 < d < c$; then for $d \geq 1$,
$\begin{pmatrix} a & b \\ c & d \end{pmatrix} = \begin{pmatrix} b & a-xb \\ d & c-xd \end{pmatrix}\begin{pmatrix} x & 1 \\ 1 & 0 \end{pmatrix}$ for all x; for $d = 1$,
$\begin{pmatrix} a & b \\ c & d \end{pmatrix} = \begin{pmatrix} b & a-bc \\ 1 & 0 \end{pmatrix}\begin{pmatrix} c & 1 \\ 1 & 0 \end{pmatrix} = \begin{pmatrix} a-(c-1)b & bc-a \\ 1 & 0 \end{pmatrix}\begin{pmatrix} 1 & 1 \\ 1 & 0 \end{pmatrix}\begin{pmatrix} c-1 & 1 \\ 1 & 0 \end{pmatrix}$;

x) under the conditions of (ix) there is an integer $a_0$ and positive integers $n, a_1, \cdots, a_n$ such that

$$\begin{pmatrix} a & b \\ c & d \end{pmatrix} = \begin{pmatrix} a_0 & 1 \\ 1 & 0 \end{pmatrix}\begin{pmatrix} a_1 & 1 \\ 1 & 0 \end{pmatrix}\cdots\begin{pmatrix} a_n & 1 \\ 1 & 0 \end{pmatrix};$$

II. Put $A_m = \begin{pmatrix} 2m-1+x & 2m-1 \\ 2m-1 & 2m-1-x \end{pmatrix}$ for $m = 1, 2, \cdots$ and suppose $\prod_{m=1}^{n} A_m = \begin{pmatrix} f_n(x) & g_n(x) \\ h_n(x) & k_n(x) \end{pmatrix}$. Then

i) $h_n(x) = g_n(-x)$ and $k_n(x) = f_n(-x)$;

ii) a) $f_n(x) = \sum_{k=0}^{n} \frac{n(2n-k-1)!}{(n-k)!\,k!} x^k$;

b) $g_n(x) = \sum_{k=0}^{n} \frac{(n-k)(2n-k-1)!}{(n-k)!\,k!} x^k$;

iii) a) $\frac{f_n(x)}{n(n+1)\cdots(2n-1)} \to e^{x/2}$;

b) $\frac{g_n(x)}{n(n+1)\cdots(2n-1)} \to e^{x/2}$;

iv) $K(A_1A_2\cdots) = e^x$ for all x;

v-a) $\begin{pmatrix} a+1 & a \\ a & a-1 \end{pmatrix} = \begin{pmatrix} 1 & 1 \\ 1 & 0 \end{pmatrix}\begin{pmatrix} a-1 & 1 \\ 1 & 0 \end{pmatrix}\begin{pmatrix} 1 & 1 \\ 1 & 0 \end{pmatrix}$ ;

b) for $k > 0$,

$[1, k-1, 1, 1, 3k-1, 1, 1, 5k-1, 1, \cdots] = e^{1/k}$ ;

c) $e = [2, 1, 2, 1, 1, 4, 1, 1, 6, 1, 1, 8, \cdots]$ ;

(The method of this problem is due to Walters [1968].)

24. I We write $\frac{b_1}{a_1-}\frac{b_2}{a_2-}\frac{b_3}{a_3-}\cdots$ for the continued fraction

$$\cfrac{b_1}{a_1 - \cfrac{b_2}{a_2 - \cfrac{b_3}{a_3 - \cfrac{b_4}{a_4 - \cdots}}}} .$$

Further, let $c_n = \frac{b_1}{a_1-}\frac{b_2}{a_2-}\frac{b_3}{a_3-}\cdots\frac{b_n}{a_n}$ and define

$b_0 = 1$

$$\left.\begin{array}{lll} p_{-1} = -1 & p_0 = 0 & p_n = a_n p_{n-1} - b_n p_{n-2} \\ q_{-1} = 0 & q_0 = 1 & q_n = a_n q_{n-1} - b_n q_{n-2} \end{array}\right\} \text{ for } n \geq 1 .$$

Finally, in all parts below we assume all $a_j$ and $b_j$ positive and $a_n \geq b_n + 1$ for $n \geq 0$.

i) $c_n = \frac{p_n}{q_n}$ for $n \geq 1$ ;

ii) for $n \geq 0$

$p_n \geq p_{n-1} + b_0 \cdots b_n$ , $q_n \geq q_{n-1} + b_0 \cdots b_n$ so
$p_n \geq b_0 + b_0 b_1 + \cdots + b_0 \cdots b_n$ , $q_n \geq b_0 + b_0 b_1 + \cdots + b_0 \cdots b_n$
and strict inequality holds unless $a_j = b_j + 1$
for $0 \leq j \leq n$, in which case equality holds ;

iii) $p_n q_{n-1} - p_{n-1} q_n = b_0 \cdots b_n$ for $n \geq 0$ so

$$\frac{p_n}{q_n} = \frac{p_{n-1}}{q_{n-1}} + \frac{b_0 \cdots b_n}{q_n q_{n-1}} ;$$

iv) $q_n - p_n \geq q_{n-1} - p_{n-1} \geq 1$ for $n \geq 0$; further, for a given $n$, if $a_j = b_j + 1$ for $j \leq n$ then both inequalities are equalities, otherwise, at least one of the inequalities is strict ;

v) $\lim\limits_n \frac{p_n}{q_n}$ exists and is always $\leq 1$ ;
if $a_n > b_n + 1$ for some $n$ then $\lim \frac{p_n}{q_n} < 1$ ;

vi) when $a_n = b_n + 1$ for all $n$ then

$b_0 + b_0 b_1 + \cdots + b_0 \cdots b_n$ diverges

implies $\frac{p_n}{q_n} \to 1$ ;

$\alpha = b_0 + b_0 b_1 + \cdots + b_0 \cdots b_n + \cdots$ implies

$$\frac{p_n}{q_n} \to 1 - \frac{1}{\alpha} ;$$

II. Let all $a_i$, $b_i$ be positive integers and suppose $a_n \geq b_n + 1$ for all $n \geq 1$, with strict inequality holding infinitely often.

Write $\alpha_n = \frac{b_n}{a_n -} \frac{b_{n+1}}{a_{n+1} -} \cdots$ and prove:

i) $0 < \alpha_n < 1$ for all $n \geq 1$;

ii) if any $\alpha_j$ is rational so also are all others;

iii) if $r$ and $s$ are positive integers and $\alpha_j = \frac{r}{s}$ then there is a positive integer $t < r$ such that $\alpha_{j+1} = \frac{t}{r}$;

iv) $\alpha_1$ is irrational;

v) if $\alpha = \frac{b_1}{a_1 -} \frac{b_2}{a_2 -} \frac{b_3}{a_3 -} \cdots$, where the $a_i$ and $b_i$ are positive integers, and if $a_n \geq b_n + 1$ for all sufficiently large $n$, with strict inequality infinitely often, then $\alpha$ is irrational.

25. i) $\sum_{k=1}^{n} \frac{1}{c_k} = \frac{1}{c_1 -} \frac{c_1^2}{c_1 + c_2 -} \frac{c_2^2}{c_2 + c_3 -} \cdots \frac{c_{n-1}^2}{c_{n-1} + c_n}$;

ii) $\sum_{k=1}^{n} c_k = \frac{c_1}{1 -} \frac{c_2}{c_1 + c_2 -} \frac{c_1 c_3}{c_2 + c_3 -} \cdots \frac{c_{n-2} c_n}{c_{n-1} + c_n}$;

iii) $\sum_{k=1}^{\infty} \frac{1}{c_k}$, $\sum_{k=1}^{\infty} c_k$ converge to $\alpha, \beta$ respectively if and only if the right sides in (i), (ii) converge to $\alpha, \beta$;

iv) from $\frac{1}{e} = \frac{1}{2!} - \frac{1}{3!} + \frac{1}{4!} - \cdots$ deduce

$$e = 2 + \frac{1}{1+}\,\frac{1}{2+}\,\frac{2}{3+}\,\frac{3}{4+}\,\frac{4}{5+}\cdots$$

$$= 2 + \cfrac{1}{1+\cfrac{1}{2+\cfrac{2}{3+\cfrac{3}{4+\cfrac{4}{5+\cdots}}}}} \quad ;$$

v) $$\frac{\pi}{4} = \cfrac{1}{1+\cfrac{1}{2+\cfrac{3^2}{2+\cfrac{5^2}{2+\cfrac{7^2}{2+\cfrac{9^2}{2+\cdots}}}}}} \quad .$$

26. (Lambert's 1761 proof of the irrationality of $\pi$) Define $f_m(x)$, for each real positive $m$, by:

$$f_m(x) = 1 - \frac{x^2}{2^2 m} + \frac{x^4}{2^4 2!\, m(m+1)} - \frac{x^6}{2^6 3!\, m(m+1)(m+2)} + \frac{x^8}{2^8 4!\, m(m+1)(m+2)(m+3)} - \cdots$$

$$= 1 + \sum_{k=1}^{\infty} \frac{(-1)^k x^{2k}}{2^{2k} k!\, m(m+1)\cdots(m+k-1)} \; .$$

Then:

i) $f_m(x)$ exists for all $x$;

ii) $\frac{f_{m+1}(x)}{f_m(x)} = \left(1 - \frac{x^2}{2^2 m(m+1)} \frac{f_{m+2}(x)}{f_{m+1}(x)}\right)^{-1} =$

$$\frac{1}{1-} \; \frac{x^2/2^2 m(m+1)}{1-} \; \frac{x^2/2^2(m+1)(m+2)}{1-} \; \frac{x^2/2^2(m+2)(m+3)}{1-} \cdots;$$

iii) $\frac{f_{3/2}(x)}{f_{1/2}(x)} = \frac{1}{1-} \; \frac{x^2}{3-} \; \frac{x^2}{5-} \; \frac{x^2}{7-} \cdots = \frac{\tan x}{x}$;

iv) $\pi$ is not rational, since if $\pi$ were rational, say $\frac{\pi}{4} = \frac{m}{n}$ for positive integers $m$ and $n$, then

$$1 = \frac{m}{n-} \; \frac{m^2}{3n-} \; \frac{m^2}{5n-} \; \frac{m^2}{7n-} \cdots,$$

which is not possible since the right hand side is irrational.

27. (Some irrational & transcendental numbers)

I. If $\alpha$ is a real number which is a zero of an integral polynomial of degree $n$ but of no such polynomial of smaller degree then $\alpha$ is said to be *algebraic of degree $n$*.

All numbers algebraic of degree $>1$ are thus irrational. A number which is irrational but not algebraic of any degree is said to be *transcendental*. A simple cardinality argument may be used to prove the existence of transcendental numbers.

II. Let $\alpha$ be algebraic of degree $n$ and let $f$ be an integral polynomial of degree $n$ having $\alpha$ as a zero. Then

i) there is an integral polynomial $g$ such that $f(x) = (x-\alpha)g(x)$, $g(\alpha) \neq 0$;

ii) there is a positive $\delta$ such that if $\alpha - \delta \leq x \leq \alpha + \delta$ then $g(x) \neq 0$;

iii) there is a constant $M$ and integers $a, b$ with $b > 0$ such that

$$\left|\alpha - \frac{a}{b}\right| = \left|\frac{f(\frac{a}{b})}{g(\frac{a}{b})}\right| \geq \frac{1}{Mb^n} \quad ;$$

iv) (Liouville 1844)

If $\alpha$ is algebraic of degree $n$ then there is a positive constant $c$ such that $|\alpha - \frac{a}{b}| > \frac{c}{b^n}$ for all integral $a, b$ with $b > 0$.

III i) $\alpha = \sum_{m=0}^{\infty} 10^{-2^m}$ is irrational;

ii) an irrational number is called a *Liouville number* if for each positive integer $n$ and constant $c > 0$ there is a rational number $\frac{a}{b}$, $b > 0$, such that $|\alpha - \frac{a}{b}| < \frac{c}{b^n}$. Every Liouville number is transcendental;

iii) $\alpha = \sum_{m=0}^{\infty} a_m 10^{-m!}$ is transcendental when the $a_m$ are integers satisfying $|a_m| \leq M$ for some $M$.

IV i) let $\alpha = [a_0, a_1, \cdots]$ be an irrational number with convergents $\frac{p_n}{q_n}$; then if $a_{k+1} > q_k^{k-1}$ for all $k \geq 1$, $\alpha$ is transcendental;

ii) using (i) exhibit a transcendental number;

iii) if $a_{k+1} \geq (2^k a_1 \cdots a_k)^{k-1}$ for $k \geq 1$ and $a_0, a_1$, are arbitrary then $[a_0, a_1, a_2, \cdots]$ is transcendental.

28. In this problem $x$ ranges over the irrational numbers in the interval $[0,1]$. We write $x=[0, a_1(x), a_2(x), \cdots]$ so that each $a_j(x)$ is a positive integer. We denote the probability that $a_n(x)=k$ by $P_{nk}$ and the probability that $(a_1(x), \cdots, a_t(x)) = (a_1, \cdots, a_t)$ by $P(a_1, \cdots, a_t)$.

I i) $\sum_{k=1}^{\infty} P_{nk} = 1$ for $n = 1, 2, \cdots$

and

$\sum_{k=1}^{\infty} P(a_1, \cdots a_{n-1}, k) = P(a_1, \cdots, a_{n-1})$ for $n=2,3\cdots$;

ii) $P_{1k} = \frac{1}{k(k+1)}$ ;

iii) $P_{2k} = P_{1k} \sum_{n=1}^{\infty} \frac{1}{(n+\frac{1}{k})(n+\frac{1}{k+1})} = \frac{\pi^2}{6k(k+1)}(1-\epsilon_k)$,

where $\epsilon_k \longrightarrow 0$ as $k \longrightarrow \infty$ ;

iv) $P_{2k} \geq P_{1k}$ for $k \geq 2$ but $P_{21} < P_{11}$.

II Let $\frac{p_j}{q_j}$, $0 \leq j \leq n$, be the convergents to $[0, a_1, \cdots, a_{n-1}, k]$. Then

i) $P(a_1, \cdots, a_{n-1}, k) = 1/q_{n-1}^2 (k + \frac{q_{n-2}}{q_{n-1}})(k+1+\frac{q_{n-2}}{q_{n-1}})$;

ii) $\frac{k}{k+2} < \frac{P(a_1,\cdots,a_{n-1},k+1)}{P(a_1,\cdots,a_{n-1},k)} = \frac{k+\frac{q_{n-2}}{q_{n-1}}}{k+2+\frac{q_{n-2}}{q_{n-1}}} < \frac{k+1}{k+3}$;

iii) $\frac{2}{k(k+1)} < \frac{P(a_1,\cdots,a_{n-1},k)}{P(a_1,\cdots,a_{n-1},1)} < \frac{6}{(k+1)(k+2)}$ for $k \geq 2$;

iv) $2 = \sum_{k=1}^{\infty} \frac{2}{k(k+1)} < \frac{P(a_1,\cdots,a_{n-1})}{P(a_1,\cdots,a_{n-1},1)} < \sum_{k=1}^{\infty} \frac{6}{(k+1)(k+2)} = 3$;

v) $\frac{2}{3k(k+1)} < \frac{P(a_1,\cdots,a_{n-1},k)}{P(a_1,\cdots,a_{n-1})} = P_{nk} < \frac{3}{(k+1)(k+2)}$.

III i) $\sum_{k=1}^{M} P_{1k} = \frac{M}{M+1}$;

ii) $\sum_{k=1}^{M} P(a_1,\cdots,a_{n-1},k) < \alpha P(a_1,\cdots,a_{n-1})$,

for $n>2$ and where $0 < \alpha = 1 - \frac{2}{3(M+1)} < 1$;

iii) $\sum_{\substack{1 \leq a_j \leq M \\ 1 \leq j \leq n}} P(a_1,\cdots,a_n) < \alpha^{n-1} \frac{M}{M+1}$;

iv) if $[a_0, a_1, a_2, \cdots]$ is the scf expansion of a number picked at random from $[0,1]$ then the set of $a_1, a_2, \cdots$ is, with probability 1, unbounded; i.e. almost all real numbers fail to have bounded partial quotients.

IV Let $\varphi$ be an arbitrary positive valued function defined over the positive integers and suppose $N \geq 1$. Then

i) $\frac{2}{3(\varphi(t)+1)} < \sum_{k \geq \varphi(t)} \frac{P(a_1,\cdots,a_{t-1},k)}{P(a_1,\cdots,a_{t-1})} < \frac{3}{\varphi(t)+1}$ for $t>N$;

ii) $1 - \frac{3}{\varphi(t)+1} < \sum_{1 \le k < \varphi(t)} \frac{P(a_1, \cdots, a_{t-1}, k)}{P(a_1, \cdots, a_{t-1})} < 1 - \frac{2}{3(\varphi(t)+1)}$ for $t > N$;

iii) $P(a_1, \cdots, a_N) \prod_{j=N+1}^{t} \left(1 - \frac{3}{\varphi(j)+1}\right) < \sum P(a_1, \cdots, a_t)$

$< P(a_1, \cdots, a_N) \prod_{j=N+1}^{t} \left(1 - \frac{2}{3(\varphi(j)+1)}\right)$ for $t > N$, where the middle sum is over all $(t-N)$-tuples $(a_{N+1}, \cdots, a_t)$ such that $1 \le a_j < \varphi(j)$, $N < j \le t$;

iv) if $\sum_{n=1}^{\infty} \frac{1}{\varphi(n)}$ diverges, respectively converges, then the probability that a random $x$ in $[0,1]$ satisfies $a_n(x) < \varphi(n)$ for all sufficiently large $n$ is 0, respectively 1.

Remarks.

1. The particular approach to continued fractions, via Euler brackets, as presented here is somewhat unusual. Chrystal [1904] gives a more complete discussion of Euler brackets than we have given. A particularly nice geometrical introduction to continued fractions is given by Stark [1970].

2. The treatment of Farey fractions, see #9, follows Бухштаб [1966] and, in fact, our exposition in many parts of this chapter owes much to the same source.

3. For #21 see Klein [1924] or Davenport [1952].

4. In connection with #23 one might consult Matthews, Walters [1970].

5. The details in #25 will be found in Cheney [1966].

6. The natural continuation of #28 would be to prove Kuzmin's theorem which states that for almost all real numbers $x$,

$$\sqrt[n]{a_1(x)\cdots a_n(x)} \rightarrow \prod_{k=1}^{\infty}\left(1+\frac{1}{k(k+2)}\right)^{\frac{\ln k}{\ln 2}},$$

where it will be noted that the limit is an absolute constant. For good expositions of this theorem, along entirely separate lines, the reader might consult Khinchin [1964] and/or Kac [1959]. The second of these gives a very interesting account of the theorem connecting it with statistical mechanics and the ergodic theorem.

7. The most complete treatment of all aspects of continued fractions will be found in Perron [1954].

## XIV More on Primes

1. (Bonse inequality) Let $p_1, p_2, \cdots$ be the primes in their natural order and suppose $n \geq 10$. Further, let $j$ satisfy $2 \leq j \leq n-1$ and set

$$N_1 = p_1 \cdots p_{j-1} - 1,\ N_2 = 2p_1 \cdots p_{j-1} - 1,\ N_3 = 3p_1 \cdots p_{j-1} - 1,$$
$$\cdots,\ N_{p_j} = p_j p_1 \cdots p_{j-1} - 1.$$

Then:

i) each of $p_j, \cdots, p_n$ divides at most one of

$$N_1, \cdots, N_{p_j};$$

ii) there is a $j$, $2 \leq j \leq n-1$, for which

$$p_j > n - j + 1;$$

iii) letting $i$ be the smallest $j$ for which $p_j > n - j + 1$, there is a $k$, $1 \leq k \leq p_i$, such that $p_1, \cdots, p_n$ all fail to divide $kp_1 \cdots p_{i-1} - 1$ and, therefore, $p_{n+1} < p_1 \cdots p_i$;

iv) the $i$ in (iii) exceeds 4 so $p_{i-1} - 2 \geq i$

$$\text{and } p_1 \cdots p_i < p_{i+1} \cdots p_n;$$

v) for $n \geq 4$, $p_{n+1}^2 < p_1 \cdots p_n$.

2. (A property of 30)

Suppose $p_k \le \sqrt{n} < p_{k+1}$, where $p_k$ is again the $k^{th}$ prime number. Then:

*i*) if for some $j \le k$, $p_j$ does not divide $n$ then $(p_j^2, n) = 1$ and $p_j^2 < n$;

*ii*) if no composite integer $< n$ is prime to $n$ then $p_1 \cdots p_k \mid n$ and, therefore, $p_1 \cdots p_k \le n$;

*iii*) if $n \ge 49$ there is a composite integer $< n$ and prime to $n$;

*iv*) all positive integers $> 1$ and $< 30$ which are prime to 30 are themselves prime and no integer larger than 30 has this property.

3. (A property of 24)

The number 24 is the largest integer which is divisible by every positive integer smaller than its square root.

4. (Erdös) We write $\pi(x)$ for the number of primes $\leq x$ and $N_j(x)$ for the number of positive integers not exceeding $x$ having no prime factor $>p_j$, the $j^{th}$ prime number.

Then:

*i*) every integer having no prime factor $>p_j$ is of the form $m^2 p_1^{\epsilon_1} \cdots p_j^{\epsilon_j}$, where each $\epsilon_i$ is 0 or 1;

*ii*) $N_j(n) \leq \sqrt{n} \cdot 2^j$;

*iii*) $\pi(n) \geq \ln n / 2 \ln 2$ and, therefore, the number of primes is not finite;

*iv*) $p_n < 4^n$;

*v*) the series $\sum_{n=1}^{\infty} \frac{1}{p_n}$ diverges, since otherwise there is a $j$ such that

$$2^j \sqrt{x} \geq N_j(x) \geq x - \sum_{n>j} \left[ \frac{x}{p_n} \right] \geq x - \sum_{n>j} \frac{x}{p_n} > \frac{x}{2}.$$

5. (Sierpinski 1953)

$$\pi(x) = 1 + \sum_{k=3}^{[x]} \left\{ 1 - \lim_{m \to \infty} \left( 1 - \prod_{j=2}^{k-1} \left( \sin \frac{k\pi}{j} \right)^2 \right)^m \right\}.$$

6. (Hardy 1906) Let $\vartheta(n)$ be the largest prime factor of $n$. Then

$$\vartheta(n) = \lim_{s\to\infty}\lim_{m\to\infty}\lim_{k\to\infty}\sum_{j=0}^{m}\left\{1-\left(\cos\frac{(j!)^s\pi}{n}\right)^{2k}\right\}.$$

7. (Moser [1950]) Let $p_n$ be the $n^{th}$ prime. Then:

i) $\sum_{m=k+1}^{\infty} p_m 10^{-\frac{m(m+1)}{2}}$ converges to $\beta_k$, $k \geq 0$, where $0 < \beta_k < 10^{-\frac{k(k+1)}{2}}$;

ii) $p_n = \left[10^{\frac{n(n+1)}{2}}\beta_0\right] - 10^n\left[10^{\frac{n(n-1)}{2}}\beta_0\right]$.

8. (Sierpinski [1952]) Let $p_n$ be the $n^{th}$ prime. Then:

i) $\sum_{m=k+1}^{\infty} p_m 10^{-2^m}$ converges to $\alpha_k$, $k > 0$, where $0 < \alpha_k < 10^{-2^k}$;

ii) $p_n = \left[10^{2^n}\alpha_0\right] - 10^{2^{n-1}}\left[10^{2^{n-1}}\alpha_0\right]$.

9. (Härtter [1961]) Let $A = \{a_1, a_2, \cdots\}$ be an arbitrary monotone increasing sequence, $a_n \leq a_{n+1}$, of positive integers. Then:

i) there is a numerical function $f$ satisfying:

a) $f(n)/f(i)$ is an integer for all $i \le n$ ;

b) $\sum_{\nu=n+1}^{\infty} \frac{a_\nu}{f(\nu)}$ converges to a value $< \frac{1}{f(n)}$ for $n \ge 0$;

ii) for any function $f$ satisfying (a), (b) of (i)

$$a_n = [f(n)\alpha] - \frac{f(n)}{f(n-1)}[f(n-1)\alpha],$$

where $\alpha = \sum_{\nu=1}^{\infty} \frac{a_\nu}{f(\nu)}$ ;

iii) the results in #7, #8 are special cases of (ii).

10. (Chebyshev) Let $n \ge 2$ be given. Then:

i) $2^n < \binom{2n}{n} < 2^{2n}$ ;

ii) $\prod_{n<p\le 2n} p$ divides $\binom{2n}{n}$ which, in turn, divides $\prod_{p<2n} p^{t_p}$, where $t_p = \left[\frac{\ln 2n}{\ln p}\right]$ ;

iii) a) $2^n < \prod_{p<2n} p^{t_p}$;

b) $\prod_{n<p\le 2n} p < 2^{2n}$ ;

iv) $\sum_{p\le x} \ln p \ge \frac{1}{2}(\pi(x) - \sqrt{x})\ln x$ and, therefore, $\pi(x) \le \frac{2}{\ln x}\sum_{p\le x} \ln p + \sqrt{x}$ ;

v) from (iii-a), $\pi(2n) > \frac{n \ln 2}{\ln 2n}$ and, therefore, there exists a constant A such that $\pi(x) > A\frac{x}{\ln x}$ for all $x \ge 2$ ;

*vi)* from (*iii*-*b*), $\sum_{p \le 2n} \ln p < 2n \ln 2 + \sum_{p \le n} \ln p$, and, therefore, $\sum_{p \le 2^k} \ln p < 2^{k+1}$;

*vii)* there exists a positive constant A such that $\sum_{p \le x} \ln p < Ax$ for all $x \ge 2$;

*viii)* there exists a positive constant A such that $\pi(x) < A \frac{x}{\ln x}$ for all $x \ge 2$;

*ix)* (Chebyshev 1852) for $x \ge 2$ and suitable positive constants A and B,

$$A \frac{x}{\ln x} < \pi(x) < B \frac{x}{\ln x}.$$

11. (Approximate order of $p_n$)

Let $\epsilon > 0$ be given. Then:

*i)* from Chebyshev's inequality there is an $x_0$ such that $1 - \epsilon < \frac{\ln \pi(x)}{\ln x} < 1 + \epsilon$ for $x > x_0$;

*ii)* for suitable positive constants A and B and for $x > x_0$,

$$A(1-\epsilon) < \frac{\pi(x) \ln \pi(x)}{x} < B(1+\epsilon);$$

*iii)* for $n$ sufficiently large

$$A(1-\epsilon) < \frac{n \ln n}{p_n} < B(1+\epsilon);$$

iv) (Theorem) There exist positive constants A and B such that

$$A n \ln n < p_n < B n \ln n ;$$

v) from (iv) one easily deduces that the series $\sum_{j=1}^{\infty} \frac{1}{p_j^{\alpha}}$ is divergent for $\alpha \le 1$ and convergent for $\alpha > 1$ .

12. (Bertrand) Let $P_n = \prod_{n<p<2n} p$, where $P_n = 1$ when there are no primes between $n$ and $2n$.

Then:

i) $P_n < \binom{2n-1}{n} < 4^{n-1}$ for $n \ge 2$ ;

ii) $\prod_{p \le x} p < 4^x$ for all real $x \ge 2$ ;

iii) if $2 \le \frac{2n}{3} < p \le n$ then $p$ does not divide $\binom{2n}{n}$;

iv) $\frac{4^n}{2\sqrt{n}} < \binom{2n}{n} \le (2n)^{\frac{1}{2}\sqrt{2n}} 4^{\frac{2n}{3}} P_n$ for $n \ge 3$ ;

v) $P_n > 1$ if $4^{2n} > 8(2n)^{3\sqrt{2n}+3}$ ;

vi) $P_n > 1$ if $n > 500$ ;

vii) (Bertrand) if $n \ge 2$ there is a prime strictly between $n$ and $2n$ ;

viii) $p_{n+1} < 2p_n$ if $n \ge 1$ .

13. (Finsler) Let $P_n$ be as in #12. Then from #12 (iv), $P_n > \frac{4^{n/3}}{2\sqrt{n}\,(2n)^{1/2\sqrt{2n}}}$. Setting the right side of this inequality equal to $(2n)^x$ we have

i) $x < \pi(2n) - \pi(n)$ if $n \geq 3$;

ii) $x \ln 2n = \frac{n}{3}\left(\ln 4 - \frac{3\ln 4n}{2n} - \frac{3\ln 2n}{\sqrt{2n}}\right) > \frac{n}{3}$ if $n \geq 2500$;

iii) $\frac{n}{3\ln 2n} < \pi(2n) - \pi(n)$ if $n \geq 2500$;

iv) $\pi(2n) - \pi(n) < \frac{7n}{5\ln n}$ if $n \geq 2$;

v) (Finsler) if $n \geq 2$ then

$$\frac{n}{3\ln 2n} < \pi(2n) - \pi(n) < \frac{7n}{5\ln n}\ ;$$

vi) $\pi(2n) - \pi(n) \geq 2$ if $n \geq 6$;

vii) $p_{n+2} < 2p_n$ if $n \geq 4$;

viii) there is a prime $p$ satisfying $n < p < 2n-2$ if $n \geq 4$; though not equivalent to #12 (vii) this is sometimes referred to as Bertrand's postulate;

ix) $\pi(2n) - \pi(n)$ is an unboundedly increasing function of $n$.

14. Starting with Finsler's theorem (#13-v) one finds:

i) $\pi(2^k) < \frac{2^{k+1}}{k \ln 2}$ if $k \geq 1$;

ii) $\pi(n) < 4 \frac{n}{\ln n}$ if $n \geq 2$;

iii) $\frac{1}{12} \frac{n}{\ln n} < \pi(n) - 1$ if $n \geq 2$;

iv) (Chebyshev's theorem again)

$$\frac{1}{12} \frac{x}{\ln x} < \pi(x) < 4 \frac{x}{\ln x} .$$

(These values of the positive constants in Chebyshev's theorem are far from "best possible".)

15. Consider

$$(*) \quad \pi(mn) > \pi(m) + \pi(n) .$$

i) Using #14 (iv) (Chebyshev) it is easy to prove (*) for $192 \leq n \leq m$;

ii) using #13 (v) (Finsler) one obtains (*) for $2 \leq n \leq 192$, $4000 \leq m$;

(using tables and a computer one gets (*) for $2 \leq n \leq m$, $6 \leq m$; see Trost [1968].)

16. (Assuming the result in the parenthetical remark of # 15.)

i) $\pi(p_m p_n) > \pi(p_{m+n})$ for $2 \le n \le m$, $4 \le m$;

ii) $p_m p_n > p_{m+n}$ for $1 \le n$, $1 \le m$;

iii) $p_{n+1-j} p_j > p_{n+1}$ for $1 \le j \le n$;

iv) $p_{n+1}^{n} < (p_1 \cdots p_n)^2$ when $n \ge 1$.

17. (A special case of Dirichlet's theorem on primes in arithmetic progressions.)

Put $F_n(x) = \prod_{\substack{k=1 \\ (k,n)=1}}^{n} (x - e^{\frac{2\pi i k}{n}})$, $n \ge 1$.

Then:

i) $x^n - 1 = \prod_{d \mid n} F_d(x)$;

ii) $F_n(x)$ is an integral polynomial of degree $\varphi(n)$;

iii) $F_n(0) = 1$ for $n > 1$;

iv) if $p$ is a prime factor of $F_n(a)$, $n > 1$, then $(a, p) = 1$;

*v*) if $p \mid F_n(a)$ and $t$ is the smallest positive $n$ for which $p \mid a^n - 1$ then:

*a*) $t \mid n$;

*b*) $t < n$ implies $a^n - 1 \equiv (a+p)^n - 1 \equiv 0 \pmod{p^2}$;

*c*) $t < n$ implies $p \mid n$;

*d*) $p \nmid n$ implies $t = n$ implies $p \equiv 1 \pmod{n}$;

*vi*) let $p_1, \cdots, p_k$ be any finite set of primes; then for $y$ sufficiently large and $n > 1$,

$$F_n(nyp_1 \cdots p_k) > 1 \text{ and}$$

$$F_n(nyp_1 \cdots p_k) \equiv F_n(0) \equiv 1 \pmod{np_1 \cdots p_k};$$

thus there is a prime $p \neq p_j$, $1 \leq j \leq k$, satisfying $p \equiv 1 \pmod{n}$;

*vii*) for $n > 1$ there are infinitely many primes in the arithmetic progression

$$1, 1+n, 1+2n, 1+3n, \cdots.$$

18. Let $F_n(x)$ be as in #17 and let $p$ be a prime. Further, suppose $n$ has exactly $r$ distinct prime factors and that $A_j$, $1 \leq j \leq r$, is the

set of products of $j$ of the $r$ primes entering into $n$. We put

$$g(x) = F_n(x) \prod_{\alpha \in A_1} (x^{\frac{n}{\alpha}} - 1) \prod_{\alpha \in A_3} (x^{\frac{n}{\alpha}} - 1) \cdots$$

$$f(x) = (x^n - 1) \prod_{\alpha \in A_2} (x^{\frac{n}{\alpha}} - 1) \prod_{\alpha \in A_4} (x^{\frac{n}{\alpha}} - 1) \cdots$$

and $\epsilon_m = e^{\frac{2\pi i m}{n}}$, with $(m,n) = d$.

*i*) $\epsilon_m$ is a zero of order 1 of each $x^{\frac{n}{\alpha}} - 1$ for which $\alpha | m$ and is not a zero of all other $x^{\frac{n}{\alpha}} - 1$;

*ii*) if $d > 1$ and has $s$ distinct prime factors then the highest power of $x - \epsilon_m$ in $g(x)$ (respectively $f(x)$) is $\binom{s}{1} + \binom{s}{3} + \binom{s}{5} + \cdots$ (respectively $1 + \binom{s}{2} + \binom{s}{4} + \cdots$);

*iii*)
$$F_n(x) = \frac{(x^n - 1) \prod\limits_{\substack{1 \le j \le r \\ j \text{ even}}} \prod\limits_{\alpha \in A_j} (x^{\frac{n}{\alpha}} - 1)}{\prod\limits_{\substack{1 \le j \le r \\ j \text{ odd}}} \prod\limits_{\alpha \in A_j} (x^{\frac{n}{\alpha}} - 1)};$$

*iv*) if $p | n$ then $F_{np}(x) = F_n(x^p)$;

*v*) if $p \nmid n$ then
$$F_{np}(x) = \frac{F_n(x^p)}{F_n(x)};$$

*vi*) $F_p(x) = \dfrac{x^p - 1}{x - 1} = x^{p-1} + x^{p-2} + \cdots + x + 1$;

vii) $F_n(1) = \begin{cases} 0 \text{ if } n=1; \\ p \text{ if } n \text{ is a power of } p; \\ 1 \text{ if } n \text{ has at least 2 distinct prime factors;} \end{cases}$

viii) $F_n(x) = \prod_{d|n} (x^{\frac{n}{d}} - 1)^{\mu(d)}$, where $\mu$ is the number theoretic function defined by $\mu(1) = 1$, $\mu(n) = 0$ if $n$ contains as a factor the square of any prime, $\mu(n) = (-1)^k$ if $n$ is a product of $k$ distinct primes;

ix) let $n = p_1 \cdots p_t$, where $t$ is odd and the $p_i$ are primes satisfying $p_1 < p_2 < \cdots < p_t$ and $p_1 + p_2 > p_t$ (see #13-ix); then

$$F_n(x) \equiv \frac{1}{1-x} \prod_{i=1}^{t} (1 - x^{p_i})$$

$$\equiv (1 + x + \cdots + x^{p_t - 1})(1 - x^{p_1}) \cdots (1 - x^{p_{t-1}})$$

$$\equiv (1 + x + \cdots + x^{p_t - 1})(1 - x^{p_1} - x^{p_2} - \cdots - x^{p_{t-1}}) \pmod{x^{p_t + 1}};$$

x) (Schur 1931) for $n$ as in (ix) the coefficient of $x^{p_t}$ in $F_n(x)$ has absolute value $t-1$.

19. (Richert) Let $S_n$ be the set of sums of 2 or more of the $1^{st}$ $n$ primes, no repetitions permitted. Then:

*i*) all integers between 12 and 29 (note $29 = 12 + p_7$), inclusive, are in $S_6$;

*ii*) if $n \geq 7$ all integers between 12 and $29 + p_7 + \cdots + p_n$ are in $S_n$;

*iii*) all integers $\geq 12$ are sums of 2 or more distinct primes;

*iv*) all integers $\geq 7$ are either prime or a sum of two or more distinct primes;

*v*) 6 is the largest positive integer which is neither a prime nor the sum of two or more distinct primes and 11 is the largest positive integer not the sum of two or more distinct primes.

20. (Furstenberg) Let S be the set of all integers. Take as a basis for a topology in

S the collection of all two way infinite arithmetic progressions. Thus a set of S is open if it is the union of such arithmetic progressions.

*i*) S with the specified basis is a topological space;

*ii*) each arithmetic progression is both open and closed;

*iii*) each finite union of arithmetic progressions is closed;

*iv*) if $A_p = \{0, \pm p, \pm 2p, \cdots\}$, where $p$ is a prime, and if $A = \bigcup_p A_p$, then the complement of A is $\{-1, 1\}$;

*v*) that there are infinitely many primes follows from (*iv*).

21. (Nicol) For each positive real number $x$ let $\pi_2(x)$ denote the number of twin primes $(p, p+2)$ with $p \leq x$. Then

$$\pi_2(x) = 2 + \sum_{7 \le n \le x} \sin\left\{\frac{\pi}{2}(n+2)\left[\frac{n!}{n+2}\right]\right\} \sin\left\{\frac{\pi}{2} n \left[\frac{(n-2)!}{n}\right]\right\},$$

where [...] is the largest integer not exceeding ... .

22. ( Willans [1964] )

Define $F(n)$ by $F(n) = \left[\cos^2 \pi \frac{(n-1)!+1}{n}\right]$. Then:

i) $F(n) = \begin{cases} 1 & \text{for } n \text{ prime or } n = 1; \\ 0 & \text{for } n \text{ composite}; \end{cases}$

ii) the $n^{th}$ prime $p_n$ is given by

$$p_n = 1 + \sum_{m=1}^{2^{2n}} \left[\frac{n}{\sum_{k=1}^{m} F(k)}\right].$$

Remarks.

1. The Bonse inequality in #1 is not very strong and its main interest is in the simplicity of its proof and its application to problems #2,3. The property of 30 given in #2 goes back to Schatunovsky who proved it in 1893.

Generalizations have been given ~ see e.g. Dickson I [1952] pp. 132, 133, 137, 138 and Landau I [1909] pp. 229-234.

2. For other work on prime representing functions, #7-9, one might consult Nambooodiripad [1971], Willans [1964], Sato & Straus [1970], Dudley [1969], Mills [1947], and the references therein.

3. Problem #4 is due to Erdös [1938], #13 to Finsler [1945], #19 to Richert [1941], #20 to Furstenberg [1955], #21 to Nicol [1974], #22 to Willans [1964]. For an exposition of the results in #12-16 see Trost [1968]. Expositions of the special case of Dirichlet's theorem may be found in Nagell [1951] and Landau [1909]. A simple, even more elementary, proof is given in Niven, Powell [1976]. A complete proof of Dirichlet's theorem is given in XXIV.

4. Taking $n \geq 4$ #16(iv) yields $p_{n+1}^{2} \leq p_{n+1}^{n/2} < p_1 \cdots p_n$, which is the Bonse inequality of #1. As with the property of 30 one can use #16(iv) to prove that when $n > p_{2k}^{k}$ then there is a $t$, $1 \leq t \leq 2k$, such that $p_t, p_t^2, \cdots, p_t^k$ are all prime to $n$ and smaller than $n$. With $k=2$ we see all numbers $n \geq p_4^2 = 49$ are prime to a smaller composite number.

5. The polynomials $F_n(x)$ introduced in #17 are called *cyclotomic polynomials*. This particular use of these polynomials goes back to Bang and Sylvester in the late 1880's. The general Dirichlet theorem was first proved by "non-elementary" means by Dirichlet in 1837. The cyclotomic polynomials offer another of those curious instances where intuitive induction may lead one astray.

The 1st ten cyclotomic polynomials are:

$F_1(x) = x - 1$ $F_6(x) = x^2 - x + 1$

$F_2(x) = x + 1$ $F_7(x) = x^6 + x^5 + x^4 + x^3 + x^2 + x + 1$

$F_3(x) = x^2 + x + 1$ $F_8(x) = x^4 + 1$

$F_4(x) = x^2 + 1$ $F_9(x) = x^6 + x^3 + 1$

$F_5(x) = x^4 + x^3 + x^2 + x + 1$ $F_{10}(x) = x^4 - x^3 + x^2 - x + 1$.

Even this small sample of data might lead one to conjecture that all non-zero coefficients of $F_n(x)$ are 1 or -1. Further data up to $n = 104$ would support this conjecture. However, for $n = 105$ a coefficient 2 appears. If one lets $A_n$ be the largest modulus of a coefficient of $F_n(x)$ then as Schur proved in 1931 (see #18(x)), $\limsup A_n = \infty$. (See I. Schur, *Gesammelte Abhandlungen* III p. 460-1.) This was sharpened by Emma Lehmer [1936] and then later Erdös [1946] proved that for every $k$ there are infinitely many $n$ for which

$$A_n > n^k.$$

There is a continuing interest in these matters and the reader might see Zeitlin [1968] and/or Beiter [1971].

6. Similar results to that of #19 may be found in Dressler [1972,3] and the references contained therein.

# XV Quaternions, Complex Numbers, & Sums of 4 and 2 Squares

1. (Quaternions)

Let $\mathbb{R}$ and $\mathbb{C}$ stand for the fields of real and complex numbers respectively and let

$$\mathbb{R}' = \left\{ \begin{pmatrix} a & b \\ -b & a \end{pmatrix} \;\middle|\; a \text{ and } b \text{ are in } \mathbb{R} \right\} ;$$

$$\mathbb{R}'' = \left\{ \begin{pmatrix} a & -b & -c & -d \\ b & a & -d & c \\ c & d & a & -b \\ d & -c & b & a \end{pmatrix} \;\middle|\; a, b, c, \text{ and } d \text{ are in } \mathbb{R} \right\} ;$$

$$\mathbb{C}' = \left\{ \begin{pmatrix} a & b \\ -\bar{b} & \bar{a} \end{pmatrix} \;\middle|\; a \text{ and } b \text{ are in } \mathbb{C} \right\} , \text{ where}$$

$\bar{c}$ is the ordinary complex conjugate of $c$. In $\mathbb{R}''$ we write $1, i, j, k$ for the elements

$$\begin{pmatrix} 1&0&0&0 \\ 0&1&0&0 \\ 0&0&1&0 \\ 0&0&0&1 \end{pmatrix}, \begin{pmatrix} 0&-1&0&0 \\ 1&0&0&0 \\ 0&0&0&-1 \\ 0&0&1&0 \end{pmatrix}, \begin{pmatrix} 0&0&-1&0 \\ 0&0&0&1 \\ 1&0&0&0 \\ 0&-1&0&0 \end{pmatrix}, \begin{pmatrix} 0&0&0&-1 \\ 0&0&-1&0 \\ 0&1&0&0 \\ 1&0&0&0 \end{pmatrix}$$

respectively.

with the usual operations of matrix addition and multiplication:

i) $\mathbb{C}'$ is a non-commutative field with $\mathbb{R}'$ a subfield;

ii) $\mathbb{R}'$ is isomorphic to $\mathbb{C}$;

iii) $\mathbb{C}'$ is isomorphic to $\mathbb{R}''$;

iv) $\mathbb{R}''$ is a 4 dimensional vector space over $\mathbb{R}$;

v) the set $\mathbb{Q}$ of all rational linear combinations of $1, i, j, k$ is a non-commutative subfield of $\mathbb{R}''$;

vi) multiplication in $\mathbb{R}''$ and in $\mathbb{Q}$ (see (v)) is characterized by the equations

$$i^2 = j^2 = k^2 = -1, \quad ij = k, \quad jk = i, \quad ki = j$$

and the fact that $1$ is a multiplicative identity.

Elements of $\mathbb{Q}$ are called *rational quaternions* while elements of $\mathbb{R}''$ are called *real quaternions*.

2. (Conjugate, Trace, Norm)

If $\alpha = a + ib + jc + kd$ is a quaternion (real or rational) we put $\bar{\alpha} = a - ib - jc - kd$, $T\alpha = 2a$, $N\alpha = a^2 + b^2 + c^2 + d^2$ and call these the conjugate, trace, and norm of $\alpha$, respectively.

i) $\alpha = \bar{\alpha}$ if and only if $b = c = d = 0$ ;

ii) $N\alpha = N\bar{\alpha}$ and $T\alpha = \alpha + \bar{\alpha}$ ;

iii) $N\alpha = 0$ if and only if $\alpha = 0$ ;

iv) $\overline{\alpha + \beta} = \bar{\alpha} + \bar{\beta}$ ;

v) $\overline{\alpha\beta} = \bar{\beta}\,\bar{\alpha}$ ;

vi) $N\alpha = \alpha\bar{\alpha}$ ;

vii) $N(\alpha\beta) = (N\alpha)(N\beta)$ ;

viii) each of $\alpha, \bar{\alpha}$ satisfies the equation $x^2 - xT\alpha + N\alpha = 0$ ; this equation is called the principal equation for $\alpha$.

3. (Integral Quaternions)

In this problem we deal exclusively with rational quaternions, i.e. with elements of $\mathcal{Q}$ (see #1(v)). We use $\mathbb{Z}$ for the set of rational integers and call an element of $\mathcal{Q}$ *integral* if it is a zero of a monic polynomial with rational integer coefficients. We put $\rho = \frac{1}{2}(1 + i + j + k)$ and define $\mathcal{L}, \mathcal{H}, I$ by:

$\mathcal{L} = \{a + ib + jc + kd \mid a, b, c, d \text{ are in } \mathbb{Z}\}$;

$\mathcal{H} = \mathcal{L} \cup \{\rho + \alpha \mid \alpha \text{ is in } \mathcal{L}\}$:

$I$ = set of all integral quaternions.

i) $\alpha$ is in $I$ if and only if $T\alpha$ and $N\alpha$ are in $\mathbb{Z}$;

ii) $\mathcal{L} \subsetneqq \mathcal{H} \subsetneqq I$;

iii) if $A$ is either $\mathcal{L}$ or $\mathcal{H}$ and if $\alpha \in A$ then $\bar{\alpha}$, $i\alpha$, $j\alpha$, $k\alpha$ are in $A$;

iv) $I$ is not closed under multiplication, not even under left multiplication by $i$;

v) $\mathcal{L}$ and $\mathcal{H}$ are integral domains while $I$ is not ;

vi) $\mathcal{H}$ is a maximal integral domain in $I$ ;

vii) $\mathcal{H} = \{a\rho + ib + jc + kd \mid a, b, c, d \text{ are in } \mathbb{Z}\}$.

We call $\mathcal{L}$ the *Lipschitz'* and $\mathcal{H}$ the *Hurwitz'* integral subdomain of $Q$. In the following problems all quaternions unless stated specifically to the contrary are to be taken from $\mathcal{H}$. When noted they may often be even more restricted and be taken from $\mathcal{L}$. We systematically use small Greek letters for quaternions.

A quaternion $\alpha$ is called a *unit* if $N\alpha = 1$. If $\alpha = \beta\gamma$ we call $\beta$ a *left divisor* of $\alpha$ and if $\alpha = \gamma\beta$ we call $\beta$ a *right divisor* of $\alpha$. If there are units $\nu$, $\mu$ such that $\alpha = \mu\beta\nu$ we call $\alpha$ and $\beta$ *associates*. If $\mu = 1$ we call $\beta$ a *left associate* of $\alpha$.

A common left divisor of $\alpha$ and $\beta$ which is left divisible by every common left divisor of $\alpha$ and $\beta$ is called a *left gcd* of $\alpha$ and $\beta$. In #4(v) below it is shown that left gcd's always exist. For each $\alpha, \beta$ we let $(\alpha, \beta)$ denote one such gcd of $\alpha$ and $\beta$.

4. i) Every element $\alpha$ has a left associate in $\mathcal{L}$;

ii) given $m \in \mathbb{Z}$ and $\alpha \in \mathcal{H}$ there is a $\beta \in \mathcal{H}$ such that $N(\alpha - m\beta) < m^2$ ; i.e. every "open circle" of radius $m^2$ contains an element of $\mathcal{H}$ which is divisible by $m$;

iii) (analogue of the Euclidean algorithm) if $\alpha$ and $\beta$ are in $\mathcal{H}$, $\beta \neq 0$, then there exist $\gamma_1, \delta_1, \gamma_2, \delta_2$ in $\mathcal{H}$ such that

$$\alpha = \beta\gamma_1 + \delta_1, \quad N\delta_1 < N\beta \quad ;$$

$$\alpha = \gamma_2\beta + \delta_2, \quad N\delta_2 < N\beta \quad ;$$

iv) if $\delta$ is a left gcd of two quaternions then $\delta_1$ is a left gcd of these quaternions if and only if $\delta$ and $\delta_1$ are left associates ;

v) for fixed $\alpha, \beta$ in $\mathcal{H}$ if $\delta$ is an element of $A = \{\alpha\mu + \beta\nu \mid \mu, \nu \in \mathcal{H}\}$ such that

$$0 < N\delta \leq N\gamma \text{ for all } \gamma \in A$$

then $\delta$ is a left gcd of $\alpha$ and $\beta$ ;

vi) there are exactly 24 units in $\mathcal{H}$ and each of them is a two sided divisor of all elements of $\mathcal{H}$ ; there are exactly 8 units in $\mathcal{L}$ and they are $\pm 1, \pm i, \pm j, \pm k$ ;

vii) if $n$ is a rational integer $> 1$ and $n$ divides $N\alpha$ then $(\alpha, n)$ is not a unit.

5. (Primes)

An element $\alpha$ in $\mathcal{H}$ is called *composite* if it may be written as the product of two elements of $\mathcal{H}$ each having norm $> 1$. Non-zero elements of $\mathcal{H}$ which are neither units nor composite are called *prime*. If $\alpha$ in $\mathcal{H}$ has no rational integer divisor other than $\pm 1$ then $\alpha$ is said to be *primitive*.

i) All associates of a prime are prime;

ii) if $N\alpha$ is a rational prime then $\alpha$ is a prime in $\mathcal{H}$;

iii) if $p$ is a rational prime dividing the norm of a primitive $\alpha$ then $(\alpha, p)$ is prime in $\mathcal{H}$ and $N((\alpha, p)) = p$;

iv) every rational prime is the norm of a prime in $\mathcal{H}$ and, therefore, is not a prime in $\mathcal{H}$;

v) $N\alpha$ is a rational prime if and only if $\alpha$ is prime in $\mathcal{H}$;

vi) if 2 divides $N\alpha$ then $\alpha = (1+i)\beta$ for suitable $\beta$ in $\mathcal{H}$;

vii) every element $\alpha$ in $\mathcal{H}$ may be written in the form $\alpha = (1+i)^r m\beta\nu$, where $r, m$ are non-negative rational integers, $r = 0$ or $1$, $\nu$ is a unit in $\mathcal{H}$, and $\beta$ is a primitive element of $\mathcal{L}$ of odd norm;

viii) suppose $\alpha$ and $\gamma$ are primitive and $N\alpha = 2^r N\gamma$, $N\gamma = p_1 \cdots p_s$, where $p_1, \cdots, p_s$ are odd primes in $\mathbb{Z}$ (not necessarily distinct); then there exist primes $\pi_1, \cdots, \pi_s$ in $\mathcal{H}$ such that $N\pi_t = p_t$, $1 \le t \le s$, and $\gamma = \pi_1 \cdots \pi_s$,

$$\alpha = (1+i)^r \pi_1 \cdots \pi_s ;$$

ix) for $\alpha$, $N\alpha$ as in (viii), if $\alpha = (1+i)^r \tau_1 \cdots \tau_s$, where the $\tau_s$ are primes in $\mathcal{H}$ and $N\tau_t = p_t$, $1 \le t \le s$, then for each $t$, $\tau_t$ and $\pi_t$ are associates ;

x) if $\alpha$ is a primitive element of $\mathcal{H}$ and $N\alpha = 2^r p_1 \cdots p_s$, where the $p_j$ are odd primes and $r = 0$ or $1$, then there exist unique, up to associates, primes $\pi_1, \cdots, \pi_s$ in $\mathcal{H}$ such that

$$\alpha = (1+i)^r \pi_1 \cdots \pi_s , N\pi_j = p_j \text{ for } 1 \le j \le s ;$$

xi) non-primitive elements of $\mathcal{H}$ may have distinct prime factorizations.

6. (Number of quaternions with given norm)

i) The number of quaternions with norm 2 is exactly 24 and they are all in $\mathcal{L}$ ;

ii) in this part let $p$ be an odd prime in $\mathbb{Z}$ and suppose $A, B, C, D$ are integers (mod $p$) in $\mathbb{Z}$. Then :

a) given $a, b$ in $\mathbb{Z}$ the congruences

1) $A^2 + B^2 + C^2 + D^2 \equiv 0 \pmod p$ and

2) $A^2 + (-aA + B)^2 + (-bA + C)^2 + D^2 \equiv 0 \pmod p$

have the same number of solutions $A, B, C, D$;

b) $a, b$ may be chosen so that $p$ divides $1 + a^2 + b^2$ and (2) of (a) becomes

3) $B^2 + C^2 + D^2 \equiv 2A(aB + bC) \pmod p$;

c) if in (3), $aB + bC \equiv 0 \pmod p$ then there are $p$ solutions when $B \equiv C \equiv 0 \pmod p$ and $2p(p-1)$ solutions otherwise;

d) if in (3), $aB + bC \not\equiv 0 \pmod p$ then there are $p(p^2 - p)$ solutions;

e) the number of solutions of (1), and therefore the number (modulo $p$) of $\alpha$'s in $\mathcal{L}$ such that $N\alpha \equiv 0 \pmod{p}$ is $(p^2-1)(p+1)+1$.

iii) In this part we write $i = i_1$, $j = i_2$, $k = i_3$ and suppose all subscripts larger than 3 to be reduced modulo 3. Thus $i_4 = i_1$, $a_5 = a_2$, $x_4 = x_1$, etc. Further, $p$ is an odd prime in $Z$, $\alpha = a_0 + i_1 a_1 + i_2 a_2 + i_3 a_3$ is in $\mathcal{L}$, and $p$ divides $N\alpha$ but does not divide $\alpha$.

a) There is a $\nu$, $1 \leq \nu \leq 3$, such that $p$ does not divide $a_0^2 + a_\nu^2$;

b) for $\nu$ as in (a) and $x = x_0 + i_1 x_1 + i_2 x_2 + i_3 x_3$ in $\mathcal{L}$ define $\beta = a_0 + a_\nu i_\nu$ $\quad \eta = x_0 + x_\nu i_\nu$

$$\gamma = a_{\nu+1} + a_{\nu+2} i_\nu \quad \delta = x_{\nu+1} + x_{\nu+2} i_\nu ;$$

then 1) $\alpha = \beta + \gamma i_{\nu+1}$, $\quad x = \eta + \delta i_{\nu+1}$;

2) if $a, b, c, d$ are scalars then $a + b i_\nu$ and $c + d i_\nu$ commute;

3) if $\omega$ is any of $\beta, \gamma, \eta, \delta$ then

$$\omega i_{\nu+1} = i_{\nu+1} \bar{\omega} ;$$

c) $p$ divides $\alpha x$ if and only if $p$ divides

$$\beta\eta - \gamma\xi \,;$$

d) $\alpha x \equiv 0 \pmod{p}$ has, modulo $p$, exactly $p^2$ solutions in $\mathcal{L}$.

iv) If $p$ is an odd prime in $\mathbb{Z}$ then, aside from associates, there are exactly $p+1$ primes in $\mathcal{L}$ of norm $p$.

7. (Jacobi's Theorem)

Let $m$ be an odd number with the ordered factorization

$$m = \underbrace{p_1 \cdots p_1}_{\alpha_1} \underbrace{p_2 \cdots p_2}_{\alpha_2} \cdots \underbrace{p_t \cdots p_t}_{\alpha_t} \,,$$

where $p_1, \cdots, p_t$ are distinct primes in $\mathbb{Z}$ and $\alpha_1, \cdots, \alpha_t$ are positive integers. Then for each $\alpha$ in $\mathcal{L}$ with $N\alpha = m$ there is a prime factorization $\alpha = \pi_{11} \cdots \pi_{1\alpha_1} \cdots \pi_{t1} \cdots \pi_{t\alpha_t}$ such that $N\pi_{\nu\eta} = p_\nu$ for $1 \le \nu \le t$, $1 \le \eta \le \alpha_\nu$. Note that the primes $\pi_{\nu\eta}$ may be in $\mathcal{H} \setminus \mathcal{L}$.

i) If for a given $\nu$ there is an $\eta$ such that $\Pi_{\nu,\eta+1}$ and $\bar{\Pi}_{\nu\eta}$ are associates then $\alpha$ is not primitive;

ii) if $\alpha$ is not primitive there exists a pair $\nu, \eta$ such that $\Pi_{\nu,\eta+1}$ and $\bar{\Pi}_{\nu\eta}$ are associates;

iii) the number of primitive $\alpha$ in $\mathcal{L}$ with $\mathcal{N}\alpha = m$ is $8(p_1+1)p_1^{\alpha_1-1}\cdots(p_t+1)p_t^{\alpha_t-1}$

$$= 8m\prod_{p|m}\left(1+\frac{1}{p}\right) \quad ;$$

iv) the number of $\alpha$ in $\mathcal{L}$ with $\mathcal{N}\alpha = m$ is

$$8\sum_{d^2|m}\frac{m}{d^2}\prod_{p|\frac{m}{d^2}}\left(1+\frac{1}{p}\right) = 8\sigma(m) \quad ;$$

v) the number of $\alpha$ in $\mathcal{L}$ with $\mathcal{N}\alpha = n$, $n$ an even integer in $\mathbb{Z}$, is $24\sigma^\circ(n)$, where $\sigma^\circ(n)$ is the sum of the odd divisors of $n$;

vi) (Jacobi's Theorem) the number of representations of a positive integer $n$ as a sum of 4 squares, representations which differ only in order or sign being counted as distinct, is 8 times the sum of the divisors of $n$ which are not divisible by 4.

8. Write $i = i_1$, $j = i_2$, $k = i_3$ and put $G_t = \{a + i_t b \mid a, b \text{ in } \mathbb{Z}\}$, $1 \le t \le 3$. Then $G_t \subset \mathcal{H}$ and, under the induced operations from $\mathcal{H}$, is a commutative subring of $\mathcal{H}$. It is easy to verify that $G_1, G_2, G_3$ are isomorphic. We write G for any of these and shall call the elements of G *Gaussian integers*. It is clear that if $\mathbb{C}$ is the set of all complex numbers one can write (where equality here indicates isomorphism) $G = \mathcal{L} \cap \mathbb{C}$. Further,

$$\mathcal{L} = \{\alpha + i_2\beta \mid \alpha, \beta \text{ in } G_1\} = \{\alpha + i_3\beta \mid \alpha, \beta \text{ in } G_2\}$$
$$= \{\alpha + i_3\beta \mid \alpha, \beta \text{ in } G_3\}.$$

If one replaces $\mathcal{H}$ by G in #4-7 much of that theory carries over, in a simplified way (since we now have commutativity) to G. Carrying out the details leads to the theorem:

if the canonical prime factorization of the rational integer $n$ contains a $4k+3$ prime to an odd power then $n$ is not expressible as the sum of two rational integer squares; otherwise the number of representations as such a sum is $4(d_1(n)-d_3(n))$, where $d_j(n)$ is the number of odd divisors of $n$ which are congruent to $j$ modulo 4.

9. Let $r_s(n)$ be the number of representations of $n$ as a sum of $s$ squares and define $f_s(n)$ by $f_s(n)=(2s)^{-1}r_s(n)$. Then

i) if $n=ab$, $(a,b)=1$, then

a) $d_1(n)=d_1(a)d_1(b)+d_3(a)d_3(b)$;

b) $d_3(n)=d_3(a)d_1(b)+d_1(a)d_3(b)$;

c) $f_2$ is multiplicative;

here $d_1$ and $d_2$ are as in #8;

ii-a) $r_4(n) = \begin{cases} 8\sigma(n) & \text{for } n \text{ odd;} \\ 24\sigma^\circ(n) & \text{for } n \text{ even,} \end{cases}$

where $\sigma^\circ(n)$ is the sum of the odd divisors of $n$;

b) $f_4$ is multiplicative;

iii-a) $r_s(2) = 4\binom{s}{2}$;

b) $r_s(3) = 8\binom{s}{3}$;

c) $r_s(6) = 64\binom{s}{6} + 24\binom{s}{3}$;

d) $f_s(6) - f_s(2)f_s(3)$

$= \frac{2}{45}s(s-1)(s-2)(s-4)(s-8)$;

e) (Bateman [1969]) the only possible positive integers $s$ for which $f_s(n)$ is multiplicative are 1, 2, 4, 8.

Remarks.

1. Clearly $f_1$ is multiplicative and in view of #9 (i & ii) each of $f_2, f_4$ is multiplicative. Using the expression $f_8(n) = (-1)^{n-1}16\sum_{d|n}(-1)^{d-1}d^3$ (see Dickson's *History* II p. 315) it is easy to

see that $f_8$ is multiplicative. Thus $f_s$ is multiplicative precisely for $s = 1, 2, 4, 8$.

2. If $A$ is any associative algebra over a field $F$ with basis $e_1, \cdots, e_n$ one can define a function $N : A \to F$ by the equation

$$N(a_1e_1 + \cdots + a_ne_n) = a_1^2 + \cdots + a_n^2 .$$

When this function satisfies the equation $N(ab) = N(a)N(b)$ for all $a, b$ in $A$ one calls the algebra a *normed algebra*. If $A$ is a normed algebra with identity $1$ (we identify $a \in F$ with $a \cdot 1 \in A$) one may define a *conjugate function* $\bar{a}$ satisfying $\bar{a} \in A$, $\overline{ab} = \bar{b}\,\bar{a}$, $a\bar{a} = Na$ for all $a, b$ in $A$. Further, putting $Ta = a + \bar{a}$ we see that each $a \in A$ satisfies its monic second degree equation $x^2 - x \cdot Ta + Na = 0$. The real numbers, the complex numbers, and the real quaternions afford three examples of normed algebras with identity over the reals.

Suppose, now, $A$ is an arbitrary normed algebra with identity over the reals and suppose $\bar{a}$ is the conjugate of $a$ mentioned above. Put $B = A \times A$ and define multiplication in $B$ by $(a, b)(c, d) = (ac - \bar{d}b, da + b\bar{c})$. Identifying $1, j$ with $(1, 0), (0, 1)$ of $B$ each element $(a, b)$ of $B$ may be written $a + jb$. Putting $\bar{x} = \bar{a} - jb$ when $x = a + jb$ (the bar on the right of the expression for $\bar{x}$ is the conjugate in $A$) and defining $Nx = x\bar{x}$ one can prove, when $A$ is associative, that $B$ is a normed algebra with identity over the reals. Starting with $A$ the real numbers, $B$ turns out to be the complex numbers. Taking $A$ to be the complex numbers, $B$ is the set of real quaternions. Finally, starting with $A$ as the real quaternions one obtains for $B$ the so-called algebra of Cayley numbers. At this point, since the algebra of Cayley numbers is not associative, no new normed algebras over the reals arise. In fact, it can be shown that only the four normed algebras mentioned exist (see Curtis [1963]). Consideration of the norm function in the system of integral Cayley numbers leads to the 8 square theorem (see the Remarks in XI p. 86, Coxeter [1946], Curtis [1963], and Dickson [1927]). For the general arithmetic properties of quaternions, not only the above references but also Redei [1967], Dickson [1919, 1923], MacDuffee [1940], Hurwitz [1896, 1919] might be consulted. Finally we mention the paper by Линник [1949] and the references therein.

3. An integer is a sum of 3 squares when it is not of the form $4^s(8t+7)$, $s \geq 0$, $t \geq 0$. For an elementary discussion see Weil [1974].

# XVI Brun's Theorem

Pairs of primes of the form $p, p+2$ are called *twin primes*. The number of primes $p$, $p \le x$, for which $p+2$ is prime is denoted by $\pi_2(x)$. It is not known if $\pi_2(x)$ increases without bound as $x$ increases. Nevertheless, in 1919 Viggo Brun found an upper bound for $\pi_2(x)$ which, though increasing without bound as $x$ increases, was sufficiently small to show that the sum $\Sigma \frac{1}{p}$, taken over those primes $p$ for which $p+2$ is prime, converges. In this chapter a proof of this result is given.

Throughout, $x$ and $z$ are positive real numbers with $2 < z < \sqrt{x}$, $R$ is the product of the distinct primes not exceeding $z$, $a_n = n(n+2)$ for $1 \le n \le [x]$, and $\nu(n)$ is the number of distinct prime divisors of $n$.

1. Let $S = \sum_{\substack{n=1 \\ (a_n, R)=1}}^{[x]} 1$ . Then $\pi_2(x) \le z + S$ .

2. Let $S_\delta$ be the number of $n$, $1 \le n \le [x]$, for which $\delta \mid a_n$ and suppose $2k \le \nu(R)$. Then $S$, as in #1, satisfies $S = \sum_{\delta \mid R, \nu(\delta) \le 2k} (-1)^{\nu(\delta)} S_\delta$ .

3. Putting $\rho(\delta)$ for the number of $n$ in a complete system of residues mod $\delta$ for which $\delta \mid a_n$ then for $p$ any prime divisor of $R$ we have

$$\rho(p) = \begin{cases} 2 & \text{for } p \text{ odd} ; \\ 1 & \text{for } p = 2 , \end{cases}$$

and, therefore, for $\delta$ any divisor of $R$

$$\rho(\delta) = \begin{cases} 2^{\nu(\delta)} & \text{for } \delta \text{ odd} ; \\ 2^{\nu(\delta)-1} & \text{for } \delta \text{ even} . \end{cases}$$

4. For each divisor $\delta$ of $R$ there is a number $\vartheta$, $|\vartheta| \le 1$, such that

$$S_\delta = \left( \frac{x}{\delta} + \vartheta \right) \rho(\delta) ,$$

where $S_\delta$, $\rho(\delta)$ are as in #2, #3 .

5. The $S$ of #1 and #2 satisfies the inequality

$$S \le x(T_1 + T_2) + T_3, \text{ where}$$

a) $T_1 = \sum_{\delta | R} (-1)^{\nu(\delta)} \frac{\rho(\delta)}{\delta} = \frac{1}{2} \prod_{2 < p \le z} \left(1 - \frac{2}{p}\right)$;

b) $T_2 = \sum_{\delta | R, \nu(\delta) > 2k} \frac{\rho(\delta)}{\delta} \le \sum_{j=2k+1}^{\infty} \sum_{\nu(\delta)=j} \frac{2^j}{\delta}$;

c) $T_3 = \sum_{\delta | R, \nu(\delta) \le 2k} \rho(\delta) \le \sum_{j=0}^{2k} \binom{\pi(z)}{j} 2^j$.

6. It is possible to choose a positive constant $A$ so that $eA \ln 2 > 1$ and $\sum_{p \le t} \frac{1}{p} < A \ln \ln t$, for all $t$.

7. Taking $A$ as in #6 and putting $z = x^{\alpha}$, $\alpha = (6eA \ln \ln x)^{-1}$, there is an $x_0$ such that for $x \ge x_0$

$$2[2eA \ln \ln z] + 2 < \pi(z);$$

$$\frac{\ln x}{\ln \ln x} > 24eA;$$

$$\frac{1}{eA \ln \ln x} < \frac{1}{3}.$$

8. For $A$ as in #6 and $z$, $x_0$ as in #7 and with

$$k = [2eA \ln\ln z] + 1$$

we have, for $x \geq x_0$, $2 < z < \sqrt{x}$ and $2k < \nu(R)$.

9. Let $A, z, x_0, k$ be as in #6-8; then for $x \geq x_0$ and all sufficiently large positive constants $B$ we have:

a) $$T_1 \leq \frac{1}{2}\prod_{p|R}\left(1-\frac{1}{p}\right)^2 = \frac{1}{2}\,\frac{1}{\left(\prod_{p|R}\left(1-\frac{1}{p}\right)^{-1}\right)^2}$$

$$< \frac{1}{2}\,\frac{1}{\left(\sum_{m=1}^{z}\frac{1}{m}\right)^2} < \frac{B}{(\ln z)^2};$$

b) $$T_2 \leq \sum_{j=2k+1}^{\infty}\left(\sum_{p\leq z}\frac{1}{p}\right)^j \frac{2^j}{j!} \leq \sum_{j=2k+1}^{\infty}\left(\frac{2eA\ln\ln z}{j}\right)^j$$

$$< 2^{-2k} < \frac{1}{(\ln z)^2};$$

c) $$T_3 \leq (\pi(z))^{2k}\sum_{j=0}^{2k}\frac{2^j}{j!} < e^2(\pi(z))^{2k} < 9z^{2k}.$$

10. For a suitable positive constant $C$ and all sufficiently large $x$ we have

$$\pi_2(x) < Cx\left(\frac{\ln\ln x}{\ln x}\right)^2.$$

11. (Brun 1919)

a) $\sum \frac{1}{p}$, where the sum is over all primes $p$ such that $p+2$ is a prime, converges.

b) $\sum \frac{1}{p}$, where the sum is over all primes occurring in a twin prime pair, converges.

Remarks.

1. The theorem of this chapter was first proved by Brun [1919]. Expositions will be found in Gelfond, Linnik [1965], Rademacher [1964], Landau [1958]. Extensions of the method of Brun may be found in Trost [1967], Prachar [1957], and Halberstam, Roth [1966].

2. The conjecture that $\pi_2(x)$ tends to infinity with $x$ is called the *twin prime conjecture*. This is the special case $k=2$, $b_1=0$, $b_2=2$ of the conjecture: if $b_1, \cdots, b_k$ are non-negative integers such that for each prime $p$ there is an integer

$n$ with $b_1+n, \cdots, b_n+n$ all non-divisible by $p$ then for infinitely many values of $n$ the latter $k$ numbers are all prime. Though this conjecture has not been proved Richards [1974] has shown that it is not compatible with the following conjecture made by Hardy and Littlewood in 1923: $\pi(x+y) \leq \pi(x)+\pi(y)$ for $x, y$ integers $\geq 2$. An even more general conjecture, of which the twin prime conjecture is a very special case, goes under the name of *hypothesis H*. For details see Sierpinski [1964a] p 127 ff.

## XVII Quadratic Residues

1. i) Suppose $a > 0$, $D = b^2 - 4ac$, $f(x) = ax^2 + bx + c$. Then $f(x) \equiv 0 \pmod m$ if and only if

$$(2ax+b)^2 \equiv D \pmod{4am};$$

ii) when $(a, m) = d = t\delta^2$ where $t$ is the square free part of $d$, then the following two statements are equivalent:

a) $x^2 \equiv a \pmod m$ is solvable;

b) $(t, \frac{m}{d}) = 1$ and $x^2 \equiv \frac{ta}{d} \pmod{\frac{m}{d}}$ is solvable.

(Thus, solvability of a general quadratic congruence is reducible to that of a pure quadratic congruence of the form $x^2 \equiv a \pmod m$ which, in turn, may be reduced to a similar congruence in which $(a, m) = 1$.)

Definition: $n$ is a *quadratic residue* (*qr*) *of*, or *modulo*, $m$ if $x^2 \equiv n \pmod m$ is solvable; otherwise, $n$ is a *quadratic non-residue* (*qnr*) *of*, or *modulo*, $m$.

In the following whenever we speak about quadratic residues we exclude 0.

2. i) $x^2 \equiv 12 \pmod{45}$ is not solvable; i.e. 12 is a qnr of 45;

ii) $x^2 \equiv 252 \pmod{1575}$ is solvable if and only if $x^2 \equiv 3 \pmod{25}$ is solvable; i.e. 252 is a qr of 1575 if and only if 3 is a qr of 25.

3. In this problem all congruences $x^2 \equiv a \pmod{m}$ are to satisfy $(a, m) = 1$. This congruence, where $m = p^\alpha$, with $p$ a prime will be denoted by $(*_\alpha)$. Thus

$(*_\alpha)$ $x^2 \equiv a \pmod{p^\alpha}$, $(a, p) = 1$, $p$ prime.

i) If $p = 2$ then

a) $(*_1)$ is solvable and has 1 solution;

b) $(*_2)$ is solvable if and only if $a \equiv 1 \pmod{4}$;

c) $(*_2)$, when solvable, has exactly 2 solutions;

d) $(*_3)$ is solvable if and only if $a \equiv 1 \pmod 8$;

e) if $\alpha \geq 3$ then $(*_{\alpha+1})$ is solvable if and only if $(*_\alpha)$ is solvable;

f) for $\alpha \geq 3$, $(*_\alpha)$, when solvable, has exactly 4 solutions;

g) the number of solutions of $(*_\alpha)$, $\alpha \geq 1$, is:

$$\begin{cases} 1 \text{ for } \alpha = 1,\ a \text{ odd}; \\ 2 \text{ for } \alpha = 2,\ a \equiv 1 \pmod 4; \\ 4 \text{ for } \alpha \geq 3,\ a \equiv 1 \pmod 8; \\ 0 \text{ otherwise.} \end{cases}$$

ii) if $p > 2$ then

a) $(*_1)$ has 0 or 2 solutions;

b) if $x_0$ is a solution of $(*_\alpha)$ then there is a unique $t$, $0 \leq t < p$, such that $x_0 + tp^\alpha$ is a solution of $(*_{\alpha+1})$;

c) $(*_{\alpha+1})$ is solvable if and only if $(*_\alpha)$ is solvable;

d) $(*_\alpha)$, when solvable, has exactly 2 solutions;

iii) (Helping lemma from the theory of congruences) let $f$ be an integral polynomial and suppose $m_1, \cdots, m_r$ are pairwise relatively prime; let $N_j$, $1 \le j \le r$, be the number, modulo $m_j$, of solutions of $f(x) \equiv 0 \pmod{m_j}$ and let $N$ be the number, modulo $m_1 \cdots m_r$, of solutions of the system

$$f(x) \equiv 0 \pmod{m_1}, \cdots, f(x) \equiv 0 \pmod{m_r};$$

then $N = N_1 \cdots N_r$;

iv) a) $a$ is a qr of $m$ if and only if

$a$ is a qr of every prime divisor of $m$;

$a \equiv 1 \pmod 4$ when $4 | m$ and $8 \nmid m$;

$a \equiv 1 \pmod 8$ when $8 | m$;

b) when $a$ is a qr of $m$ the number of solutions of $x^2 \equiv a \pmod m$ is:

$$\begin{cases} 2^k & \text{if } 4 \nmid m; \\ 2^{k+1} & \text{if } 4 | m \text{ and } 8 \nmid m; \\ 2^{k+2} & \text{if } 8 | m, \end{cases}$$

where $k$ is the number of distinct odd prime factors of $m$.

(Thus, solvability of general quadratic congruences reduces to solvability of quadratic congruences of the form $x^2 \equiv a \pmod{p}$, where $p$ is an odd prime not dividing $a$.)

4. Let $p$ be an odd prime and suppose $p$ does not divide $a$. Then:

i) exactly one of the following congruences is valid: $a^{\frac{p-1}{2}} \equiv 1 \pmod{p}$, $a^{\frac{p-1}{2}} \equiv -1 \pmod{p}$;

ii) if $a$ is a qr of $p$ then $a^{\frac{p-1}{2}} \equiv 1 \pmod{p}$;

iii-a) $x^{p-1} - 1 = (x^2 - a)q(x^2) + a^{\frac{p-1}{2}} - 1$, where $q$ is a polynomial of degree $\frac{p-3}{2}$;

b) if $a^{\frac{p-1}{2}} \equiv 1 \pmod{p}$ then $(x^2-a)q(x^2) \equiv 0 \pmod{p}$ is satisfied for all $x$ not divisible by $p$;

iv) (Euler's criterion)

a is a qr of p if and only if
$a^{\frac{p-1}{2}} \equiv 1 \pmod{p}$;

a is a qnr of p if and only if
$a^{\frac{p-1}{2}} \equiv -1 \pmod{p}$.

Definition: For $p$ an odd prime not dividing $a$

we write $\left(\frac{a}{p}\right) = \begin{cases} 1 & \text{if } a \text{ is a qr of } p\,; \\ -1 & \text{if } a \text{ is a qnr of } p. \end{cases}$

This symbol $\left(\frac{a}{p}\right)$ is called the *Legendre symbol*.

5. Let $p$ be an odd prime not dividing $ab$. Then:

i) $\left(\frac{a^2}{p}\right) = 1$ ;

ii) $\left(\frac{a}{p}\right) \equiv a^{\frac{p-1}{2}} \pmod p$ (this is also called Euler's criterion);

iii) if $a \equiv b \pmod p$ then $\left(\frac{a}{p}\right) = \left(\frac{b}{p}\right)$ ;

iv) if $p \nmid ab$ then $\left(\frac{ab}{p}\right) = \left(\frac{a}{p}\right)\left(\frac{b}{p}\right)$ ;

v- a) if $0 \le a < b \le \frac{p-1}{2}$ then $a^2 \not\equiv b^2 \pmod p$ ;

b) if $0 < a \le p-1$ then there is a $b$, $0 < b < \frac{p-1}{2}$ such that $a^2 \equiv b^2 \pmod p$ ;

c) if $\left(\frac{a}{p}\right) = 1$ then $a \equiv b^2 \pmod p$ for some $b$, $0 < b \le \frac{p-1}{2}$;

vi) $\left(\frac{-1}{p}\right) = (-1)^{\frac{p-1}{2}}$ .

(Referring to the result in (vi) Gauss wrote in the introduction to his monumental *Disquisitiones Arithmeticae* the following: "Engaged in other work I chanced upon an extraordinary arithmetic truth .... Since I considered it to be so beautiful in itself and since I suspected its connection with even more profound results, I concentrated on it all my efforts in order to understand the principles on which it depended and to obtain a rigorous proof.")

6\. *i*) Every square is a qr of $p$;

*ii*) numbers congruent modulo $p$ are either both qr or both qnr of $p$;

*iii*) the product of two numbers with the same quadratic character modulo $p$ is a qr of $p$, while the product of two numbers with different quadratic character modulo $p$ is a qnr of $p$;

iv) half of the numbers $1,2,\cdots,p-1$ are qr and half are qnr of $p$.

7. Let $p$ be an odd prime.

i) if $\left(\frac{n}{p}\right)=-1$ then $\sum_{d|n} d^{\frac{p-1}{2}} \equiv 0 \pmod p$;

ii) $\sum_{k=1}^{p-1}\left(\frac{k}{p}\right)=0$;

iii-a) at least one of $a, b, ab$ is a qr of $p$ when $p \nmid ab$;

b) every prime divides at least one value of $x^6-11x^4+36x^2-36$;

iv) if $(ax_0, by_0)=1$ then $ax_0^2+by_0^2 \equiv 0 \pmod p$ implies $\left(\frac{a}{p}\right)=\left(\frac{-b}{p}\right)$.

8. Let $p$ be an odd prime. Then:

i) $\prod_{\substack{a \\ \left(\frac{a}{p}\right)=1}} a \equiv -\left(\frac{-1}{p}\right) \pmod p$ and $\prod_{\substack{a \\ \left(\frac{a}{p}\right)=-1}} a \equiv \left(\frac{-1}{p}\right) \pmod p$;

ii) if $a_1,\cdots,a_s$ are the qr of $p$ among $1,2,\cdots,\frac{p-1}{2}$ then

a) if $p \equiv 1 \pmod 4$ then $p-a_1,\cdots,p-a_s$ are the qr of $p$ among $\frac{p+1}{2},\cdots,p-1$;

b) if $p \equiv 3 \pmod 4$ then $p-a_1, \cdots, p-a_s$ are the qnr of $p$ among $\frac{p+1}{2}, \cdots, p-1$;

c) if $p \equiv 1 \pmod 4$ then the qr of $p$ are symmetrically distributed about $\frac{p}{2}$;

d) $(-1)^{s+1} \equiv \begin{cases} (a_1 \cdots a_s)^2 \pmod p & \text{if } p \equiv 1 \pmod 4; \\ (\frac{p-1}{2})! \pmod p & \text{if } p \equiv 3 \pmod 4. \end{cases}$

9. The mod $p$ residues of $a, 2a, \cdots, \frac{p-1}{2}a$ which lie between $\frac{-p}{2}$ and $\frac{p}{2}$ will be denoted by $a_1, \cdots, a_{\frac{p-1}{2}}$. Again $p$ is an odd prime and we assume $p$ does not divide $a$. Then:

i) the $|a_j|$, $1 \le j \le \frac{p-1}{2}$, are pairwise unequal;

ii) if $\nu$ is the number of $a_1, \cdots, a_{\frac{p-1}{2}}$ which are negative then

$$(-1)^{\nu}\left(\tfrac{p-1}{2}\right)! = a_1 \cdots a_{\frac{p-1}{2}} \equiv a^{\frac{p-1}{2}}\left(\tfrac{p-1}{2}\right)! \pmod p;$$

iii) (Lemma of Gauss)

$$\left(\frac{a}{p}\right) = (-1)^{\nu}, \text{ where } \nu \text{ is as in (ii)}.$$

In problems #10-13 the $\nu$ is always that introduced in #9(ii).

10. Let $p$ be an odd prime and suppose $a = 2$. Then:

i) $\nu$ is the number of $j$, $1 \le j \le \frac{p-1}{2}$, for which $\frac{p}{2} < 2j < p$;

ii) $$\nu = \left[\frac{p}{2}\right] - \left[\frac{p}{4}\right] \equiv \begin{cases} 0 \pmod 2 & \text{if } p \equiv \pm 1 \pmod 8; \\ 1 \pmod 2 & \text{if } p \equiv \pm 3 \pmod 8; \end{cases}$$

iii) $\left(\frac{2}{p}\right) = (-1)^{\frac{p^2-1}{8}}$;

iv) 2 is a qr of all $8k \pm 1$ primes and a qnr of all $8k \pm 3$ primes.

(Fermat claimed to have a proof of #10(iv) but did not publish it. Euler tried to prove it but without success. Lagrange gave the first correct proof in 1775.)

11. Let $p$ be an odd prime other than 3 and suppose $a = 3$. Then:

i) $\nu$ is the number of $j$, $1 \le j \le \frac{p-1}{2}$, for which $\frac{p}{2} < 3j < p$;

ii) $\nu = [\frac{p}{3}] - [\frac{p}{6}] \equiv \begin{cases} 0 \pmod 2 & \text{if } p \equiv \pm 1 \pmod{12}; \\ 1 \pmod 2 & \text{if } p \equiv \pm 5 \pmod{12}; \end{cases}$

iii) 3 is a qr of all $12k \pm 1$ primes and a qnr of all $12k \pm 5$ primes.

(Fermat knew these results but they were first proved by Euler. It is interesting to note, as does Gauss in his *Disquisitiones Arithmeticae*, that even after Euler proved #11 (iii) he was unable to prove #10 (iv).)

12. Let $p$ be an odd prime other than 5 and suppose $a = 5$. Then:

i) $\nu$ is the number of $j$, $1 \le j \le \frac{p-1}{2}$, for which $\frac{p}{2} < 5j < p$ or $\frac{3p}{2} < 5j < 2p$;

ii) $\nu = [\frac{p}{5}] - [\frac{p}{10}] + [\frac{2p}{5}] - [\frac{3p}{10}]$;

iii) 5 is a qr of all $20k \pm 1$ and $20k \pm 9$ primes and is a qnr of all $20k \pm 3$ and $20k \pm 7$ primes.

13. ( The Quadratic Reciprocity Law )

Let $p$ and $q$ be different odd primes . Then :

i) if $p \equiv \pm q \pmod{4a}$ then $\left(\frac{a}{p}\right) = \left(\frac{a}{q}\right)$ ;

ii) there is an $a$ such that $(a, pq) = 1$ and

$$p = \pm q + 4a \ ;$$

iii) $$\left(\frac{p}{q}\right) = \begin{cases} \left(\frac{-q}{p}\right) & \text{if } p \equiv q \pmod 4 \ ; \\ \left(\frac{q}{p}\right) & \text{if } p \not\equiv q \pmod 4 \ ; \end{cases}$$

iv) ( the reciprocity law )

$$\left(\frac{p}{q}\right)\left(\frac{q}{p}\right) = (-1)^{\frac{p-1}{2} \cdot \frac{q-1}{2}} \ .$$

( Bachmann in his *Niedere Zahlentheorie* (v. 1 p. 202) calls this theorem one of the most beautiful and important in all of number theory. Euler discovered the law by induction in 1738 . Legendre invented his symbol in 1785 and stated the law in the form given in #13 (iv) . Gauss discovered the law independently in 1795 . He claimed that his first proof , which was the first known, eluded his most strenuous efforts for more than a year . Altogether Gauss gave

8 proofs of the law before he died. Bachmann catalogues over 45 proofs as of 1901 and there have been many more discovered since that time. In the 2nd volume of Gauss' collected works one finds a table, computed by Gauss, listing the values of $\left(\frac{p}{q}\right)$ for $2 \le p \le 997$, $3 \le q \le 503$.)

14. Consider the pair of congruences
$$x^2 \equiv p \pmod{q}, \quad x^2 \equiv q \pmod{p},$$
where $p$ and $q$ are odd primes. If at least one of $p$ and $q$ is a $4k+1$ prime then the two congruences are either both solvable or both not solvable while if both $p$ and $q$ are $4k+3$ primes then exactly one of the congruences is solvable.

Definition (The Jacobi Symbol): Let $m = p_1^{\alpha_1} \cdots p_k^{\alpha_k}$ be a positive odd integer. For $(n,m)=1$ put
$$\left(\frac{n}{m}\right) = \left(\frac{n}{p_1}\right)^{\alpha_1} \cdots \left(\frac{n}{p_k}\right)^{\alpha_k}.$$
This symbol $\left(\frac{n}{m}\right)$ is called the Jacobi symbol.

15. Let $m$ and $n$ be odd positive integers and suppose $(a,m) = (b,m) = 1$. Then:

i) for $m$ a prime the Jacobi symbol $\left(\frac{n}{m}\right)$ is the Legendre symbol $\left(\frac{n}{m}\right)$;

ii) $\left(\frac{1}{m}\right) = 1 = \left(\frac{a^2}{m}\right)$;

iii) if $a$ is a qr of $m$ then $\left(\frac{a}{m}\right) = 1$;

iv) $\left(\frac{a}{m}\right)$ may equal 1 without $a$ being a qr of $m$;

v) if $a \equiv b \pmod{m}$ then $\left(\frac{a}{m}\right) = \left(\frac{b}{m}\right)$;

vi) $\left(\frac{a}{mn}\right) = \left(\frac{a}{m}\right)\left(\frac{a}{n}\right)$;

vii) $\left(\frac{ab}{m}\right) = \left(\frac{a}{m}\right)\left(\frac{b}{m}\right)$;

viii) $\left(\frac{-1}{m}\right) = (-1)^{\frac{m-1}{2}}$;

ix) $\left(\frac{2}{m}\right) = (-1)^{\frac{m^2-1}{8}}$;

x) $\left(\frac{n}{m}\right)\left(\frac{m}{n}\right) = (-1)^{\frac{n-1}{2}\cdot\frac{m-1}{2}}$.

16. i) Calculate

a) $\left(\frac{89}{197}\right)$; b) $\left(\frac{1050}{1573}\right)$; c) $\left(\frac{12345}{6789}\right)$;

ii) which of the following congruences are solvable;

a) $x^2 \equiv 89 \pmod{197}$; b) $x^2 \equiv 197 \pmod{89}$;

c) $x^2 \equiv 1050 \pmod{1573}$; d) $x^2 \equiv 1573 \pmod{1050}$;

e) $x^2 \equiv 111 \pmod{219}$; f) $x^2 \equiv 219 \pmod{111}$.

17. Let $f(x) = x^2 + x + 41$. Then:

i) no prime $< 41$ divides any value of $f(x)$;

ii) all values of $f(x)$ in absolute value $< 41^2$ are prime;

iii) $f(x)$ is prime for 80 consecutive integers.

(The polynomial in #17 represents a prime 4506 times in the first 11000 values of $x$. It was discovered in 1772 by Euler. A somewhat "better" prime representing polynomial is $x^2 + x + 72491$ which represents a prime 4923 times in the first 11000 values of $x$. See Beeger [1939], Szekeres [1974].)

18. (Theorem of Zolotareff)

Let $p$ be an odd prime and suppose $A$ is a reduced residue system modulo $p$. For each integer $a$ not divisible by $p$ there is a modulo $p$ reciprocal $a^{-1}$ and an element $\tilde{a}$ in $A$ congruent to $a$. Thus

$$aa^{-1} \equiv 1 \pmod{p}, \quad a \equiv \tilde{a} \pmod{p}, \quad \tilde{a} \in A.$$

For each integer $D$ not divisible by $p$ we define mappings $Z_D, T_D$ of $A$ into $A$ by

$$Z_D a = \widetilde{Da} \qquad , \qquad T_D a = \widetilde{Da^{-1}} \ .$$

i) $Z_D, T_D$ are permutations ;

ii) $T_D$ is an involution ; i.e. $T_D^{-1} = T_D$ ;

iii) $Z_D = T_D T_1$ ;

iv) if $\alpha_D$ is the number of elements kept fixed by $T_D$ then

$$\alpha_D = \begin{cases} 2 & \text{if } D \text{ is a qr of } p \ ; \\ 0 & \text{otherwise} \ ; \end{cases}$$

v) defining the signature of a permutation $\pi$, $\operatorname{sgn} \pi$, to be $1$ or $-1$ depending on whether the permutation is even or odd we see that

$$\operatorname{sgn} T_D = (-1)^{\frac{\varphi(p) - \alpha_D}{2}} \ , \ \operatorname{sgn} Z_D = (-1)^{\frac{\alpha_D + \alpha_1}{2}} \ ;$$

vi) $\alpha_1 = 2$ ;

vii) $\operatorname{sgn} Z_D = \left(\frac{D}{p}\right)$ ;

viii) (Theorem of Zolotareff)

D is a qr of p if and only if the least positive residues of $D, 2D, \cdots, (p-1)D$ constitute an even permutation of $1, 2, \cdots, p-1$;

ix) noting that A can be any reduced system of residues modulo p we can use the above to give independent evaluations of $\left(\frac{-1}{p}\right)$ and $\left(\frac{2}{p}\right)$.

19. (Quadratic reciprocity theorem from Zolotareff's theorem) Let $A^+ = \left\{1, 2, \cdots, \frac{p-1}{2}\right\}$, $A^- = \left\{-\frac{p-1}{2}, \cdots, -1\right\}$, and put $A = A^- \cup A^+$. Define $Z_D$ as in #18 and call $a', a''$ an *inversion* if $a' < a''$ and $Z_D a' > Z_D a''$. Then:

i) if $a', a''$ is an inversion so also is $-a'', -a'$;

ii) inversions occur in pairs except for those of the form $-a, a$;

iii) (Lemma of Gauss) $\operatorname{sgn} Z_D = (-1)^{\nu}$, where $\nu$ is the number of elements in $\left\{D, 2D, \cdots, \frac{p-1}{2}D\right\}$ with least absolute remainders, modulo p, in $A^-$;

iv) if $\mu$ is the number of elements in $\{q, 2q, \cdots, \frac{p-1}{2}q\}$ with negative least absolute residues modulo $p$ and $\nu$ is the number of elements in $\{p, 2p, \cdots, \frac{q-1}{2}p\}$ with negative least absolute residues modulo $q$, $q$ an odd prime distinct from $p$, then

$$\left(\frac{p}{q}\right)\left(\frac{q}{p}\right) = (-1)^{\mu+\nu} \quad ;$$

v) let $x, y$ satisfy $1 \leq x \leq \frac{p-1}{2}$, $1 \leq y \leq \frac{q-1}{2}$; then each pair $x, y$ leads to exactly one of the following four inequalities

$$qx - py < -\frac{p}{2}$$
$$-\frac{p}{2} < qx - py < 0$$
$$0 < qx - py < \frac{q}{2}$$
$$\frac{q}{2} < qx - py \quad ;$$

vi) the number of pairs $x, y$ in (v) leading to the first of the inequalities is the same as the number leading to the last of the inequalities and the number of pairs satisfying the second (third) inequalities is just $\mu$ ($\nu$) ;

vii) (the Quadratic reciprocity law)

$$\left(\frac{p}{q}\right)\left(\frac{q}{p}\right) = (-1)^{\frac{p-1}{2}\cdot\frac{q-1}{2}} .$$

(The theorem of Zolotareff given in #18 was first proved by Zolotareff [1872]. In our discussion in #18, #19 we have followed Riesz [1953], Cartier [1970], and Frobenius [1914] (see Gesammelte Abh. 1968 pp 628-649). For other recent treatments of Zolotareff's theorem along other lines see Rademacher, Grosswald [1972], Brenner [1973], and Roberts [1969].)

# XVIII Exponents, Primitive Roots, & Power Residues

When $a$ and $m$ are relatively prime positive integers Euler's theorem assures us of the existence of positive integers $t$ for which $a^t \equiv 1 \pmod{m}$. The smallest such $t$ is called the *exponent* of $a$ mod $m$ and will be denoted by $P_m(a)$ (or just $P(a)$ when the modulus is understood). The number of mod $m$ solutions of $P_m(x) = t$ is denoted by $\psi_m(t)$ (or just $\psi(t)$). If $P_m(a) = \varphi(m)$ one says that $a$ is a *primitive root* of $m$. As we shall see, not all $m$ have primitive roots. Throughout we assume $(a, m) = 1$.

1. In this problem $P(a)$ and $\psi(t)$ are used for $P_m(a)$, $\psi_m(t)$. Then:

i) $a^s \equiv a^t \pmod{m}$ if and only if

$$s \equiv t \pmod{P(a)};$$

ii) from (i):

a) $a^s \equiv 1 \pmod{m}$ if and only if

$$s \equiv 0 \pmod{P(a)};$$

b) $P(a) \mid \varphi(m)$;

c) $a, a^2, \cdots, a^{P(a)}$ are incongruent, modulo m, solutions of $x^{P(a)} \equiv 1 \pmod{m}$;

iii) $P(a^k) = \frac{P(a)}{(k, P(a))}$;

iv-a) $\psi(t) = 0$ when $t \nmid \varphi(m)$;

b) $\sum_{t \mid \varphi(m)} \psi(t) = \varphi(m)$;

c) if $\psi(t) \neq 0$ then $\varphi(t) \leq \psi(t)$;

v) if $m = p$, p a prime, then

a) if $\psi_p(t) \neq 0$ then $\varphi(t) = \psi(t)$;

b) $\psi_p(t) = \varphi(t)$ for all t such that $t \mid p-1$;

vi) every prime p has exactly $\varphi(p-1)$ primitive roots.

2. Let p be an odd prime and $\alpha$ be a positive integer. Then:

i) if g is a primitive root of p and if $g^{\frac{\varphi(p^\alpha)}{p}} \not\equiv 1 \pmod{p^\alpha}$ then g is a primitive root of $p^\alpha$;

ii) if $q$ is a primitive root of $p$ then one of $q$, $q+p$ is a primitive root of $p^2$;

iii) every primitive root of $p^2$ is also a primitive root of $p^\alpha$ for $\alpha > 2$;

iv) if $q$ is a primitive root of $p^\alpha$ the odd one of $q$, $q+p^\alpha$ is a primitive root of $2p^\alpha$ while the even one is not;

v) every number of the form $p^\alpha$ or $2p^\alpha$ has a primitive root (recall that $p$ here is odd);

vi) only 2, 4 and the numbers specified in (v) have primitive roots.

3. Let $p$ be an odd prime. Then:

i) every primitive root of any positive integral power of $p$ is a primitive root of all smaller positive integral powers of $p$;

ii) when $q$ is a primitive root of $p$ the numbers $(1+sp)q$, $0 \leq s < p$, with one exception, are primitive roots of $p^2$;

iii) every primitive root of $p$ is congruent modulo $p$ to exactly $p-1$ primitive roots of $p^2$ and, consequently, there are exactly $\varphi(\varphi(p^2))$ primitive roots of $p^2$;

iv) every primitive root of $p^\alpha$, $\alpha \geq 2$, is congruent modulo $p^\alpha$ to exactly $p$ primitive roots of $p^{\alpha+1}$;

v) there are exactly $\varphi(\varphi(p^\alpha))$ primitive roots of $p^\alpha$;

vi) if $m$ has a primitive root it has exactly $\varphi(\varphi(m))$ of them.

( In the above we have proved:
the only integers having primitive roots are 2, 4, powers of odd primes, and twice such powers; when $m$ has a primitive root it has, modulo $m$, exactly $\varphi(\varphi(m))$ of them. )

4. In this problem we discuss the exponent function in a little greater detail and the results lead to another proof that every prime has exactly $\varphi(p-1)$ primitive roots. The only earlier results we use are #1 (i ~ iii). Throughout we use $P(a)$ for $P_m(a)$.

i) If $P(a) = uv$ then $P(a^u) = v$;

ii) if $(P(a), P(b)) = 1$ then $P(ab) = P(a)P(b)$;

iii) it is false that $P(ab)$ is the least common multiple of $P(a)$ and $P(b)$ in all instances;

iv) for given $a, b$ there is a $c$ such that $P(c)$ is the largest common multiple of $P(a)$ and $P(b)$;

v) all exponents modulo a given $m$ divide the largest exponent;

vi) every prime has a primitive root;

vii) every prime $p$ has $\varphi(p-1)$ primitive roots.

5. Some of the ideas in #4 are useful in the study of Abelian groups. We illustrate this here. Once more the results lead to a proof of #1 (vi). We let G be an Abelian group and write $A_G$ for the set of positive integers which are orders of elements of G. Then:

i) if $u, v$ are in $A_G$ then $[u, v] \in A_G$ ;

ii) if $A_G$ has a largest element P then all elements of $A_G$ divide P ;

iii) every finite subgroup of the multiplicative group of a field is cyclic ;

iv) the multiplicative group of a finite field is cyclic ;

v) every prime has $\varphi(p-1)$ primitive roots.

Definition: If p is a prime and $x^n \equiv a \pmod{p}$ is solvable we call a an $n^{th}$ power residue modulo p.

6. Let a be an integer not divisible by the prime p and suppose $(n, p-1) = d$. Then:

i) a is an $n^{th}$ power residue modulo p if and only if $a^{\frac{p-1}{d}} \equiv 1 \pmod{p}$;

ii) let $\delta$ be a divisor of $p-1$; then a is a $\delta^{th}$ power residue modulo p if and only if

$$a^{\frac{p-1}{\delta}} \equiv 1 \pmod{p};$$

iii) if $\delta \mid p-1$, $\delta > 1$, then there is an integer which is not a $\delta^{th}$ power residue modulo p;

iv) if $p-1 = p_1^{\alpha_1} \cdots p_k^{\alpha_k}$ and $A_i$, $1 \le i \le k$, is not a $p_i^{th}$ power residue modulo p and $B_i = A_i^{\frac{p-1}{p_i^{\alpha_i}}}$, $1 \le i \le k$, then

a) the exponent of $B_i$ modulo p is $p_i^{\alpha_i}$;

b) $B_1 \cdots B_k$ is a primitive root modulo p.

7. i) Using #6 it is easy to find a primitive root modulo 43;

ii) using the primitive root found in (i) enables us to construct a table of exponents modulo 43 and use it to find all primitive roots, the least positive primitive root, and the primitive root with least absolute value.

8. In the following, in each instance, "all integers" refers to all integers not divisible by the primes in question.

i) All integers are cubic residues modulo 5 and 11;

ii) all integers are quintic residues modulo 7;

iii) all integers are $n^{th}$ power residues, for all odd $n$, modulo 5 and 17;

iv) a necessary and sufficient condition that all integers are $n^{th}$ power residues modulo an odd prime $p$, for all odd $n$, is that $p$ be a Fermat prime;

v) if $n$ is a fixed odd integer there are infinitely many primes for which not all integers are $n^{th}$ power residues.

9. For p a prime

$$1^n + 2^n + \cdots + (p-1)^n \equiv \begin{cases} 0 \pmod{p} \text{ if } p-1 \nmid n ; \\ -1 \pmod{p} \text{ if } p-1 \mid n . \end{cases}$$

10. When a and n are positive integers

n divides $\varphi(a^n - 1)$.

11. The product of all primitive roots of an odd prime p is congruent to 1 modulo p.

12. The following generalization of Wilson's theorem is true.

$$\prod_{\substack{n=1 \\ (n,m)=1}}^{m} n \equiv \begin{cases} -1 \pmod{m} \text{ if m has a primitive root;} \\ 1 \pmod{m} \text{ otherwise.} \end{cases}$$

13. (L. Marx)

The arithmetic progression $(x > 0)$

$$x,\ 3x+1,\ 5x+2,\ 7x+3, \cdots$$

always contains a power of 2 or a number 1 smaller than a power of 2; i.e. it always contains a term of the form $2^a$ or $2^a - 1$ $(Q \geq 0)$; further, the smallest such $Q$ is $< x$ and, if $t$ is the exponent of 2 modulo $2x+1$, is given by

$$Q = \begin{cases} \frac{1}{2}t - 1 \text{ when } t \text{ is even and } 2^{\frac{1}{2}t} \equiv -1 \pmod{2x+1}; \\ t - 1 \text{ otherwise.} \end{cases}$$

14. The sequence 5, 12, 19, 26, 33, 40, 47, ⋯ contains no term of the form $2^a$ or $2^a - 1$.

15. i) For each positive integer $n$

$$5^{2^n} - 1 = 4 \prod_{j=0}^{n-1} (5^{2^j} + 1) \begin{cases} \equiv 0 \pmod{2^{n+2}} \\ \not\equiv 0 \pmod{2^{n+3}}; \end{cases}$$

ii) if $\alpha > 2$ then the exponent of 5 modulo $2^\alpha$ is $2^{\alpha-2}$;

iii) for $\alpha > 2$ the set of numbers

$$\pm 1, \pm 5, \pm 5^2, \cdots, \pm 5^{2^{\alpha-2}-1}$$

is a reduced system of residues modulo $2^\alpha$.

(N.B. When $\alpha > 2$ we know $2^\alpha$ has no primitive root; however, as this problem shows, 5 is the "next best thing" to a primitive root for $2^\alpha$.)

Remarks.

With respect to #2 (ii) we note that there are cases where $g$ may be a primitive root of $p$ but not of $p^2$. However up to 1,001,321 all primes, with the single exception of 40 487, have least primitive roots which are also primitive roots of all higher powers. For 40 487 one finds the least primitive root 5 but also observes that $5^{40\,486} \equiv 1 \pmod{40\,487^2}$. (See Лурбер and Юдина [1971] and Riesel [1964].) It is interesting to note that 10 is a primitive root of 487 but not of $487^2$; in fact, $10^{486} \equiv 1 \pmod{487^2}$. This does not contradict the above since 3 and not 10 is the least primitive root of 487.

Again, the exposition in this chapter owes much to Бухштаб [1966].

# XIX Special Primes and the Lucas-Lehmer Theorem

1. As earlier, we let $F_n$ be the $n^{th}$ Fermat number. (see III #9.) Thus $F_n = 2^{2^n} + 1$, $n \geq 0$. Then, when $n \geq 1$,

i) if $F_n$ is a prime then $\left(\frac{3}{F_n}\right) = -1$, where $\left(\frac{3}{F_n}\right)$ is the Legendre symbol;

ii) if $3^{\frac{F_n - 1}{2}} \equiv -1 \pmod{F_n}$ then $P_{F(n)}(3) = F_n - 1$ and, therefore, $F_n$ is prime (here $P_{F(n)}(3)$ is the exponent of 3 modulo $F_n$);

iii) $F_n$ is prime if and only if $3^{\frac{F_n - 1}{2}} \equiv -1 \pmod{F_n}$.

2. Let $p$ be a prime dividing $F_n = 2^{2^n} + 1$, $n > 1$. Then:

i) $P_p(2) = 2^{n+1}$;

ii) $2^{n+1} \mid p - 1$ and $\left(\frac{2}{p}\right) = 1$;

iii) $2^{n+1} \mid \frac{p-1}{2}$ and, therefore, $p = 1 + 2^{n+2} \cdot t$;

iv) every prime divisor of $F_n$, $n > 1$, must be of the form $1 + 2^{n+2} \cdot t$ ;

v) for each fixed $k$, $k \geq 1$, the sequence

$$2^k + 1,\ 2 \cdot 2^k + 1,\ 3 \cdot 2^k + 1,\ 4 \cdot 2^k + 1,\ \ldots$$

contains infinitely many primes.

3. i) Using #2 (iv) one may factor $F_5$ rather quickly (compare with III #9 (iii)) ;

ii) (If computor is available) use #1 (iii) to show $F_7$ is composite ; (though Morehead found this result in 1905 it was not until 1971 that the factorization of $F_7$ was determined; see also the remarks at the end of III).

4. A number of the form $2^n - 1$ is called a Mersenne number and is denoted by $M_n$. A Mersenne prime is a prime Mersenne number.

i) If $M_n$ is prime then $n$ is prime ;

ii) if $p$ and $q$ are primes and $q = 2p+1$, $p \equiv 3 \pmod 4$ then:

a) $\left(\frac{2}{q}\right) = 1$ ; b) $q \nmid 2^p + 1$; c) $q \mid M_p$ ;

iii) $23 \mid M_{11}$, $47 \mid M_{23}$, $503 \mid M_{251}$ ;

iv) the converse of (i) is false.

5. (The Lucas-Lehmer Theorem)

As in #4 we write $M_n = 2^n - 1$. Further, for $n = 1, 2, \cdots$ let

$$U_n = \frac{1}{2\sqrt{3}}\{(1+\sqrt{3})^n - (1-\sqrt{3})^n\}, \quad V_n = (1+\sqrt{3})^n + (1-\sqrt{3})^n.$$

Then:

i) $U_n, V_n$ are integers, $V_n$ is even, and

$$U_{n+1} = U_n + \tfrac{1}{2}V_n, \quad V_{n+1} = 6U_n + V_n ;$$

ii) for all $m \geq 1$, $n \geq 1$

a) $2U_{m+n} = U_m V_n + V_m U_n$ ; d) $U_{2n} = U_n V_n$ ;

b) $(-2)^{m+1} U_n = U_m V_{m+n} - V_m U_{m+n}$; e) $V_{2n} = V_n^2 + (-2)^{n+1}$ ;

c) $2V_{m+n} = V_m V_n + 12 U_m U_n$; f) $V_n^2 - 12U_n^2 = (-2)^{n+2}$ ;

iii) if $p$ is a prime larger than 3 then

a) $U_p \equiv \left(\frac{3}{p}\right) \pmod p$ ; b) $V_p \equiv 2 \pmod p$ ; c) $p \mid U_{p-1} U_{p+1}$ ;

iv) if $p$ is a prime larger than 3 and $S_p$ is the set of integers $n$ for which $p \mid U_n$ then :

a) $m, n$ in $S_p$ imply $m+n$ is in $S_p$ ;

b) $m, n$ in $S_p$ and $n < m$ imply $m-n$ is in $S_p$ ;

c) if $\omega_p$ is the smallest element of $S_p$ then

1) $\omega_p \leq p+1$ ;

2) $n$ is in $S_p$ if and only if $\omega_p \mid n$ ;

v) if $M_p = 2^p - 1$ is prime then

a) $2V_{2^p} = 2V_{2^{p-1}} + 12U_{2^{p-1}} \equiv -8 \pmod{M_p}$ ;

b) $V_{2^p} = V_{2^{p-1}}^2 - 4 \cdot 2^{2^{p-1}-1}$ ;

c) $V_{2^{p-1}}^2 \equiv 4\left(2^{\frac{M_p-1}{2}} - 1\right) \pmod{M_p}$ ;

d) $M_p \mid V_{2^{p-1}}$ ;

vi) if $q$ is an odd prime and $p$ is a prime divisor of $M_q$, which, in turn, divides $V_{2^{q-1}}$ then :

a) $p > 3$ ;

b) $p \mid U_{2^q}$ ;

c) $\omega_p \mid 2^q$ ;

d) $\omega_p \nmid 2^{q-1}$ ;

e) $\omega_p = 2^q$ ;

f) $p = M_q$ ;

vii) there are integers $s_1, s_2, \cdots$ such that

$$s_1 = 4, \quad s_{k+1} = s_k^2 - 2 \text{ for } k \geq 1$$

and $$V_{2^k} = 2^{2^{k-1}} \cdot s_k \text{ for } k \geq 1 \;;$$

viii) ( the Lucas-Lehmer theorem )

if $p$ is an odd prime larger than 3 then $M_p$ is prime if and only if $M_p \mid s_{p-1}$, where

$$s_1 = 4, \quad s_{k+1} = s_k^2 - 2 \text{ for } k \geq 1 \;;$$

ix) ( the theorem of Lucas )

if $p$ is an odd prime then $M_p$ is prime if and only if $M_p \mid t_{p-1}$, where

$$t = 2, \quad t_{k+1} = 2t_k^2 - 1 \text{ for } k \geq 1 .$$

*Remarks*

Much of the information about large prime numbers $M_p$ has been deduced from computations made possible by the Lucas-Lehmer theorem. See, for example, Sierpinski's book *Elementary Theory of Numbers*. For a complete list of all known Mersenne primes see the remarks in VIII. For an early history of such primes see Uhler [1952]. For more recent information see recent issues of the journal Mathematics of Computation.

## XX Pell Equation

1. Let $\alpha$ be an irrational real number and $D$ be a positive non-square integer. Then:

i-a) if $y$ is a non-zero integer then there is an integer $x$ satisfying $0 < x - \alpha y < 1$;

b) for each positive integer $m$ there are integers $x, y$, $0 < y \leq m$, such that

$$|x - \alpha y| < \frac{1}{m};$$

c) there exist infinitely many distinct pairs $x, y$ such that $|x - \alpha y| < \frac{1}{y}$;

ii-a) if $\alpha = \sqrt{D}$ and $x, y$ is a pair of integers satisfying the inequality in (i-c) then

$$|x^2 - Dy^2| < 1 + 2\sqrt{D};$$

b) for some integer $k$ there are infinitely many integer pairs $x, y$, $y > 0$, such that $x^2 - Dy^2 = k$;

c) there are distinct integer pairs $x_1, y_1$ and $x_2, y_2$ satisfying the conditions of (b) such that $x_1 \equiv x_2$, $y_1 \equiv y_2 \pmod{k}$;

d) if $x_1, x_2, y_1, y_2$ are as in (c) and $x_3 + y_3\sqrt{D} = (x_1 + y_1\sqrt{D})(x_2 - y_2\sqrt{D})$ then

1) $k \mid x_3$, $k \mid y_3$;

2) $y_3 \neq 0$;

3) $x_3^2 - Dy_3^2 = k^2$;

e) the equation $x^2 - Dy^2 = 1$ has a solution in integers $x, y$.

2. Consider the equation

(1) $x^2 - Dy^2 = 1$, $D$ a positive non-square integer.

when $x', y'$ satisfy (1) we call $x' + y'\sqrt{D}$ a solution of (1). If $x', y'$ are positive we call $x' + y'\sqrt{D}$ a positive solution of (1) when it is a solution. Among all positive solutions $x + y\sqrt{D}$ we call the one which is smallest the fundamental solution of (1). (Note that because of the irrationality of $\sqrt{D}$ there can be only one smallest positive solution.) We denote the fundamental solution by $x_0 + y_0\sqrt{D}$. Then:

i) if $x_1 + y_1\sqrt{D}$ and $x_2 + y_2\sqrt{D}$ are solutions of (1) so also is $x_3 + y_3\sqrt{D}$ where

$$x_3 + y_3\sqrt{D} = (x_1 + y_1\sqrt{D})(x_2 + y_2\sqrt{D})\ ;$$

ii) all non-positive solutions of (1) other than $1 + 0\cdot\sqrt{D}$, $-1 + 0\cdot\sqrt{D}$ are obtainable from the positive solutions by making one or both of $x, y$ negative ;

iii) if $x + y\sqrt{D}$ is a solution of (1) and if

$$1 < x + y\sqrt{D} \text{ then } x > 0,\ y > 0\ ;$$

iv) every positive solution of (1) is a positive integral power of the fundamental solution of (1);

v) as $k$ runs over all integers (positive, negative, zero) then $\pm(x_0 + y_0\sqrt{D})^k$ runs over all solutions of (1).

3. Consider the equation

(2) $x^2 - Dy^2 = -1$, $D$ a positive non-square integer.

Then :

i) (2) is not always solvable ;

ii) when (2) is solvable there is a smallest positive solution all odd integral powers of which are solutions and if taken with ± signs exhaust all solutions ;

iii) if (2) is solvable and $x' + y'\sqrt{D}$ is the fundamental solution then $(x' + y'\sqrt{D})^2$ is the fundamental solution of (1) .

4. Consider the equation

(3) $x^2 - Dy^2 = \sigma^2$, D a positive non-square integer.

Then :

i) for each integer $\sigma$ the equation (3) has infinitely many solutions ;

ii) if $x_1 + y_1\sqrt{D}$ and $x_2 + y_2\sqrt{D}$ are solutions of (3) and $x_3, y_3$ are defined by

$$\frac{x_3 + y_3\sqrt{D}}{\sigma} = \frac{x_1 + y_1\sqrt{D}}{\sigma} \cdot \frac{x_2 + y_2\sqrt{D}}{\sigma}$$

then $x_3 + y_3\sqrt{D}$ is a rational solution of (3) (i.e. $x_3, y_3$ are rational numbers and $x_3^2 - Dy_3^2 = \sigma^2$) ;

iii) the $x_3, y_3$ in (ii) are not necessarily integers ;

iv) the $x_3, y_3$ in (ii) are integers if $D \equiv 0 \pmod{\sigma^2}$ ;

v) if $4D \equiv \sigma^2 \pmod{4\sigma^2}$ then

a) $\sigma$ is even, say $\sigma = 2\rho$ ;

b) $D = D'\rho^2$, where $D' \equiv 1 \pmod 4$ ;

c) if $x + y\sqrt{D}$ is an integral solution of (3) then $\rho \mid x$ and $\frac{x}{\rho}$, $y$ have the same parity;

d) $x_3$ and $y_3$ in (ii) are integers ;

vi) if $D \equiv 0 \pmod{\sigma^2}$ or $4D \equiv \sigma^2 \pmod{4\sigma^2}$ then $x^2 - Dy^2 = \sigma^2$ has integral solutions and if $x_1 + y_1\sqrt{D}$, $x_2 + y_2\sqrt{D}$ are such solutions so also is $x_3 + y_3\sqrt{D}$, where $x_3, y_3$ are as defined in (ii) ; further, if $x_0 + y_0\sqrt{D}$ is the smallest integral solution with $x_0 > 0, y_0 > 0$ then all solutions $x + y\sqrt{D}$ are obtained by allowing $k$ to run over all the integers in the equation

$$\frac{x + y\sqrt{D}}{\sigma} = \pm\left(\frac{x_0 + y_0\sqrt{D}}{\sigma}\right)^k .$$

5. (Miscellaneous)

i) For every rational number $r$,

$\frac{D+r^2}{D-r^2} - \frac{2r}{D-r^2}\sqrt{D}$ is a rational solution of (1);

ii) every rational solution of (1) is of the form indicated in (i);

iii) the formula in (i) provides an integral solution for $x^2 - 7y^2 = 1$ when $r = -\frac{7}{3}$;

iv) a parametric solution of (3) is given by

$x = m^2 + Dn^2$, $y = 2mn$, $\sigma = m^2 - Dn^2$.

6. (A small application)

i) Give complete solutions to

$$x^2 + 1 = 2y^2 \text{ and } x^2 - 1 = 2y^2;$$

ii) let $s_n$ be the sum of the lengths of the legs and $h_n$ be the length of the hypotenuse of a Pythagorean triangle (a right triangle with integer length sides) whose legs are consecutive integers; show that the pair $s_n, h_n$ is such a pair if and only if

$$s_n + h_n\sqrt{2} = (1+\sqrt{2})(3+2\sqrt{2})^n$$

for some positive integer $n$ ;

iii) if $(x, x+1, z)$ is a Pythagorean triple show $f(x, x+1, z) = (3x+2z+1, 3x+2z+2, 4x+3z+2)$ is also such a triple ;

iv) all Pythagorean triples with consecutive integer legs appear in the sequence

$(3,4,5)$, $f(3,4,5)$, $ff(3,4,5)$, $fff(3,4,5)$, ⋯

where $f$ is as in (iii) ;

v) compute the 1st four terms of the sequence in (iv) .

*Remarks*

Equations of the form $x^2 - Dy^2 = a$ are called Pell equations, though some authors feel the reference to be sufficiently unreliable as to refuse to call them by this name. An elementary exposition of these equations will be found in Gelfond [1961]. The result in #6(iv) is proved in quite a different manner in Sierpinski's delightful little book *Pythagorean Triangles*. The method used here derives from Carmichael [1915]. The Pell equation arose earlier, in our chapter on continued fractions, see XIII #17 (vii). If one examines the diagram in XI #15 one finds all the Pythagorean triangles with consecutive integer legs on the horizontal line through (3,4,5). This may easily be confirmed by comparing the diagram with #6 (iv) above.

## XXI Weyl's Theorem on Uniform Distribution

In the following all functions have domain $[0,1]$ and the sequences $\{s_n\}$ are to satisfy
$$0 \le s \le 1 \text{ for all } n.$$

**Definition**: $\{s_n\}$ is *uniformly distributed* if for every pair of $a, b$, $0 \le a < b \le 1$ the number, $n(a,b)$, of $s_1, \cdots, s_n$ lying in $[a,b]$ satisfies $\lim_{n \to \infty} \frac{n(a,b)}{n} = b - a$.

1. Define the characteristic function $\chi_I$ of a subinterval $I$ of $[0,1]$ by
$$\chi_I(x) = \begin{cases} 1 & \text{for } x \text{ in } I; \\ 0 & \text{otherwise.} \end{cases}$$

Then:

i-a) $\sum_{m=1}^{n} \chi_{[a,b]}(s_m) = n(a,b)$;

b) $\int_0^1 \chi_{[a,b]}(x)\,dx = b - a$;

ii) if for every $f$ which is a characteristic function of a subinterval of $[0,1]$ it is true that $\lim_{n\to\infty} \frac{1}{n}\sum_{m=1}^{n} f(s_m) = \int_0^1 f(x)\,dx$ then $\{s_n\}$ is uniformly distributed;

iii) the converse of (ii) is also true.

Definition: When the limit expression in #1 (ii) holds we write $f(s_n) \rightsquigarrow \int f$.

2. Let $f$ be Riemann integrable and suppose that for each $\epsilon > 0$ there are Riemann integrable functions $g$ and $h$ such that $g \le f \le h$, $0 \le \int_0^1 (h(x) - g(x))\,dx < \epsilon$, $g(s_n) \rightsquigarrow \int g$, $h(s_n) \rightsquigarrow \int h$.

Then $f(s_n) \rightsquigarrow \int f$.

3. If $\{s_n\}$ is uniformly distributed then for every Riemann integrable function $f$ it is true that $f(s_n) \rightsquigarrow \int f$.

4. *i*) By treating real and complex parts separately we see that the result of #3 remains valid for $f$ a Riemann integrable complex valued function of a real variable;

*ii*) if $\{s_n\}$ is uniformly distributed and $f(x) = e^{2\pi i k x}$, $k$ a positive integer, then $f(s_n) \rightsquigarrow 0$.

5. Let P be the proposition "$e^{2\pi i k s_n} \rightsquigarrow 0$ for all $k \geq 0$" and let $T$ be an arbitrary trigonometric polynomial with zero constant term; i.e. $T(x) = \sum_{k=1}^{q} (a_k \cos 2\pi k x + b_k \sin 2\pi k x)$; then:

*i*) if P then $T(s_n) \rightsquigarrow 0$;

*ii*) if for all such $T$ as described $T(s_n) \rightsquigarrow 0$ then P;

*iii*) P if and only if $T(s_n) \rightsquigarrow 0$ for all trigonometric polynomials with zero constant term;

*iv*) the proposition in (*iii*) is true even if on the right we eliminate the condition "with zero constant term";

v) if P then $f(s_n) \sim \int f$ for all continuous $f$;

vi) if P then $\{s_n\}$ is uniformly distributed;

vii) (Weyl's theorem)

$\{s_n\}$ is uniformly distributed if and only if

$\lim_{n\to\infty} \frac{1}{n}\sum_{m=1}^{n} e^{2\pi i k s_m} = 0$ for all $k \geq 0$.

6. i) If $\alpha$ is irrational and $s_n$ is the fractional part of $n\alpha$, i.e. $s_n = n\alpha - [n\alpha] = (n\alpha)$, then $\{s_n\}$ is uniformly distributed;

ii) if $\alpha$ is irrational and $\beta$ is arbitrary and $s = (n\alpha+\beta)$ then $\{s_n\}$ is uniformly distributed.

7. i) The sequence $\{s_n\}$ is uniformly distributed if and only if for all $a$, $0 \leq a \leq 1$, $\lim_{n\to\infty} \frac{n(0,a)}{n} = a$;

ii) let $\{\beta_n\}$ be uniformly distributed and suppose $|\alpha_n - \beta_n| < \frac{1}{n}$ for all $n$; then $\{\alpha_n\}$ is uniformly distributed;

iii) let $S$ be a countable subset of $[0,1]$; then $S$ is dense in $[0,1]$ if and only if some enumeration of $S$ is uniformly distributed.

8. Let $\alpha_1, \cdots, \alpha_Q$ be complex numbers and set $\alpha_q = 0$ for $q \leq 0$ and for $q > Q$.

Then, for $1 \leq H \leq Q$,

i) $H \sum_{1 \leq q \leq Q} \alpha_q = \sum_{0 < p < H+Q} \left\{ \sum_{0 \leq r < H} \alpha_{p-r} \right\}$;

ii) $\sum_{\substack{p, r, s \\ 0 < p < H+Q \\ 0 \leq r < H, 0 \leq s < H}} \alpha_{p-r} \bar{\alpha}_{p-s} =$

$$H \sum_{1 \leq q \leq Q} \alpha_q \bar{\alpha}_q + \sum_{1 \leq h < H} (H-h) \sum_{1 \leq q \leq Q-h} (\alpha_q \bar{\alpha}_{q+h} + \bar{\alpha}_q \alpha_{q+h}).$$

9. Let $\alpha_1, \cdots, \alpha_Q$ be complex numbers and define $\alpha_j$ for $j \leq 0$, $j > Q$ as in #8. Further, suppose $1 \leq H \leq Q$. Then:

$$H^2 \left| \sum_{1 \leq q \leq Q} \alpha_q \right|^2 \leq$$

$$(H+Q-1) \left\{ H \sum_{1 \leq q \leq Q} |\alpha_q|^2 + 2 \sum_{0 < h < H} (H-h) \left| \sum_{1 \leq q \leq Q-h} \bar{\alpha}_q \alpha_{q+h} \right| \right\}.$$

10. i) If $e^{2\pi i (s_{n+h} - s_n)} \rightsquigarrow 0$ for all positive integers $h$ then $e^{2\pi i s_n} \rightsquigarrow 0$;

ii) if $\{s_{n+h} - s_n\}$ is uniformly distributed for each positive integer $h$ then $\{s_n\}$ is uniformly distributed.

11. (Weyl) If $f(x) = a_r x^r + \cdots + a_0$ and for some $j$, $a_j$ is irrational, then the fractional parts $(f(n))$ of $f$ are uniformly distributed.

Remarks

The work of this chapter follows the expositions given by Hardy [1949] and Cassels [1957]. Vinogradoff proved in 1937 that if one replaces $n$ by $p_n$, the $n^{th}$ prime, in #6 (i) then the resulting sequence $\{(\alpha p_n)\}$ is uniformly distributed. By using Vinogradoff's method Rhin has recently proved #11 with $n$ replaced by $p_n$. The interested reader should consult the review of Rhin's paper:

MR 48 (1974) # 2087.

For a proof of #6(i) not dependent on Weyl's theorem but making use of continued fractions one might consult Niven's *Irrational Numbers*. Weyl originally proved his theorem in 1916 and it has long been considered an outstanding contribution to the theory. The result in #7(iii) is a very special case of a general theorem proved by John von Neumann [1925]. The reader might consult Koksma [1936] for this and many other aspects of the material of this chapter.

## XXII Möbius Functions

1. Let $f(x) = x + x^2 + x^3 + \cdots$ and define $a_1, a_2, a_3, \cdots$ to be that sequence of integers for which

$$x = a_1 f(x) + a_2 f(x^2) + a_3 f(x^3) + \cdots$$

when one carries out the operations on the right in a purely formal manner. Then:

i-a) $a_1 = 1$, $\sum_{d \mid m} a_d = 0$ for $m > 1$;

b) if $(s, t) = 1$ then $a_{st} = a_s a_t$;

c) $a_{p^k} = \begin{cases} 1 \text{ for } k = 0, \\ -1 \text{ for } k = 1, \\ 0 \text{ for } k > 1, \end{cases}$ where $p$ is a prime;

d) $a_n = \begin{cases} 1 \text{ for } n = 1, \\ (-1)^k \text{ for } n \text{ the product of } k \text{ distinct primes}, \\ 0 \text{ for } n \text{ divisible by a square} > 1; \end{cases}$

ii-a) $x = \sum_{m=1}^{\infty} \frac{a_m x^m}{1 - x^m}$;

b) $\frac{1}{10} = \frac{1}{9} - \frac{1}{99} - \frac{1}{999} - \frac{1}{99999}$

$+ \frac{1}{999999} - \frac{1}{9999999} + \frac{1}{9999999999} - \cdots;$

c) $\frac{1}{10} = \frac{1}{11} + \frac{1}{111} + \frac{1}{11111} - \frac{1}{111111}$

$+ \frac{1}{1111111} - \frac{1}{1111111111} + \cdots.$

(This function $a_n$ was first introduced by A. F. Möbius [1831] in just this way. Nowadays one writes $\mu(n)$ rather than $a_n$ and defines the function by (i-d) above. Quite recently Gian-Carlo Rota [1963-4] has shown how the Möbius function arises quite naturally in a considerably wider setting and with many applications in combinatorial analysis. Rota's work has been extended and generalized in a great proliferation of papers in the last 12 years.)

Definition. Define the function $\mu$ by:

$$\mu(n) = \begin{cases} 1 \text{ for } n = 1; \\ (-1)^k \text{ for } n \text{ the product of } k \text{ distinct primes}; \\ 0 \quad \text{for } n \text{ divisible by a square} > 1; \end{cases}$$

i.e. define $\mu$ by $\mu(n) = a_n$ for $n \geq 1$, $a_n$ as in #1.

2\. i) $\mu$ is a multiplicative function; i.e. if $(s,t) = 1$ then $\mu(st) = \mu(s)\mu(t)$;

ii) if $f$ is multiplicative then

$$\sum_{d|n} \mu(d) f(d) = \prod_{p|n} (1 - f(p)),$$

where $p$ in the index denotes a prime number;

a) $\sum_{d|n} \mu(d) = 0$ for $n > 1$;

b) $\sum_{d|n} \mu(d) d = \prod_{p|n} (1-p)$;

c) $\sum_{d|n} \frac{\mu(d)}{d} = \prod_{p|n} (1 - \frac{1}{p}) = \frac{1}{n} \varphi(n) = \frac{1}{n} \sum_{d|n} \mu(\frac{n}{d}) d$;

d) $\sum_{d|n} \mu(d)^2 = 2^t$, where $t$ is the number of distinct prime factors of $n$.

3\. i) (Möbius inversion)

$f(n) = \sum_{d|n} g(d)$ if and only if $g(n) = \sum_{d|n} \mu(d) f(\frac{n}{d})$;

a) define $\Lambda(n)$ to be $\ln p$ when $n$ is a power of the prime number $p$ and to be $0$ otherwise; then

1) $\ln n = \sum_{d|n} \Lambda(d)$;

a) $\Lambda(n) = -\sum_{d|n} \mu(d) \ln d$ ;

( $\Lambda(n)$ is called the Mangoldt function. )

b) $n = \sum_{d|n} \varphi(d)$ ;

c) $F(n) = \prod_{d|n} f(d)$ if and only if

$$f(n) = \prod_{d|n} F\left(\frac{n}{d}\right)^{\mu(d)};$$

d) $F_n(x) = \prod_{d|n} \left(x^{\frac{n}{d}} - 1\right)^{\mu(d)}$, where $F_n(x)$ is as in XIV# 17 ;

e) define $\bar{\Psi}(n)$ by :

$$\bar{\Psi}(1) = 1,\quad \mu(n) = \sum_{d|n} \bar{\Psi}(d) \text{ for } n>1;$$

then $\bar{\Psi}(n) = \begin{cases} 0 \text{ if } n \text{ is divisible by a cube} > 1; \\ (-2)^t \text{ if } n \text{ is cubefree and the squarefree part of } n \text{ is divisible by exactly } t \text{ different primes}; \end{cases}$

ii) (Shapiro)

if $\psi$ is a real function defined on $[0,1]$ then if $f(n) = \sum_{\substack{(r,n)=1 \\ r\le n}} \psi\left(\frac{r}{n}\right)$, $g(n) = \sum_{r\le n} \psi\left(\frac{r}{n}\right)$ then

$$f(n) = \sum_{d|n} \mu(d) g\left(\frac{n}{d}\right);$$

a) $\mu(n) = \sum_{\substack{(r,n)=1 \\ r\le n}} e^{\frac{2\pi i r}{n}}$ ;

b) let $S_k(n)$ be the sum of the $k^{th}$ powers of the positive integers prime to n; then

$$S_k(n) = n^k \sum_{d|n} \mu\left(\frac{n}{d}\right) \frac{1^k + \cdots + d^k}{d^k};$$

1) $S_1(n) = \frac{1}{2} n \varphi(n),\ n > 1;$

2) $S_2(n) = \frac{1}{3} n^2 \varphi(n) + \frac{1}{6} n \prod_{p|n} (1-p),\ n > 1;$

3) $S_3(n) = \frac{1}{4} n^3 \varphi(n) + \frac{1}{4} n^2 \prod_{p|n} (1-p),\ n > 1;$

4) $1^k + 2^k + \cdots + n^k = n^k \sum_{d|n} \frac{S_k(d)}{d^k} = \sum_{d|n} d^k S_k\left(\frac{n}{d}\right);$

c) let $w(n)$ be the product of those integers prime to n and not exceeding n; then

$$w(n) = n^{\varphi(n)} \prod_{d|n} \left(\frac{d!}{d^d}\right)^{\mu\left(\frac{n}{d}\right)};$$

iii) (Prachar)

let $k_1, \cdots, k_N$ be N numbers of which $\alpha$ are equal to 1 and suppose $f$ is defined for each $k_i$; then

if $S_d = \sum_{d|k_i} f(k_i)$ then $\sum_d \mu(d) S_d = \alpha f(1);$

a) let $\varphi(x,y)$ be the number of integers not exceeding x which are divisible by no prime not exceeding y and let $P_y = \prod_{p \le y} p$; then

$$\varphi(x,y) = \sum_{d|P_y} \mu(d) \left[\frac{x}{d}\right];$$

1) $\sum_{m\le n} \mu(m)\left[\frac{n}{m}\right] = 1$ ;

2) $\left|\sum_{m\le n} \frac{\mu(m)}{m}\right| \le 1$ ;

3) $\pi(x)-\pi(\sqrt{x})+1 = [x] - \sum_{i=1}^{r}\left[\frac{x}{p_i}\right] + \sum_{\substack{i,j=1\\ i\ne j}}^{r}\left[\frac{x}{p_ip_j}\right] - \cdots$,

where $p_1, \cdots, p_r$ are the primes not exceeding $\sqrt{x}$.

4. i) For $h$ a function of two variables

$$\sum_{mn\le x} \mu(n)h(x, mn) = h(x, 1) ;$$

ii) (Shapiro) let $P$ be completely multiplicative; i.e. $P(ab) = P(a)P(b)$ for all $a, b$ ; then $g(x) = \sum_{n\le x} P(n)f\left(\frac{x}{n}\right)$ if and only if $f(x) = \sum_{n\le x} \mu(n)P(n)g\left(\frac{x}{n}\right)$;

iii) given $\sum_{m,n=1}^{\infty} |f(mnx)| = \sum_{\nu=1}^{\infty} \tau(\nu)|f(\nu x)|$ we have $g(x) = \sum_{m=1}^{\infty} f(mx)$ if and only if

$$f(x) = \sum_{n=1}^{\infty} \mu(n)g(nx) ;$$

here $\tau(n)$ is the number of divisors of $n$ ;

iv) (Halberstam and Roth)

let $\mathcal{D}$ be *divisor closed*; i.e. $\mathcal{D}$ is a set such that $\mathcal{D}$ contains all integral divisors of any of its elements ; then $F(d) = \sum_{\substack{\delta\\ d|\delta, \delta\in\mathcal{D}}} G(\delta)$ if and only if

$$G(d) = \sum_{\substack{t\\ td\in\mathcal{D}}} \mu(t)F(td).$$

## XXIII Some Analytic Methods

In the following we shall often write expressions like $O(f(x))$, where $f$ is a positive real function. Whenever we write this we intend it to stand for an unspecified complex valued function of a real variable, say $g(x)$, with the following property:

there exist constants $x_0$, $A$ such that $g(x)$ and $f(x)$ are defined for all $x \geq x_0$ and, for such values of $x$,

$$|g(x)| \leq Af(x).$$

1. Let $f$ be a complex valued function of a real variable and suppose $M, N$ are integers with $M < N$. Further, put $F(m) = \sum_{k=M+1}^{m} f(k)$, $F(M) = 0$. Then, for $g$ any real function,

i) (Abel partial summation formula)

$$\sum_{m=M+1}^{N} f(m)g(m) = F(N)g(N+1) - \sum_{m=M+1}^{N} F(m)(g(m+1) - g(m))$$
$$= F(N)g(N) - \sum_{m=M+1}^{N-1} F(m)(g(m+1) - g(m));$$

( It will be noted that this formula is a finite analogue of the integration by parts formula

$$\int_a^b u'v = uv\Big|_a^b - \int_a^b uv'.$$

Indeed, if we put $\Delta h(n) = h(n) - h(n-1)$, we may write the formula $\sum_{m=M+1}^{N} (\Delta F(m))g(m) =$

$$F(m)g(m)\Big|_M^N - \sum_{m=M+1}^{N} F(m)(\Delta g(m+1)) \quad .)$$

ii) if $g$ is monotonic and non-negative,

$$\left|\sum_{m=M+1}^{N} f(m)g(m)\right| \le \begin{cases} g(M) \max\limits_{M<m\le N} |F(m)| \text{ if } \\ \qquad g \text{ is decreasing;} \\ 2g(N) \max\limits_{M<m\le N} |F(m)| \text{ if } \\ \qquad g \text{ is increasing;} \end{cases}$$

iii) if $g$ tends monotonically downward to 0 as $n \to \infty$, and if $F$ is bounded, then

a) $\sum_{n=1}^{\infty} f(n)g(n)$ converges;

b) $\sum_{n\le x} f(n)g(n) = \sum_{n=1}^{\infty} f(n)g(n) + O(g([x]))$;

iv) if $\lambda_1, \lambda_2, \cdots$ is an unboundedly increasing sequence of real numbers and $g$ has a continuous derivative for $x \ge \lambda_1$ then, putting $F(x) = \sum_{\substack{m \\ \lambda_1 \le \lambda_m \le x}} f(m)$, we have

$$\sum_{\substack{m \\ \lambda_1 \le \lambda_m \le x}} f(m) g(\lambda_m) = F(x) g(x) - \int_{\lambda_1}^{x} F(t) g'(t)\, dt ;$$

v) if $a$ is a positive integer and $g$ has a continuous derivative for $x \ge a$ then

$$\sum_{a \le m \le x} g(m) =$$
$$\int_a^x g(t)\, dt + \int_a^x (t - [t]) g'(t)\, dt + g(a) - (x - [x]) g(x);$$

vi) if $a$ is a positive integer and $g$ is a monotone function with a continuous derivative for $x \ge a$ then

$$\sum_{a \le m \le x} g(m) = \int_a^x g(t)\, dt + O(|g(a)| + |g(x)|) ;$$

vii) if $a$ is a positive integer and $g$ is a continuously differentiable function for $x \ge a$ and if $g$ tends monotonically to $0$ as $x \to \infty$ then

$$\sum_{a < m \le x} g(m) = \int_a^x g(t)\, dt + c + O(|g(x)|),$$

where $c = \int_a^\infty (t - [t]) g'(t)\, dt$ is independent of $x$.

2. (Applications)

i) $\sum_{n\leq x} \frac{1}{n^s} = \frac{1}{1-s} x^{1-s} + C + O(x^{-s})$, $s>0$, $s\neq 1$, where $C$ is a suitable constant;

ii) $\sum_{n\leq x} \frac{1}{n} = \ln x + \gamma + O(x^{-1})$, where $\gamma$ is Euler's constant (approximately equal to 0.57721) and $\gamma = 1 - \int_1^\infty \frac{t-[t]}{t^2}\,dt = \lim_{n\to\infty}\left\{\sum_{m=1}^{n} \frac{1}{m} - \ln n\right\}$;

(It is not known whether or not $\gamma$ is rational.)

iii) $\sum_{n\leq x} \ln n = x\ln x - x + O(\ln x)$;

iv) $\sum_{n\leq x} \ln \frac{x}{n} = O(x)$;

v) $\sum_{p\leq x} \frac{\ln p}{p} + O(1) = \frac{1}{x}\sum_{p\leq x} \ln p\left\{\left[\frac{x}{p}\right] + \left[\frac{x}{p^2}\right] + \cdots\right\}$
$= \frac{1}{x}\sum_{n\leq x}\left[\frac{x}{n}\right]\Lambda(n) = \frac{1}{x}\sum_{n\leq x} \ln n = \ln x + O(1)$,
where $p$ is a prime and $\Lambda(n)$ is 0 unless $n$ is a power of a prime $p$ when it is $\ln p$;

(see XXII # 3(i-a));

vi) $\sum_{p\leq x} \frac{\ln p}{p} = \ln x + O(1) = \sum_{n\leq x} \frac{\Lambda(n)}{n} + O(1)$;

vii) $\sum_{N<p\leq N^2} \frac{\ln p}{p} = \int_N^{N^2} \pi(t)\frac{\ln t - 1}{t^2}\,dt + O(1)$.

3. ( Chebyshev's theorem of 1849 )

Suppose $\lim_{x\to\infty} \frac{\pi(x)}{x/\ln x}$ exists and equals $\beta$. Then:

i) if $\beta < 1$ then

$$\ln N + O(1) = \sum_{N<p\le N^2} \frac{\ln p}{p} = \int_N^{N^2} \pi(t)\, \frac{\ln t - 1}{t^2}\, dt + O(1)$$

$$< \frac{1+\beta}{2} \int_N^{N^2} \left(\frac{1}{t} - \frac{1}{t \ln t}\right) dt + O(1) < \frac{1+\beta}{2} \ln N + O(1),$$

which is false ;

ii) if $\beta > 1$ then a similar contradiction to that in (i) arises ;

iii) if $\lim_{x\to\infty} \frac{\pi(x)}{x/\ln x}$ exists then that limit is 1 .

( This result was proved by Chebyshev in 1849 but it was not until 1896 that it was proved the limit exists. In that year the Belgian, de la Valleé Poussin, and the Frenchman, Jacques Hadamard, independently published proofs that the limit does exist. The result

$$\lim_{x\to\infty} \frac{\pi(x)}{x/\ln x} = 1$$

has come to be known as the *Prime Number Theorem*. The first proofs made heavy use of the theory of functions of a complex variable and it was not until 1948 that the 1st so called "elementary" proofs - that is, those not using complex variable theory - were given by the Swedish mathematician Atle Selberg and the Hungarian mathematician Paul Erdös. For further information with respect to this theorem and its ramifications the reader might consult Hardy & Wright [1962], Trost [1968], Specht [1956], Prachar [1957], Landau [1953], or Levinson [1969].)

4. i) $\sum_{\substack{n \leq N \\ d|n}} n = \frac{d}{2}\left(\frac{N}{d} + O(1)\right)^2$ ;

ii) if $N^*$ is the number of positive proper irreducible fractions with denominator not exceeding N then

$$N^* = \sum_{d=1}^{N} \frac{\mu(d)}{d} \sum_{\substack{n \leq N \\ d|n}} n \quad ;$$

iii) $N^*$, as in (ii), satisfies

$$N^* = \frac{N^2}{2}\left(\sum_{n=1}^{\infty} \frac{1}{n^2}\right)^{-1} + N^2 g(N) \quad ,$$

for some function $g(N)$ which tends to 0 as $N \to \infty$;

iv) if $N'$ is the total number of positive proper fractions with denominator not exceeding $N$ then

$$\lim_{N\to\infty} \frac{N^*}{N'} = \left(\sum_{n=1}^{\infty} \frac{1}{n^2}\right)^{-1};$$

v) the probability of 2 randomly chosen integers being relatively prime is $\frac{6}{\pi^2}$.

5. (Infinite products)

The infinite product $\prod_{j=1}^{\infty} a_j$ is said to converge to $\alpha$ if $\alpha \neq 0$ and $\lim_{n\to\infty} \prod_{j=1}^{n} a_j = \alpha$. When such a non-zero $\alpha$ exists we say the product converges and in the contrary case that it diverges.

i) $x \le \ln \frac{1}{1-x} = \sum_{n=1}^{\infty} \frac{x^n}{n} \le 2x$, for $0 \le x \le \frac{1}{2}$;

ii) if $0 \le x_j < 1$ for all $j$ then $\sum_{j=1}^{\infty} \ln \frac{1}{1-x_j}$ converges if and only if $\sum_{j=1}^{\infty} x_j$ converges;

iii) if $0 \le x_j < 1$ for all $j$ then $\prod_{j=1}^{\infty} \frac{1}{1-x_j}$ converges if and only if $\sum_{j=1}^{\infty} x_j$ converges.

6. Suppose $f$ is a completely multiplicative (real or complex valued) number theoretic function; i.e. $f(ab) = f(a)f(b)$ for all integers $a$ and $b$. Then if $\sum_{j=1}^{\infty} f(j)$ is absolutely convergent we have:

i) $|f(j)| \leq 1$ for all $j$;

ii) $\left|\sum_{j=1}^{\infty} f(j) - \prod_{p \leq m} (1 - f(p))^{-1}\right| \leq \sum_{j=m+1}^{\infty} |f(j)|$;

iii) $\prod_{p} (1 - f(p))^{-1} = \sum_{j=1}^{\infty} f(j)$.

7. In (i-a) below we show $\sum_{n=1}^{\infty} \frac{1}{n^s}$ converges for $s > 1$; we denote the sum of this series, in this case, by $\zeta(s)$. In (b)-(f) $s$ is to be larger than 1.

i-a) $\sum_{n=1}^{\infty} \frac{1}{n^s}$ converges for $s > 1$;

b) $\frac{1}{s-1} < \zeta(s) < 1 + \frac{1}{s-1}$;

c) $(s-1)\zeta(s) \to 1$ as $s \to 1^+$;

d) $\zeta(s) = \prod_{p} (1 - p^{-s})^{-1}$;

e) $\ln \zeta(s) = \sum_{p} \sum_{n=1}^{\infty} \frac{1}{np^{ns}}$;

f) $0 \leq \ln \zeta(s) - \sum_{p} \frac{1}{p^s} < 1$;

ii-a) one can use (i-c) and (d) to prove the existence of infinitely many primes ;

b) one can use (i-f) to show $\sum_p \frac{1}{p^s}$ converges for $s > 1$ and diverges for $s = 1$.

(The proof in (ii-a) of the infinitude of primes goes back to Euler and this proof already contains the germ of the idea developed by Dirichlet to prove the theorem concerning the infinitude of primes in an arithmetic progression. In the next two problems we extend these notions a little further and obtain some special cases of the Dirichlet theorem.)

8. Define $\chi : \mathbb{Z} \to \{0, 1, -1\}$ by

$$\chi(n) = \begin{cases} 0 & \text{if } n \equiv 0 \pmod{2} ; \\ 1 & \text{if } n \equiv 1 \pmod{4} ; \\ -1 & \text{if } n \equiv 3 \pmod{4} . \end{cases}$$

Further, put $L(s) = \sum_{n=1}^{\infty} \frac{\chi(n)}{n^s}$ .

Then:

i-a) $L(s)$ exists for $s > 0$ ;

b) $0 < L(s) < 1$ for $s > 0$ and $\frac{2}{3} < L(s)$ for $s \geq 1$ ;

c) $L(s)$ is continuous at 1 ;

d) $\zeta(s) L(s) \to \infty$ as $s \to 1^+$ ;

e) $\lim_{s \to 1^+} \prod_{p \equiv 3 \pmod 4} (1 - p^{-2s})^{-1}$ exists ;

f) $\zeta(s) = \prod_p (1 - p^{-s})^{-1} =$

$$\frac{1}{1 - 2^s} \prod_{p \equiv 1 \pmod 4} (1 - p^{-s})^{-1} \prod_{p \equiv 3 \pmod 4} (1 - p^{-s})^{-1} ;$$

and $L(s) = \prod_p \left(1 - \frac{\chi(p)}{p^s}\right)^{-1} =$

$$\prod_{p \equiv 1 \pmod 4} (1 - p^{-s})^{-1} \prod_{p \equiv 3 \pmod 4} (1 + p^{-s})^{-1} ;$$

g) there are infinitely many primes of each of the forms $4k+1$, $4k+3$ ;

ii) one can derive the result of (i-g) along the following lines :

a) $\ln L(s) = \sum_p \frac{\chi(p)}{p^s} + O(1)$ ;

b) $\ln \zeta(s) = \sum_p \frac{1}{p^s} + O(1)$ ;

c) $\ln \zeta(s) + \chi^{-1}(a) \ln L(s) =$

$$\begin{cases} 2 \sum_{p \equiv 1 \pmod 4} \frac{1}{p^s} + O(1) & \text{for } a = 4k+1\,; \\ 2 \sum_{p \equiv 3 \pmod 4} \frac{1}{p^s} + O(1) & \text{for } a = 4k+3\,; \end{cases}$$

d) there are infinitely many primes of each of the forms $4k+1$, $4k+3$.

9. Define the four number theoretic functions $\chi_j$, $0 \le j \le 3$, by the condition

$$\chi_j(n) = \chi_j(a) \text{ when } n \equiv a \pmod 5,\ 0 \le j \le 3,$$

and the table

| j \ a | 0 | 1 | 2 | 3 | 4 |
|---|---|---|---|---|---|
| 0 | 0 | 1 | 1 | 1 | 1 |
| 1 | 0 | 1 | $i$ | $-i$ | -1 |
| 2 | 0 | 1 | -1 | -1 | 1 |
| 3 | 0 | 1 | $-i$ | $i$ | -1 |

The $i$ is just $\sqrt{-1}$. Thus, for example,

$$\chi_1(17) = \chi_1(2) = i\ , \quad \chi_2(43) = \chi_2(3) = -1\,.$$

i-a) Each $\chi_j$ is completely multiplicative;

b) for $a \not\equiv 0 \pmod 5$,

$$\sum_{\chi} \chi^{-1}(a)\chi(n) = \begin{cases} 4 \text{ if } a \equiv n \pmod 5; \\ 0 \text{ otherwise}; \end{cases}$$

c) if $L_0(\chi) = \sum_{n=1}^{\infty} \frac{\chi(n)}{n}$, $L_1(\chi) = \sum_{n=1}^{\infty} \frac{\chi(n)\ln n}{n}$

then, for $\chi \neq \chi_0$, the two series converge to non-zero sums;

d) for $\chi \neq \chi_0$, $\sum_{n \le x} \frac{\chi(n)}{n} = L_0(\chi) + O\left(\frac{1}{x}\right)$

and $\sum_{n \le x} \frac{\chi(n)\ln n}{n} = L_1(\chi) + O\left(\frac{\ln x}{x}\right)$;

e) for $\chi \neq \chi_0$, $\sum_{n \le x} \frac{\mu(n)\chi(n)}{n} = O(1)$;

f) for $\chi \neq \chi_0$,

$$\sum_{n \le x} \frac{\chi(n)\Lambda(n)}{n} = \sum_{d \le x} \frac{\mu(d)\chi(d)}{d} \left\{ L_1(\chi) + O\left(\frac{\ln x/d}{x/d}\right) \right\} = O(1);$$

ii-a) for $\chi \neq \chi_0$,

$$\sum_{p \le x} \frac{\chi(p)\ln p}{p} = \sum_{n \le x} \frac{\chi(n)\Lambda(n)}{n} - \sum_{j=2}^{\infty} \sum_{p^j \le x} \frac{\chi(p^j)\ln p}{p^j} = O(1);$$

b) for $a \not\equiv 0 \pmod 5$,

$$\sum_{\substack{p \le x \\ p \equiv a \pmod 5}} \frac{\ln p}{p} = \frac{1}{4} \sum_{\chi} \frac{1}{\chi(a)} \sum_{p \le x} \frac{\chi(p)\ln p}{p} = \frac{1}{4}\ln x + O(1);$$

c) there are infinitely many primes of each of the forms

$$5n+1,\ 5n+2,\ 5n+3,\ 5n+4.$$

## XXIV Numerical Characters and the Dirichlet Theorem

In problems #8, 9 of XXIII we met functions $\chi, \chi_0, \cdots, \chi_4$. These functions are special cases of a class of functions called *characters*. The $\chi$ of #8 is a mod 4 *character* and the $\chi_j$ of #9 are mod 5 *characters*. In this chapter we introduce the notion of mod $k$ characters for arbitrary positive integers $k$ and will use them, much as was done in #8, 9 of XXIII, to prove the Dirichlet theorem on primes in arithmetic progressions.

*Definition*. A completely multiplicative complex valued number theoretic function of period $k$ which is zero precisely on those integers not prime to $k$ is called a (*numerical*) mod $k$ *character*.

1. (Elementary properties)

i) The function $\chi_0$ defined by

$$\chi_0(a) = \begin{cases} 1 & \text{if } (a,k)=1; \\ 0 & \text{otherwise,} \end{cases}$$

is a mod $k$ character; this character $\chi_0$ is called the *principal mod k character*;

ii) $\chi(1) = 1$ for all mod $k$ characters;

iii) if $\chi$ is a mod $k$ character and $(a,k)=1$ then $\chi(a)$ is a $\varphi(k)^{th}$ root of unity;

iv) the function $\chi$ defined in XXIII #8 is a mod 4 character;

v) the functions $\chi_j$, $0 \le j \le 3$, defined in XXIII #9 are mod 5 characters;

vi) there are only finitely many mod $k$ characters; in fact, no more than $\varphi(k)^{\varphi(k)}$ (we shall see in #3 (ii) that this bound is much too large);

vii) if $\chi$ is a mod $d$ character and $k = dn$ then $\chi^*$ defined by $\chi^*(n) = \begin{cases} \chi(n) \text{ for } (n,k)=1; \\ 0 \text{ otherwise}, \end{cases}$

is a mod $k$ character; $\chi^*$ is called the mod $k$ extension of $\chi$;

viii) $\sum_{n=1}^{k} \chi(n) = 0$ for $\chi$ any non-principal mod $k$ character;

ix) if $\chi_1$ and $\chi_2$ are mod $k$ characters so also are $\chi_1\chi_2$ and $\bar{\chi}_1$, where $\bar{\chi}_1(a) = \overline{\chi_1(a)}$, the bar on the right denoting the complex conjugate function;

x) $\chi\chi_1$ runs over all mod $k$ characters as $\chi$ does.

2. (Properties leading to a deeper result)

i) Let $p$ be an odd prime, $\beta$ be a positive integer, and $q$ be a primitive root of $p^\beta$; define $\chi$ by $\chi(n) = \begin{cases} 0 \text{ for } (n, p^\beta) = 1; \\ e^{2\pi i\lambda/\varphi(p^\beta)} \text{ for } n \equiv q^\lambda \pmod{p^\beta},\ 0 \leq \lambda < \varphi(p^\beta); \end{cases}$

then a) $\chi$ is a mod $p^\beta$ character ;

b) if $(d, p^\beta) = 1$ and $d \not\equiv 1 \pmod{p^\beta}$

then $\chi(d) \neq 1$ ;

c) if $(d, k) = 1$, $d \not\equiv 1 \pmod{p^\beta}$, and $p^\beta$ divides $k$ then there is a mod $k$ character $\chi^*$

such that $\chi^*(d) \neq 1$ ;

ii) let 4 be the highest power of 2 which divides $k$ and let $\chi^*$ be the mod $k$ extension of the mod 4 character $\chi$ defined in XXIII # 8 ; then

if $d \equiv -1 \pmod 4$ then $\chi^*(d) \neq 1$ ;

iii) let $2^\alpha$, $\alpha \geq 3$, be the highest power of 2 which divides $k$ and define $\chi$ by

$$\chi(n) = \begin{cases} 0 & \text{for } n \text{ even ;} \\ e^{2\pi i t / 2^{\alpha-2}} & \text{for } n \equiv (-1)^{\frac{n-1}{2}} \cdot 5^t \pmod{2^\alpha},\ 0 \leq t < 2^{\alpha-2} \text{ ;} \end{cases}$$

then

a) $\chi$ is a mod $2^\alpha$ character ;

b) if $(d, k) = 1$ and $d \not\equiv \pm 1 \pmod{2^\alpha}$ then there is a mod $k$ character $\chi^*$ such that

$\chi^*(d) \neq 1$ ;

iv) if $(d,k)=1$ and $d \not\equiv 1 \pmod{k}$ then there is a mod $k$ character such that $\chi(d) \neq 1$.

3. i) Let $c$ be the number of mod $k$ characters; then
$$\sum_{\chi} \chi(a) = \begin{cases} c & \text{if } a \equiv 1 \pmod{k}; \\ 0 & \text{if } a \not\equiv 1 \pmod{k}; \end{cases}$$

ii) there are exactly $\varphi(k)$ mod $k$ characters;

iii) when $(a,k)=1$,
$$\sum_{\chi} \chi(a)^{-1}\chi(n) = \begin{cases} \varphi(k) & \text{for } a \equiv n \pmod{k}; \\ 0 & \text{otherwise}; \end{cases}$$

iv) if $(a,k)=1$ then
$$\sum_{\chi} \bar{\chi}(a)\chi(n) = \begin{cases} \varphi(k) & \text{for } a \equiv n \pmod{k}; \\ 0 & \text{otherwise}. \end{cases}$$

4. (Miscellaneous)

Let $(a,k)=1$ and let $m$ be the exponent of $a$ modulo $k$. Then for each mod $k$ character $\chi$ it is true that $\chi(a)^m = \chi(a^m) = \chi(1) = 1$; thus $\chi(a)$ is an $m^{th}$ root of unity. In this problem we see that as $\chi$ runs over the mod $k$ characters none of the $m$ $m^{\text{th}}$ roots of unity is slighted in its number of appearances – i.e. they each appear $\frac{\varphi(k)}{m}$ times. Let $\omega$ be an arbitrary $m^{\text{th}}$ root of unity and suppose it appears $N$ times. Then

$$\text{i)}\ \sum_{j=1}^{m} \left(\frac{\chi(a)}{\omega}\right)^{j} = \begin{cases} m & \text{if } \chi(a) = \omega\ ; \\ 0 & \text{otherwise}\ ; \end{cases}$$

$$\text{ii)}\ Nm = \sum_{\chi} \sum_{j=1}^{m} \left(\frac{\chi(a)}{\omega}\right)^{j} = \varphi(k)\ .$$

In the next six problems the Dirichlet theorem is proved.

5. Each of the three series

$$\sum_{n=1}^{\infty} \frac{\chi(n)}{n}\ ,\qquad \sum_{n=1}^{\infty} \frac{\chi(n)\ln n}{n}\ ,\qquad \sum_{n=1}^{\infty} \frac{\chi(n)}{\sqrt{n}}$$

converges when $\chi$ is a non-principal mod $k$ character; denoting the sums by $L_0(\chi)$, $L_1(\chi)$, $L_2(\chi)$ respectively we have

$$\sum_{n\le x} \frac{\chi(n)}{n} = L_0(\chi) + O\left(\frac{1}{x}\right);$$

$$\sum_{n\le x} \frac{\chi(n)\ln n}{n} = L_1(\chi) + O\left(\frac{\ln x}{x}\right);$$

$$\sum_{n\le x} \frac{\chi(n)}{\sqrt{n}} = L_2(\chi) + O\left(\frac{1}{\sqrt{x}}\right).$$

6. Suppose $\chi$ is any real non-principal mod $k$ character. Define $F$ and $G$ by

$$F(n) = \sum_{d\mid n} \chi(d), \qquad G(x) = \sum_{n\le x} \frac{F(n)}{\sqrt{n}}.$$

Then:

i) $F$ is multiplicative and

$$F(p_1^{\alpha_1}\cdots p_s^{\alpha_s}) = \prod_{i=1}^{s} \sum_{j=0}^{\alpha_i} \chi(p_i^{\,j});$$

ii) for all $n$, $F(n) \ge 0$, $F(n^2) \ge 1$;

iii) $G(x) \to \infty$ as $x \to \infty$;

iv) $G(x) = \sum_{d\delta\le x} \frac{\chi(d)}{\sqrt{d}\sqrt{\delta}} =$

$$\sum_{d\le\sqrt{x}} \frac{\chi(d)}{\sqrt{d}} \sum_{\delta\le\frac{x}{d}} \frac{1}{\sqrt{\delta}} + \sum_{\delta<\sqrt{x}} \frac{1}{\sqrt{\delta}} \sum_{\sqrt{x}<d\le\frac{x}{\delta}} \frac{\chi(d)}{\sqrt{d}};$$

v) $G(x) = 2\sqrt{x}\,L_0(\chi) + O(1)$;

vi) if $\chi$ is any real non-principal mod $k$ character then

$$L_0(\chi) = \sum_{n=1}^{\infty} \frac{\chi(n)}{n} \ne 0.$$

7. i) $L_0(\chi)\sum_{n\le x}\frac{\nu(n)\chi(n)}{n} = O(1)$ for $\chi \ne \chi_0$ (the $\chi$ here need not be real);

ii) suppose $\chi$ is a non-real character and that $L_0(\chi) = 0$; then putting $g(x) = \sum_{n\le x}\chi(n)\frac{x}{n}\ln\frac{x}{n}$ we have

a) $g(x) = -x\,L_1(\chi) + O(\ln x)$;

b) $x\ln x = -x\,L_1(\chi)\sum_{n\le x}\frac{\nu(n)\chi(n)}{n} + O(x)$;

iii) for any $\chi \ne \chi_0$,

$$L_1(\chi)\sum_{n\le x}\frac{\nu(n)\chi(n)}{n} = \begin{cases} -\ln x + O(1) & \text{for } L_0(\chi) = 0; \\ O(1) & \text{for } L_0(\chi) \ne 0. \end{cases}$$

8. For $\chi \ne \chi_0$,

$$\sum_{p\le x}\frac{\chi(p)\ln p}{p} = \begin{cases} -\ln x + O(1) & \text{if } L_0(\chi) = 0; \\ O(1) & \text{if } L_0(\chi) \ne 0. \end{cases}$$

9. Let $N$ be the number of non-principal mod $k$ characters for which $L_0(\chi) = 0$. Then:

i) if $N \ne 0$ then $N \ge 2$;

ii) if $Q(x) = \varphi(k) \sum_{\substack{p \le x \\ p \equiv 1 \pmod{k}}} \frac{\ln p}{p}$ then

$$0 \le Q(x) = \sum_{\chi} \sum_{p \le x} \frac{\chi(p) \ln p}{p} = (1-N) \ln x + O(1);$$

iii - a) $0 \le N \le 1$;

b) if $\chi \ne \chi_0$ then $L_0(\chi) \ne 0$;

c) $\sum_{p \le x} \frac{\chi(p) \ln p}{p} = O(1)$ when $\chi \ne \chi_0$;

(note that (b) tells us the supposition made in #7 (ii) is in fact not realizable; i.e. that the number N of non-principal characters for which $L_0(\chi) = 0$ is itself 0).

10. (The Dirichlet Theorem)

For $(a,k) = 1$,

$$\varphi(k) \sum_{\substack{p \le x \\ p \equiv a \pmod{k}}} \frac{\ln p}{p} = \sum_{\chi} \chi(a)^{-1} \sum_{p \le x} \frac{\chi(p) \ln p}{p} = \ln x + O(1)$$

and, therefore, there are infinitely many primes in the arithmetic progression

$$a,\ a+k,\ a+2k,\ a+3k, \cdots .$$

Remarks.

1. The theory of numerical characters is but a small part of the general theory of characters in the theory of Abelian groups. It is for this reason we have used the word "numerical".

2. The interested reader might wonder if each residue class modulo $k$ gets its "fair share" of primes. Since there are $\varphi(k)$ possible residue classes for the primes this would mean that each class got $\frac{1}{\varphi(k)}$ th of the primes. The asymptotic Dirichlet theorem says

$$\Big(\sum_{\substack{p \leq x \\ p \equiv a \pmod{k}}} 1\Big)\Big(\frac{x}{\ln x}\Big)^{-1} \to \frac{1}{\varphi(k)} \quad \text{as } x \to \infty .$$

Using this one may immediately deduce the prime number theorem $\pi(x)\big(\frac{x}{\ln x}\big)^{-1} \to 1$ as $x \to \infty$ and then conclude $\Big(\sum_{\substack{p \leq x \\ p \equiv a \pmod{k}}} 1\Big)(\pi(x))^{-1} \to \frac{1}{\varphi(k)}$ as $x \to \infty$. Thus each class does equally well.

For an elementary, though quite involved, proof of the asymptotic Dirichlet theorem the reader might consult Specht [1956].

3. Our proof of the Dirichlet theorem follows Shapiro [1950] and also makes use of ideas of Rademacher [1964], Hasse [1950], and Prachar [1957].

# SOLUTIONS

## I The Game of Euclid & the Euclidean Algorithm

~ solutions ~

1\. i, ii) The derived sets are

$$\{m, n-m\}, \{m, n-2m\}, \cdots, \{m, n-tm\},$$

where $tm \le n < (t+1)m$ ;

iii) since we are assuming $m \le n$ we must have $\{a, b\} = \{m, n-sm\}$ for some positive integer $s$ ; now any common divisor of $n$ and $m$ clearly divides $m$ and $n-sm$ and conversely ; thus $(n, m) = (m, n-sm) = (a, b)$ ;

iv) since negative integers are not permitted and each move reduces one of the two elements it must happen, after a finite number of steps, that one of the elements is reduced to 0 ; the other, by (iii), must then be $(m, n)$.

2. *i*) This is clear ;

*ii*) if a player starting with $\{2,5\}$ moves to the minimal derived set $\{2,1\}$ then the other player will immediately win by moving to $\{0,1\}$ ;

*iii*) if there is but one derived set the proposition is true ; otherwise, since we are assuming there is a winning strategy for A and since one of A or B must ultimately make the move from the minimal derived set, the advantage must lie in either making or not making this move, and for A to do anything other than asserted enables B to decide who will move from the minimal derived set and thus transfers to B the winning strategy ;

iv) since $m < a < m\tau < 2m$ the only possible move from $\{a, m\}$ is to $\{a-m, m\}$; hence, $r = a - m$ and

$$\frac{m}{r} = \frac{m}{a-m} = \frac{1}{\frac{a}{m} - 1} > \frac{1}{\tau - 1} = \tau .$$

3. i) Since one is not able to win in one move from a position $\{m, n\}$, $1 < \frac{n}{m} < \tau$, it is enough to show that when A starts from $\{m, n\}$, $\frac{n}{m} > \tau$, then he may either win in one move or leave to B a position with $1 < \frac{n}{m} < \tau$, from which, by #2 (iv), B's sole move is to a position with ratio $> \tau$ from which the process is repeated; when $\frac{n}{m} > 2$ there are at least the two moves

$$\{m,n\} \begin{array}{l} \nearrow \{m, r\} \\ \searrow \{m, m+r\} \end{array} , \quad 0 \le r < m ,$$

where $r$ is the remainder obtained when one divides $n$ by $m$; if $r = 0$, A may win in one move by moving to $\{m, r\}$; otherwise, since (by an elementary calculation) $\tau$ is strictly between $\frac{m}{r}$ and $\frac{m+r}{m}$, A moves to that position for which the ratio lies strictly between 1 and $\tau$; when $\tau < \frac{n}{m} < 2$ A moves to $\{m,r\}$;

*ii*) this follows from (*i*).

4. By #1 (*iii*) we know $(r_0, r_1) = (r_1, r_2) = (r_2, r_3) = \cdots = (r_{n-1}, r_n) = (r_n, 0) = r$.

5. In #4 each $r_j$, $j \geq 2$, is a linear combination of $r_{j-1}, r_{j-2}$; thus starting at the bottom of the Euclidean algorithm and solving for $r_n$ first in terms of $r_{n-1}, r_{n-2}$ and then in terms of $r_{n-2}, r_{n-3}$ etc. we ultimately find $r_n$ expressed in the form $cr_0 + dr_1$, where $c$ and $d$ are integers; since $0 < r_n \leq \min\{r_0, r_1\}$ exactly one of $c, d$ is $\leq 0$;

if $c>0, d=0$ or $c=0, d>0$

then $r_0 = r_1 = r_2$ so $r_n = 2r_0 - 1\cdot r_1$ ;

if $c>0, d<0$ then $r_n = cr_0 - |d|\, r_1$ ;

if $c<0, d>0$ then $r_n = (sr_1 - |c|)r_0 + (d - sr_0)r_1$

and we may select $s$ so that

$$sr_1 - |c| > 0 , d - sr_0 < 0 .$$

6. *i*) The proof is by induction; since $u_6 = 13 > 10$ it is true for $n=1$; supposing it to be true for $n$ we have

$u_{5(n+1)+1} = u_{5n+5} + u_{5n+4} = 2u_{5n+4} + u_{5n+3}$

$= 3u_{5n+3} + 2u_{5n+2} = 5u_{5n+2} + 3u_{5n+1} =$

$8u_{5n+1} + 5u_{5n} > 8u_{5n+1} + 2(u_{5n} + u_{5n-1}) = 10u_{5n+1}$

$> 10^{n+1}$, and thus it is also true for $n+1$ ;

*ii*) reading the Euclidean algorithm from bottom to top we see

$$r_{n-1} \geq r_n + r_n \geq r_n + 1 \geq u_2$$

$$r_{n-2} \geq r_{n-1} + r_n \geq u_2 + 1 = u_3$$

$$r_{n-3} \geq r_{n-2} + r_{n-1} \geq u_3 + u_2 = u_4$$

$$\cdots$$

$$r_1 \geq r_2 + r_3 \geq u_{n-1} + u_{n-2} = u_n \;;$$

*iii*) if the number of divisions $n$ is 1 this is clear; otherwise suppose $0 < r_1 < r_0$ and the first step has $r_1$ as divisor and $r_1$ has $t$ base 10 digits; then, by (*ii*), $r_1 \geq u_n$ and, if $m$ is such that

$$5m + 1 \leq n \leq 5(m+1)$$

then $r_1 \geq u_n \geq u_{5m+1} > 10^m$ so $t \geq m + 1 \geq \frac{1}{5}n$

and $n \leq 5t$ ;

*iv*) direct calculation;

*v*) by (*ii*), $b \geq u_{5t}$ and since $b$ has $t$ base 10 digits $u_{5t}$ has $t$ or fewer base 10 digits ;

vi) by induction $|u_{n+1}^2 - u_n u_{n+2}| = 1$ for all $n$ and $\frac{u_{n+1}}{u_n}$ is $< \tau$ ($> \tau$) when $n$ is even (odd); hence $\left|\frac{u_{n+1}}{u_n} - \tau\right| < \left|\frac{u_{n+1}}{u_n} - \frac{u_{n+2}}{u_{n+1}}\right| = \frac{1}{u_n u_{n+1}} < \frac{1}{u_n^2}$ ; for further details see II #3, 5 ;

vii) for $n = 4$ we have $u_9 = 55 > 10 \cdot 5 = 10 u_4$ ; assuming true for $n$ we have

$u_{n+5} = 5u_{n+1} + 3u_n = u_n\left\{5\tau + 3 + 5\left(\frac{u_{n+1}}{u_n} - \tau\right)\right\}$

$> u_n\left(\frac{15}{2} + 3 - \frac{1}{5}\right) > 10 u_n$ for $n \geq 4$ (using (vi));

viii) for $t = 4$, $u_{20} = 10946 > 10^4$ ; assuming true for $t$, then, by (vii),

$u_{5(t+1)} = u_{5t+5} > 10 u_{5t} > 10 \cdot 10^t = 10^{t+1}$ ;

ix) by (viii) when $t$ is at least 4, $u_{5t}$ has more than $t$ base 10 digits while, by (v), if the process starting with $a$ and $b$, $a > b$, $b$ having $t$ base 10 digits, takes $5t$ steps then $u_{5t}$ has no more than $t$ base 10 digits ; the conclusion follows .

# II The Golden Mean ~ Solutions

1. Let $\frac{m}{r} = x$ ; then $x = 1 + \frac{1}{x}$ so $x^2 = x + 1$ and, therefore, $x = \frac{1 \pm \sqrt{5}}{2}$; since $x > 0$ we must have $\frac{m}{r} = x = \tau$.

2. By the above $\tau^2 = \tau + 1$ ; dividing by $\tau$ and then subtracting $\tau$ from both sides yields $\tau^{-1} = \tau - 1$ ; finally since $\tau\tau' = -1$ the last relationship follows.

3. If $\frac{m}{r} < \tau$ then

$$\frac{m+r}{m} = 1 + \frac{1}{m/r} > 1 + \frac{1}{\tau} = \frac{\tau + 1}{\tau} = \frac{\tau^2}{\tau} = \tau ;$$

similarly if $\frac{m}{r} > \tau$ then $\frac{m+r}{m} < \tau$. Finally, if $\alpha$ is between $\frac{m}{r}$ and $\frac{m+r}{m}$ then

$$|\alpha - \tau| \leq \left|\frac{m+r}{m} - \frac{m}{r}\right| = \frac{|mr + r^2 - m^2|}{mr}$$

$$= \frac{\left| 1 + \frac{m}{r} - \left(\frac{m}{r}\right)^2 \right|}{\frac{m}{r}} ;$$

now since we may select integers $m$ and $r$ so that $\frac{m}{r}$ is as close as we like to $\tau$ we see that $|\alpha-\tau|$ is smaller than every positive number, hence $\alpha=\tau$.

4. i) The inequality is true for the 1st two terms; supposing it to be true up to and including the $n^{th}$ term we have $\beta \geq (n-1)c$, $\alpha \geq nc$ when $\frac{\beta}{\gamma}$, $\frac{\alpha}{\beta}$ are the $n-1^{st}$ and $n^{th}$ terms respectively; now the $n+1^{st}$ term has a numerator $\alpha+\beta$ which is

$$\geq (n-1)c + nc \geq (n+1)c \;;$$

the remainder of the assertion follows from this and the fact that the typical denominator is the previous numerator;

ii) let the terms be $\frac{a}{b}$, $\frac{a+b}{a}$, $\frac{2a+b}{a+b}$; then direct calculation yields the result;

*iii*) this follows immediately from (*ii*);

*iv*) if any two consecutive terms of the sequence are equal then, by #1, all terms are equal to $\tau$ and the sequence converges to $\tau$; otherwise by #3, $\tau$ lies strictly between each consecutive pair of terms and hence, using (*i*) and (*iii*), the sequence converges to $\tau$.

5. *i*) Immediate from the definition of the sequence in #4;

*ii*) this follows by induction from (*i*) and the truth for $n = 0$; from (*i*) we see that any divisor of $u_{n+2}$ and $u_{n+1}$ is also a divisor of $u_{n+1}$ and $u_n$ ( $= u_{n+2} - u_{n+1}$ );

*iii*) this follows from #4(*ii*) and the fact that $u_1^2 - u_0 u_2 = 1 - 2 = -1 = (-1)^1$;

iv) this follows from #4(i) since $c = 1$ ;

v) $$\left|\frac{u_{n+1}}{u_n} - \tau\right| \le \left|\frac{u_{n+1}}{u_n} - \frac{u_{n+2}}{u_{n+1}}\right| = \frac{|u_{n+1}^2 - u_n u_{n+2}|}{u_n u_{n+1}}$$
$$= \frac{1}{u_n u_{n+1}} < \frac{1}{u_n^{\,2}} \ ;$$

vi) this follows from (iv) and (v) or from #4(iv) ;

6. i-a) This is true for $n = 1$ by #2 ; suppose true for $n$; then $\tau^{n+2} = \tau\tau^{n+1} = \tau(u_{n-1} + u_n\tau)$ $= u_{n-1}\tau + u_n\tau^2 = (u_{n-1} + u_n)\tau + u_n = u_n + u_{n+1}\tau$ and the induction is complete ;

b) for $n = 1$ we have $-\tau^{-2} = \tau^{-1} - 1$ which follows also from #2; suppose true for $n$; then $(-1)^{n+1}\tau^{-(n+2)} = -\tau^{-1}(-1)^n\tau^{-(n+1)} = -\tau^{-1}(u_n\tau^{-1} - u_{n-1})$ $= -u_n\tau^{-2} + u_{n-1}\tau^{-1} = u_n(\tau^{-1} - 1) + u_{n-1}\tau^{-1} = u_{n+1}\tau^{-1} - u_n$ and the induction is complete;

ii) for $n=0$ this is clear; for $n \geq 1$ we add the expressions in (i-a) and (b) to obtain

$$u_n(\tau + \tau^{-1}) = \tau^{n+1} + (-1)^n \tau^{-(n+1)};$$

since $\tau + \tau^{-1} = \tau - \tau' = \sqrt{5}$ and $\tau^{-1} = -\tau'$

we have

$$u_n = \frac{1}{\sqrt{5}} \{\tau^{n+1} - \tau'^{n+1}\}.$$

7. i) Equality of the areas implies

$$(u+v)r = mv = (m+r)u$$

and, therefore,

$$\frac{v}{u} = \frac{m+r}{m} \text{ and } \frac{u+v}{u} = 1 + \frac{v}{u} = 1 + \frac{m}{r} = \frac{m+r}{r};$$

$$\text{thus } \frac{v}{u} = \frac{m}{r} = \frac{m+r}{m} = \tau,$$

where we have used #1 for the last equality.

ii) in this case $m = u + v$ so

$$\frac{m}{r} = \frac{m+r}{m} = \frac{m+r}{u+v} = \tau$$

and the conclusion follows.

8. Consider the regular pentagon shown; then, successively, we see :

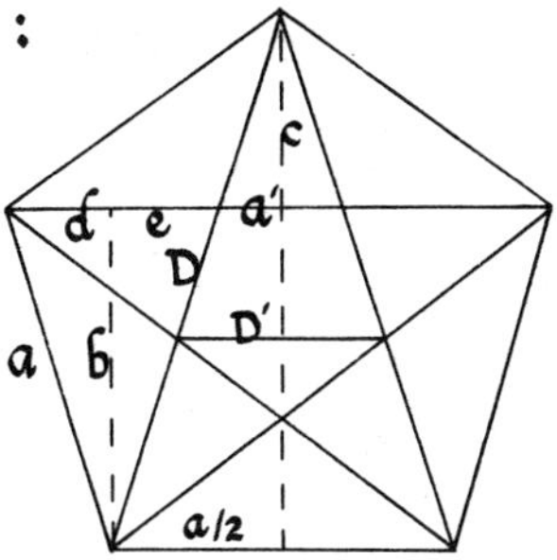


$(b+c)^2+(a/2)^2=D^2$ ;

$\frac{b+c}{a/2}=\frac{b}{d}$ ;

$b^2+d^2=a^2$ ;

$2d+a=D$ ;

$d=e$ ; thus we conclude :

$D=a\tau$ , $d+e=\frac{a}{\tau}$ , $a'=\frac{a}{\tau^2}$ , $D'=\frac{a}{\tau}$ ;

when $a=1$ this yields $D=\tau$.

9. Considering the proof as given in the solution to #8 we see that the lines are as marked at the right. This yields the result as stated. The continuation is clear.

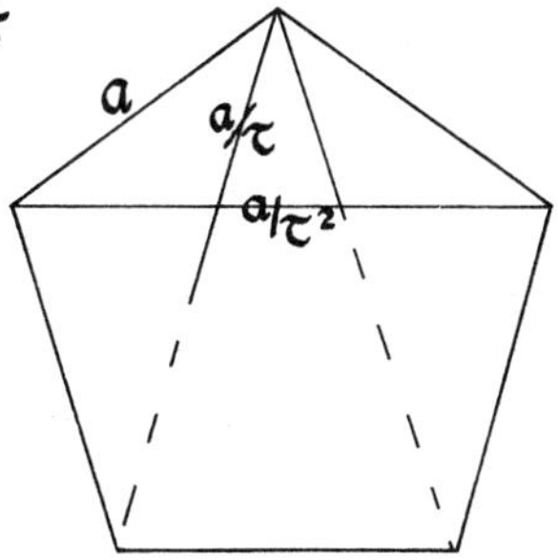

10. For any integral value of $m$ the dissections indicated below are correct.

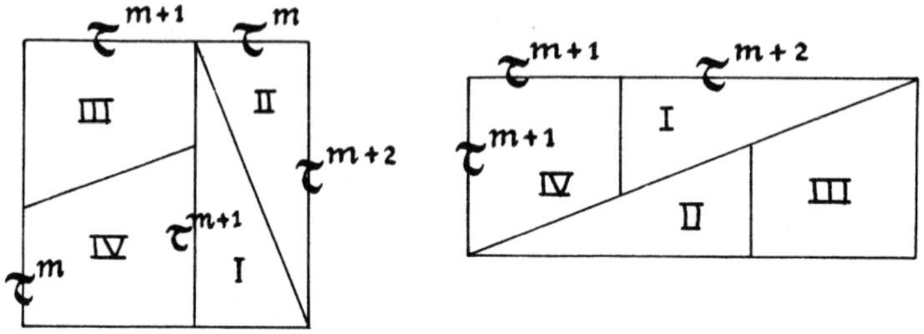


They are correct since $(\tau^{m+2})^2 = \tau^{m+1}(\tau^{m+1} + \tau^{m+2})$.

11. The model seems to form a 13 by 34 rectangle; but the area of the square is $21^2 = 441$ and the area of the rectangle is $13 \cdot 34 = 442$; the model does not indicate this discrepancy since the extra unit of area is distributed along the main diagonal and it would require an extremely accurate model to reveal the difficulty.

12. a) $\tau = 1 + \frac{1}{\tau}$ since $\tau^2 = 1 + \tau$ and successively replacing $\tau$ on the right by

$1+\frac{1}{\tau}$ yields the string of equalities; if the pieces yield $\frac{u_{n+1}}{u_n}$ up to the $n^{th}$ term then the next term is just $1+\frac{1}{\frac{u_{n+1}}{u_n}}=1+\frac{u_n}{u_{n+1}}=\frac{u_{n+2}}{u_{n+1}}$ so the expression for the general term is correct;

6) by #5 (vi) we know $\frac{u_{n+1}}{u_n}\to\tau$ so the implied limiting process does yield $\tau$.

13. i) This is immediate for $n=1,2$ and all $m$; suppose true for each of $n-1, n$ and all $m$; then $u_{m+(n+1)}=u_{m+n-1}+u_{m+n}=$
$u_{m-1}u_{n-2}+u_m u_{n-1}+u_{m-1}u_{n-1}+u_m u_n=$
$u_{m-1}(u_{n-2}+u_{n-1})+u_m(u_{n-1}+u_n)=u_{m-1}u_n+u_m u_{n+1}$;

ii) clear for $m=1$ so assume true for $m$; then $u_{n(m+1)-1}=u_{nm+n-1}=u_{nm-1}u_{n-2}+u_{nm}u_{n-1}$ and, therefore, since $u_{n-1}$ divides $u_{nm-1}$ we may conclude $u_{n-1}$ divides $u_{n(m+1)-1}$;

iii) noting $(n,m)$ divides each of $n, m$ part (ii) shows $u_{(n,m)-1}$ divides $(u_{n-1}, u_{m-1})$; now there are integers $x$ and $y$ such that $(n,m) = nx - my$ so $u_{nx-1} = u_{(n,m)+my-1} = u_{(n,m)-1}u_{my-2} + u_{(n,m)}u_{my-1}$ and, since $(u_{n-1}, u_{m-1})$ divides each of $u_{nx-1}$ and $u_{my-1}$, it divides $u_{(n,m)-1}u_{my-2}$ and is prime to $u_{my-2}$; thus $(u_{n-1}, u_{m-1})$ divides $u_{(n,m)-1}$ and the proof is complete.

14. For $n=2$ we have $\begin{pmatrix} 1 & 1 \\ 1 & 0 \end{pmatrix}^2 = \begin{pmatrix} 2 & 1 \\ 1 & 1 \end{pmatrix}$ and the result is correct;

suppose true for $n$, then

$$\begin{pmatrix} 1 & 1 \\ 1 & 0 \end{pmatrix}^{n+1} = \begin{pmatrix} 1 & 1 \\ 1 & 0 \end{pmatrix}\begin{pmatrix} u_n & u_{n-1} \\ u_{n-1} & u_{n-2} \end{pmatrix} = \begin{pmatrix} u_{n+1} & u_n \\ u_n & u_{n-1} \end{pmatrix}.$$

15. i) The contributors to $g(n)$ not containing $n$ are precisely the contributors to $g(n-1)$, while those containing $n$ are precisely the contributors to $g(n-2)$ with $n$ adjoined;

*ii*) this is true for $n = 1, 2$ by direct calculation and the recurrence of (*i*) shows that the equality continues since it is the same as the Fibonacci recurrence;

*iii*) each $k$ element subset of $\{1, 2, \cdots, n\}$ corresponds uniquely to a marking of $k$ elements of $\{1, 2, \cdots, n\}$ with a 1 and the remaining $n-k$ elements with a 0; such a subset will contribute to $f(n, k)$ precisely when no two consecutive 1's appear;

*iv*) in (*iii*) the $n-k$ 0's may be thought to define $n-k+1$ boxes; the number of ways of putting $k$ 1's into these boxes as described is just the number of strings of 0's and 1's discussed in (*iii*); i.e. it is equal to $f(n, k)$;

v) by (iv) it is just the number of ways of choosing $k$ of $n-k+1$ objects and this is $\binom{n-k+1}{k}$ when $n-k+1 \geq k$ and is 0 otherwise;

vi) $u_n = g(n-1) = \sum_{k=0}^{n-1} f(n-1,k) = \sum_{k=0}^{n-1} \binom{n-k}{k}$;

vii) this follows immediately from (vi).

16. i) By the ratio test the series converges for $\lim_{n\to\infty} \frac{u_{n+1}}{u_n}|x| < 1$; since the limit is $\tau|x|$ we see that the series converges for $|x| < \frac{1}{\tau}$; calculation shows $\mathcal{U} - x\mathcal{U} - x^2\mathcal{U} = 1$ so $\mathcal{U} = \frac{1}{1-x-x^2}$ when

$$|x| < \frac{1}{\tau};$$

ii) let $r = \tau$, $s = \tau'$; then $r+s = 1 = -rs$ and $\frac{1}{1-x-x^2} = \frac{r/(r-s)}{1-rx} - \frac{s/(r-s)}{1-sx}$;

iii) expanding $\frac{r}{1-rx}$ and $\frac{s}{1-sx}$ on the right of (ii) yields

$$\frac{1}{1-x-x^2} = \frac{1}{r-s}\sum_{n=0}^{\infty}(r^{n+1} - s^{n+1})\, x^n$$

$$= \frac{1}{\sqrt{5}}\sum_{n=0}^{\infty}(\tau^{n+1} - \tau'^{\,n+1})\, x^n$$

and comparison with $\mathcal{U}$ yields the conclusion;

iv) put $x = .01$ in $\mathcal{U}$, making use of (i);

v) by induction.

## III Prime Factorizations & Primes

~ Solutions ~

1. For the integer $n=2$ the integer 2 is itself a prime divisor of $n$. If the proposition is false let $N$ be the smallest positive integer $>1$ for which it is false. Then $N$ is not prime so $N=ab$, where $1<a<N$, $1<b<N$. This implies $a$ has a prime factor which is then a prime factor of $N$. This contradicts our assumption that the proposition is false.

2. True for $n=2$. If false for some integer let $N$ be the smallest integer for which it is false. Then $N$ is not prime so $N=ab$, $1<a<N$, $1<b<N$. Thus each of $a, b$ have prime factorizations. Putting the factorizations of $a$ and $b$ together gives a prime factorization for $N$.

This is a contradiction so the proposition must be true.

3. Suppose $p$ is a prime dividing $ab$; if $p$ does not divide $a$ then $(p, a) = 1$ so, by I#5, there are integers $x$ and $y$ such that $1 = px - ay$; multiplying this equation by $b$ yields $b = pxb - aby$; now if $p | ab$ then $p$ divides the right and, therefore, the left side of this last equation.

(Alternate proof)

Let $S$ be the set of positive integers $n$ for which, for a given prime $p$, there exists a $b$ satisfying

$p$ divides $nb$ and $p$ divides neither $n$ nor $b$.

We show $S$ is empty by an induction argument. Certainly 1 is not in $S$. Suppose no integer $< n$ is in $S$. Let $p$ divide $nb$ and suppose $n = pt + q$, $0 \leq q < p$, $b = ps + r$, $0 \leq r < p$.

Then $nb = p(pts + tr + sq) + qr$ so $p$ divides $qr$. Since $q < n$, either $p$ divides $q$ or $p$ divides $r$. In either event $p$ divides one of $n, b$ so $n$ is not in $S$. Thus $S$ is empty and the proposition is proved.

4. For $n = 2$ this is clear. Suppose true for all positive integers $>1$ and $< n$, $n > 2$. Let $n = p_1 \cdots p_k = q_1 \cdots q_s$ where the $p_i$ and $q_i$ are primes. Then by the finite extension of #3, $p_1$ divides one of $q_1, \cdots, q_s$ and hence equals one of them. Suppose, without loss of generality, $p_1 = q_1$. Then $\frac{n}{p_1}$ is an integer smaller than $n$. If $\frac{n}{p_1} = 1$ then, since $n$ is prime, the proposition is true for $n$. Otherwise $2 \le \frac{n}{p_1} < n$ so the proposition is true for $\frac{n}{p_1}$; i.e. $p_2, \cdots, p_k$ are just the $q_2, \cdots, q_s$ in some order. Thus $p_1, \cdots, p_k$ are just the $q_1, \cdots, q_s$ in some order

and the proposition is true for $n$. By induction the proposition is true for all $n \geq 2$.

5. By #2 there is a prime $p$ which divides $1+n!$. Since no prime dividing $n!$ may divide $1+n!$ and since all primes $\leq n$ divide $n!$ it must be the case that $p > n$. Hence since there can be no largest prime there must be infinitely many of them.

6. Since $1+p_1 \cdots p_k$ must have a prime factor differing from each of $p_1, \cdots, p_k$ and since the same argument shows that every finite collection of primes fails to exhaust all primes the number of primes is not finite.

7. Such a string is afforded by

$$(k+1)!+2, \cdots, (k+1)!+(k+1).$$

8. Since the product of any finite number of $4k+1$ primes is again a $4k+1$ number and since $4p_1 \cdots p_k - 1$ is not of the form $4k+1$ we conclude that among the prime factors of $4p_1 \cdots p_k - 1$ there must be a $4k+3$ prime. Since, by the argument of #6 above, no finite set of primes can exhaust all $4k+3$ primes there must be infinitely many of them.

9. i) $F_2 = 2^{2^2} + 1 = 17 \equiv 7 \pmod{10}$;

suppose $2^{2^n} + 1 = F_n \equiv 7 \pmod{10}$;

then $2^{2^n} \equiv 6 \pmod{10}$ and, therefore,

$$F_{n+1} = 2^{2^{n+1}} + 1 = (2^{2^n})^2 + 1 \equiv 6^2 + 1 \equiv 7 \pmod{10}$$

and the conclusion follows by induction.

ii) if $m = ab$, where $b$ is an odd number larger than 1 then

$$2^m + 1 = (2^a)^b + 1 \equiv 0 \pmod{(2^a+1)}$$

and, being divisible by $2^a+1$, is not prime;

iii) since $641 = 5 \cdot 2^7 + 1 = 5^4 + 2^4$ we see that $5 \cdot 2^7 \equiv -1 \pmod{641}$ and $5^4 \equiv -2^4 \pmod{641}$; raising the first congruence to the 4th power and using the second congruence we find $5^4 \cdot 2^{28} \equiv -2^{32} \equiv 1 \pmod{641}$ and the desired conclusion follows;

iv-a) since $F_0 = F_1 - 2$ the proposition is true for $m = 1$; suppose true for $m$, then

$$\prod_{0 \le n < m+1} F_n = (F_m - 2)F_m = (2^{2^m} - 1)(2^{2^m} + 1)$$
$$= 2^{2^{m+1}} - 1 = F_{m+1} - 2$$

and the proposition is true by induction;

б) without loss of generality let $n < m$; then any common factor of $F_n$ and $F_m$ would divide $\prod_{0 \le j < m} F_j = F_m - 2$ and, therefore, would divide 2; but $F_n$ and $F_m$ are odd so such a common factor may only be 1;

ν) since each $F_n$ contains a prime factor the conclusion of (ώ-б) is incompatible with the existence of only finitely many primes.

10. Without loss of generality let $P$ be an integral polynomial with positive leading coefficient; since $P(n) \longrightarrow \infty$ with $n$ we may choose an $m$ such that $P(m) > 1$; now we note that for all $n$, $P(m)$ divides $P(m + nP(m))$ and, consequently, for infinitely many integral values of $x$, $P(x)$ is composite.

11. (See Luthar [1969])

i) $p_5 = 11 > 9 = 2\cdot 4 + 1$ so the assertion is true for $n = 4$; if true for $n$ then

$$p_{n+1} \geq p_n + 2 > 2n + 1 + 2 = 2(n+1) + 1 \;;$$

ii) the assertion is true by checking for $n = 1, 2, 3, 4$; if true for $n$ $(n \geq 4)$ then

$$x_{n+1} = x_n + p_{n+1} > n^2 + 2n + 1 = (n+1)^2 \;;$$

iii) (a) follows from (i); for (b) we have, when $0 \leq j < n$, $p_{n-j} \leq p_{n-j+1} - 2 \leq p_{n-j+2} - 2\cdot 2$

$$\leq \cdots \leq p_{n+1} - 2(j+1) \leq 2(n+k) + 1 - 2(j+1)$$
$$= 2(n+k) - (2j+1) \;;$$

for (c) note $x_n = p_1 + \cdots + p_n \leq$

$$2n(n+k) - (1+3+5+\cdots+(2n-1)) = n^2 + 2nk < (n+k)^2 \;;$$

iv) follows immediately from (iii);

v) by (ii), for each $n$ there is a non-negative integer $k$ such that $(n+k)^2 \le x_n < (n+k+1)^2$; when $k=0$ we have $n^2 < x_n < (n+1)^2 < x_{n+1}$, while when $k \ne 0$ we have, by (iv),

$$x_n < (n+k+1)^2 = (n+k)^2 + 2(n+k) + 1 < x_n + p_{n+1} = x_{n+1}.$$

12. (See Grimm [1969])

i) When $q_k = k$, since $k \le n$, we have $q_k \mid n! + k$; when $q_k$ is a prime divisor of $\frac{n!}{k} + 1$ it certainly divides $n! + k = k\left(\frac{n!}{k} + 1\right)$;

ii) case 1: $\frac{n}{2} < k \le n$ and $k$ prime; then $q_k = k$ and since $q_k$ divides each of $n! + j$ and $n!$ it also divides $j$; but $2 \le j \le n < 2k$,

so $k = j$;

case 2: $q_k$ is a prime dividing $k - j$ since $k - j = k\left(\frac{n!}{k} + 1\right) - (n! + j)$; suppose now that $q_k \le n$; then $q_k \mid j$, $j = (n! + j) - n!$, and $q_k | k-j$ so $q_k | k$ ; if $k \leq \frac{n}{2}$ this means $q_k | 2k$ and $2k | \frac{n!}{k}$ which implies

$$q_k | \left(\frac{n!}{k}+1\right) - \frac{n!}{k} \quad \text{or} \quad q_k | 1 \ ;$$

thus $\frac{n}{2} < k \leq n$ and, since we are in case 2, $k$ is composite; again each prime factor of $k$, including $q_k$, divides $\frac{n!}{k}$ and hence would have to divide 1; thus $q_k > n$ and, since $q_k | k-j$ and $k$ and $j$ are positive integers, not exceeding $n$, $k = j$ ;

iii & iv) these follow immediately from (i) & (ii).

## IV Square Brackets – Solutions

1. Clearly $[\alpha] \le \alpha$ ; if $[\alpha] \le \alpha - 1$ then $[\alpha] < [\alpha] + 1 \le \alpha$ contradicting the definition of $[\alpha]$ .

2. Adding the inequalities

$$\alpha + n - 1 < [\alpha + n] \le \alpha + n, \quad -\alpha \le -[\alpha] < -\alpha + 1$$

yields $n - 1 < [\alpha + n] - [\alpha] < n + 1$

and, therefore,

$$[\alpha + n] - [\alpha] = n .$$

3. $m = \left[\frac{m}{n}\right] n + r$ , $0 \le r < n$ ; hence

$$\frac{m+1}{n} = \left[\frac{m}{n}\right] + \frac{r+1}{n} \le \left[\frac{m}{n}\right] + 1 .$$

4. $\left[\frac{[\alpha]}{n}\right] \le \frac{[\alpha]}{n} \le \frac{\alpha}{n} < \frac{[\alpha] + 1}{n} \le \left[\frac{[\alpha]}{n}\right] + 1$ ,
where we used #3 at the last inequality.

5. $-\frac{1}{2} = (\alpha + \frac{1}{2}) - 1 - \alpha < [\alpha + \frac{1}{2}] - \alpha$
$$\leq (\alpha + \tfrac{1}{2}) - \alpha = \tfrac{1}{2}.$$

6. From $-\alpha - 1 < [-\alpha] \leq -\alpha$ we see $\alpha \leq -[-\alpha] < \alpha + 1$ and the conclusion follows.

7. $[\alpha] + [\beta] = [[\alpha] + \beta] \leq [\alpha + \beta] \leq \alpha + \beta < [\alpha] + [\beta] + 2$ and the conclusion follows.

8. $[\alpha+\beta] + [\alpha] + [\beta] \leq 2[\alpha] + 2[\beta] + 1$
$$\leq [2\alpha] + [2\beta] + 1;$$
if both inequalities were equalities then
$$[\alpha+\beta] = [\alpha] + [\beta] + 1,$$
$$[2\alpha] = 2[\alpha], \quad [2\beta] = 2[\beta],$$
but then $2[\alpha+\beta] > [2\alpha] + [2\beta] + 1$
$$\geq [2(\alpha+\beta)] = [(\alpha+\beta) + (\alpha+\beta)] \geq 2[\alpha+\beta]$$
which is a contradiction; hence at least one of the inequalities is a strict inequality and the conclusion follows.

9. Let $\alpha = m+\epsilon$, $\beta = n+\nu$; then

$$[\alpha][\beta] = mn \le [\alpha\beta] = mn + [m\nu + n\epsilon + \epsilon\nu]$$
$$\le mn + m + n + [\epsilon\nu] = mn + m + n$$
$$= [\alpha][\beta] + [\alpha] + [\beta].$$

10. If $\alpha = qk + r$, $0 \le r < k$, then $q$ is the number of positive integral multiples of $k$ not exceeding $\alpha$; but $\left[\frac{\alpha}{k}\right] = q$ so the conclusion follows.

11. By #10, $[\alpha]-[\beta]$ is the number of positive integers $\le \alpha$ and not $\le \beta$.

12. $[\alpha]+[-\alpha] = [[\alpha]-\alpha]$

$$= \begin{cases} 0 & \text{if } \alpha \text{ is an integer;} \\ -1 & \text{otherwise.} \end{cases}$$

13. In #7 let $\alpha$ and $\beta$ both be $\frac{\alpha}{2}$ to obtain $2\left[\frac{\alpha}{2}\right] \le [\alpha] \le \left[\frac{\alpha}{2}\right] + \left[\frac{\alpha}{2}\right] + 1$; hence $0 \le [\alpha] - 2\left[\frac{\alpha}{2}\right] \le 1$.

14. From $\frac{n}{2}-1<\left[\frac{n}{2}\right]\le\frac{n}{2}$ and $\frac{n}{2}\le-\left[-\frac{n}{2}\right]<\frac{n}{2}+1$ we find $n-1<\left[\frac{n}{2}\right]-\left[-\frac{n}{2}\right]<n+1$, and, therefore, $\left[\frac{n}{2}\right]-\left[-\frac{n}{2}\right]=n$.

15. $\alpha-\frac{1}{n}<\frac{[n\alpha]}{n}\le\alpha$ and the conclusion follows.

16. Let $m^k\le[\alpha]\le\alpha<(m+1)^k$; then

$$\left[\sqrt[k]{\alpha}\right]=m=\left[\sqrt[k]{[\alpha]}\right].$$

17. Let $\alpha=[\alpha]+\frac{\beta}{n}$, $0\le\beta<n$, so

$$[n\alpha]=[n[\alpha]+\beta]=n[\alpha]+[\beta] \text{ and}$$

$$[\alpha]+\left[\alpha+\frac{1}{n}\right]+\cdots+\left[\alpha+\frac{n-1}{n}\right]=\sum_{j=0}^{n-1}\left[[\alpha]+\frac{\beta+j}{n}\right]$$

$$=n[\alpha]+\sum_{j=0}^{n-1}\left[\frac{\beta+j}{n}\right]=n[\alpha]+\sum_{j=1}^{n}\left[\frac{\beta+n-j}{n}\right]$$

$$=n[\alpha]+\sum_{j=1}^{[\beta]}1=n[\alpha]+[\beta].$$

18. Put $\frac{\alpha}{n}$ for $\alpha$ in #17.

19. $$\sum_{j=0}^{n-1}\left[m\alpha+\frac{jm}{n}\right]=\sum_{k=0}^{m-1}\sum_{j=0}^{n-1}\left[\alpha+\frac{k}{m}+\frac{j}{n}\right]$$

$$=\sum_{k=0}^{m-1}\left[n\alpha+\frac{kn}{m}\right],\text{ since}$$

$$\sum_{k=0}^{m-1}\left[\alpha+\frac{k}{m}+\frac{j}{n}\right]=\left[m\alpha+\frac{jm}{n}\right] \text{ and}$$

$$\sum_{j=0}^{n-1}\left[\alpha+\frac{k}{m}+\frac{j}{n}\right]=\left[n\alpha+\frac{kn}{m}\right] \text{ by \#17.}$$

20. Put $f(x)=(-1)^{[nx]}\binom{n-1}{[nx]}$,

$$g(x)=(-1)^{[mx]}\binom{m-1}{[mx]};$$

then one of $f, g$ is symmetric, the other antisymmetric about $x=\frac{1}{2}$; hence

$$\int_0^1 fg=\int_0^{1/2} fg+\int_{1/2}^1 fg=\int_0^{1/2} fg+\int_{1/2}^0 fg=0.$$

21. Let $\vartheta=\tau n-[\tau n]$; since $\tau^2=\tau+1$, $\vartheta=\tau^2 n-[\tau^2 n]$; hence $-\frac{\vartheta}{\tau}=\vartheta(1-\tau)$

$$=(\tau^2 n-[\tau^2 n])-(\tau^2 n-\tau[\tau n])$$

$$=\tau[\tau n]-[\tau^2 n];$$

since $0<\vartheta<1<\tau$, taking square brackets yields the desired result.

22. Write $n=q\sqrt{2}+\alpha$, $0<\alpha<\sqrt{2}$, $q$ integral; then

a) $[(1+\sqrt{2})n] = n + 2q + [\alpha\sqrt{2}]$ and

b) $[\sqrt{2}\,[(1+\frac{1}{\sqrt{2}})n + \frac{1}{2}]]$

$= n + 2q + [\alpha(\sqrt{2}-1) + \sqrt{2}\,[\frac{\alpha}{\sqrt{2}} + \frac{1}{2}]]$ ;

calling the last term on the right in (b) $t$

we see $\quad [\alpha\sqrt{2}] = [\frac{\alpha}{\sqrt{2}} + \frac{1}{2}]$

$$= \left\{ \begin{array}{ll} [\alpha(\sqrt{2}-1)] = 0 & \text{for } 0 < \alpha < \frac{1}{\sqrt{2}} \\ [\alpha(\sqrt{2}-1) + \sqrt{2}] = 1 & \text{for } \frac{1}{\sqrt{2}} \le \alpha < \sqrt{2} \end{array} \right\} = t,$$

where, in the second case, we use

$$1 < 1 - \frac{1}{\sqrt{2}} + \sqrt{2} \le \alpha(\sqrt{2}-1) + \sqrt{2}$$
$$< \sqrt{2}(\sqrt{2}-1) + \sqrt{2} = 2 .$$

23. We know from II #6 (ii) that

$u_n = \frac{1}{\sqrt{5}}(\tau^{n+1} - \tau'^{\,n+1})$; further $\tau\tau' = -1$ so

$$u_n = \frac{1}{\sqrt{5}}\tau^{n+1} + \frac{(-1)^n}{\sqrt{5}\,\tau^{n+1}} = [\frac{1}{\sqrt{5}}\tau^{n+1} + \frac{1}{2}] ,$$

since $\frac{1}{\sqrt{5}\,\tau^{n+1}} < \frac{1}{2}$ .

24. Among $1, 2, \cdots, n$ there are $[\frac{n}{p}]$ divisible by $p$, and, of these, there are $[\frac{[\frac{n}{p}]}{p}] = [\frac{n}{p^2}]$ divisible by $p^2$, etc; thus the highest power of $p$ in $n!$ is as stated.

25. *i*) $\binom{m}{n} = \frac{m!}{(m-n)!\,n!}$ so the highest power of p in $\binom{m}{n}$ is

$$\sum_{j=1}^{\infty} \left\{ \left[\frac{m}{p^j}\right] - \left[\frac{m-n}{p^j}\right] - \left[\frac{n}{p^j}\right] \right\} \geq 0$$

since, by #7, each term is $\geq 0$; since this is true for every prime p the conclusion follows;

*ii*) this follows exactly as does (*i*) if one uses the obvious extension of #7;

*iii*) this follows in a similar way using #8.

26. $$\left[\frac{n}{m}\right] - \left[\frac{n-1}{m}\right] = \begin{cases} 1 & \text{if } m \text{ divides } n; \\ 0 & \text{otherwise}; \end{cases}$$

thus the sum is $\geq$ the number of positive integer divisors of n and this number is 2 precisely when n is prime.

27. $$\sum_{k=2}^{n} \left[\frac{[n/k]}{n/k}\right] \geq 2 \text{ for } n \geq 2,\ n \text{ not prime};$$

for n prime this sum is 1;

thus for each composite $n$ the summand is 0 while for each prime $n$ the summand is 1.

28. For $x$ irrational $[\cos^2 m!\pi x] = 0$ while for $x$ rational, say $\frac{s}{t}$, all terms with $m \geq t$ have value 1.

29. $xy \leq n$ is equivalent, under the conditions of the system, to $x \leq \frac{n}{y}$ and, for fixed $y$, the number of such $x$ is $[\frac{n}{y}]$; now allow $y$ to vary over $1, 2, \cdots, n$; this yields the left equality; for the right equality note that the number of lattice points on the vertical line through $k$ and beneath the curve is $[\frac{n}{k}]$; the number in the shaded region is twice the number in the doubly shaded region diminished by the number in the triply shaded region.

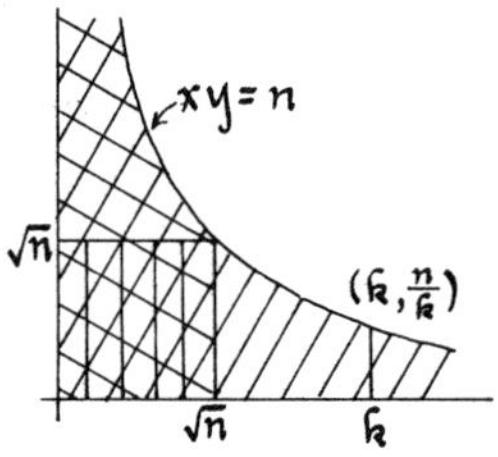

30. Suppose $2x = qb + r$, $0 \le r < b$;

i) if $q$ is even, say $q = 2s$, then $0 \le x - sb = \frac{r}{2} < \frac{b}{2}$, while if $0 \le x - sb < \frac{b}{2}$ then

$$\left[\frac{2x}{b}\right] = \left[2s + \frac{1}{2}\right] = 2s \text{ is even;}$$

ii) if $q$ is odd, say $q = 2s+1$, then $0 \le x - (s+1)b$ and $-\frac{b}{2} < x - (s+1)b < 0$, while if $-\frac{b}{2} < x - (s+1)b < 0$ then $-1 < \frac{2x}{b} - 2s - 2 < 0$ so

$$0 < \frac{2x}{b} - 2s - 1 < 1 \text{ and } \left[\frac{2x}{b}\right] \text{ is odd.}$$

31. i) $\left[\frac{an}{b}\right]$ is the number of lattice points on the half open segment $(A, B]$ in the diagram; when $d = 1$ there are no such lattice points on $OC$ and, in general, there are $d-1$ such points on $OC$; thus the indicated sum is half the number of lattice points inside the rectangle $Obca$ plus $\frac{d-1}{2}$;

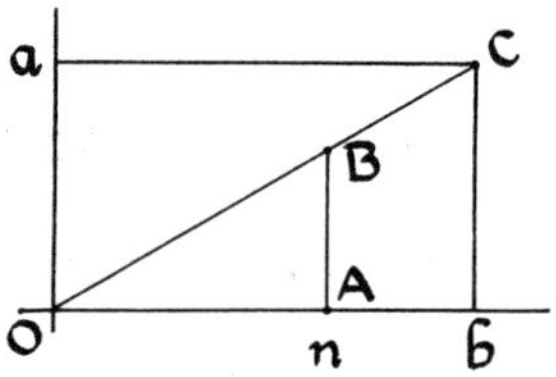

ii) the left hand side is exactly the number of lattice points in the indicated rectangle.

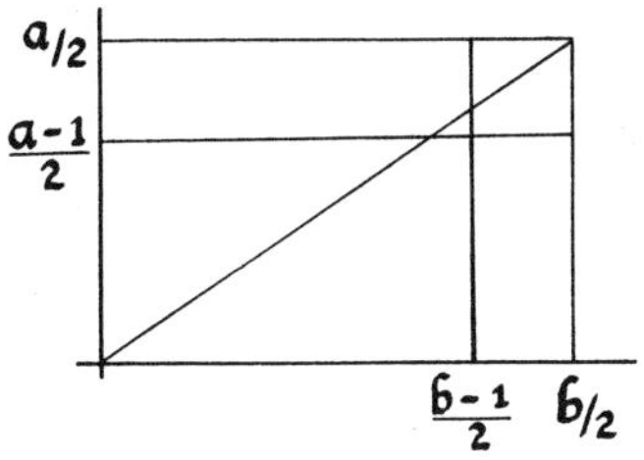


32. i) From $ax+by=k=akx_0+bky_0$ we see that $a(x-kx_0)=b(ky_0-y)$ ; since $(a,b)=1$ this means there is a $t$ such that $x-kx_0=tb$ , $ky_0-y=ta$ ;

ii) from (i) if $x,y$ is non-negative we must have $kx_0+bt\geq 0$ and $ky_0-at\geq 0$ ; i.e. $-\frac{kx_0}{b}\leq t\leq\frac{ky_0}{a}$ so the number of such solutions is $1+\left[\frac{ky_0}{a}\right]+\left[\frac{kx_0}{b}\right]$ ;

iii) if $k$ has the stated form then

$$1+\left[\frac{kx_0}{b}\right]+\left[\frac{ky_0}{a}\right]=$$

$$1+\left[\frac{r}{b}-ry_0+sx_0-ax_0\right]+\left[\frac{s}{a}-sx_0+ry_0-by_0\right]=0;$$

on the other hand, if (*) has no non-negative solution, then for suitable $r, s$, $0 \le r < b$, $0 \le s < a$,

$$0 \le \left[\frac{k}{ab}\right] = \left[\frac{kx_0}{b} + \frac{ky_0}{a}\right] \le 1 + \left[\frac{kx_0}{b}\right] + \left[\frac{ky_0}{a}\right] =$$

$$1 + \frac{kx_0 - r}{b} + \frac{ky_0 - s}{a} = \frac{k + ab - ar - bs}{ab} = 0,$$

which can happen only when $k = ar + bs - ab$;

iv) the 2nd assertion follows from (iii) by observing that $ab - a - b = a(b-1) + b(a-1) - ab$; when $k > ab - a - b$ we observe that

$$1 + \left[\frac{kx_0}{b}\right] + \left[\frac{ky_0}{a}\right] = \frac{k + ab - ar - bs}{ab} > \frac{2ab - a(1+r) - b(1+s)}{ab}$$

$\ge 0$ and the conclusion follows from (ii);

v) we need to count the number of distinct non-negative $ar + bs - ab$, $0 \le r < b$, $0 \le s < a$; if

$$ar + bs - ab = ar' + bs' - ab$$

then $a(r - r') = b(s' - s)$ and, since $(a, b) = 1$, this means $r = r'$, $s = s'$; thus they are all distinct so we need only determine how many are non-negative;

put $r = b - i$, $s = a - j$, $1 \le i \le b$, $1 \le j \le a$ and examine

$$a(b-i) + b(a-j) \ge ab$$

or, what is the same, $ab \ge ai + bj$; for fixed $i$ we have $[a - \frac{a}{b} i]$ values of $j$ so altogether we have

$$\sum_{i=1}^{b} [a - \tfrac{a}{b} i] = ab + \sum_{i=1}^{b} [-\tfrac{a}{b} i] = ab - a + \sum_{i=1}^{b-1} (-1 - [\tfrac{a}{b} i])$$

$$= ab - a - (b-1) - \frac{(a-1)(b-1)}{2} = \frac{(a-1)(b-1)}{2},$$

where we have used #31 (i).

## V Kronecker Theorems ~ Solutions

1. i) By IV #1, $0 \le n\alpha - [n\alpha] < 1$ and, since $\alpha$ is irrational, the 1st inequality is strict;

ii) $P_n - P_m = (n-m)\alpha - [n\alpha] + [m\alpha]$ and, if this were 0, $\alpha$ would be rational contrary to fact;

iii) $P_1, \cdots, P_{m+1}$ are pairwise distinct numbers between 0 and 1; thus some pair of them must be within $\frac{1}{m}$ of each other; taking $\frac{1}{m} < \epsilon$ yields the desired result;

iv) $P_r = P_{n+r} - P_n + [n\alpha + r\alpha] - [n\alpha] - [r\alpha]$; consequently when $|P_{n+r} - P_n| < \epsilon$ then since $0 < P_r < 1$ and $P_r$ is within $P_{n+r} - P_n$ of an integer, either $P_r > 1 - \epsilon$ or $P_r < \epsilon$;

v) for $r$ as in (iv) the points $P_r, P_{2r}, \cdots, P_{sr}$, where $s > \frac{1}{\epsilon}$, constitute an $\epsilon$-dense set of points (i.e. every point in the unit interval is within $\epsilon$ of one of these points); the union of such sets for $\epsilon = 1, \frac{1}{2}, \frac{1}{3}, \cdots$ is contained in $\{P_1, P_2, \cdots\}$ and, therefore, this set is dense.

2. i) If $P_n \neq P_m$, $n \neq m$, then

$$(n\alpha) = (m\alpha) \text{ and } (n\beta) = (m\beta)$$

which contradicts # 1 (ii);

ii) let $Q = (q_1, q_2)$; then $P_m Q = P_n P_{n+r}$ implies $q_1 = \{m\alpha - [m\alpha]\} + \{n\alpha + r\alpha - [n\alpha + r\alpha]\} - \{n\alpha - [n\alpha]\}$ and a similar expression for $q_2$; thus

$$f(Q) = (((m+r)\alpha), ((m+r)\beta)) = P_{m+r};$$

iii) let $Q = (q_1, q_2)$; then the given equality implies

$$q_1 = \alpha - [\alpha] + m\{\alpha + r\alpha - [\alpha + r\alpha] - \alpha + [\alpha]\}$$
$$+ n\{\alpha + s\alpha - [\alpha + s\alpha] - \alpha + [\alpha]\}$$

and a similar expression for $q_2$; thus

$$f(Q) = ((\alpha + mr\alpha + ns\alpha), (\beta + mr\beta + ns\beta))$$
$$= P_{1+mr+ns};$$

iv) as $m$ and $n$ run over the non-negative integers the points $Q$ such that

$$P_1Q = mP_1P_{1+r} + nP_1P_{1+s}$$

have $f$ images that are $\mathcal{L}$-dense in the unit square; but these $f(Q)$ are just the

$$P_{1+mr+ns}.$$

3. i) If $\alpha = \frac{a}{b}$ then $b\alpha + 0\cdot\beta - a = 0$ for $ab \neq 0$; similarly for $\beta$;

ii) by #2 (i) all $P_n$ are distinct; since they all lie in the unit square they must possess an accumulation point; consequently, there are points of the set arbitrarily close together;

iii) there is some vector $P_1 Q = P_n P_{n+r}$ with $|P_n P_{n+r}| < \epsilon$; thus we know $Q$ is in the unit square; using #2 (ii), $f(Q) = P_{1+r} = f(P_{1+r})$ and we are done since $f$ is one to one on the unit square;

iv) this follows immediately from (ii) & (iii);

v) if $P_1 P_{1+r}$ is parallel to $P_1 P_{1+s}$ then the triangle determined by $P_1, P_{1+r}, P_{1+s}$ has zero area so $0 =$

$$\begin{vmatrix} (\alpha) & (\beta) & 1 \\ (\alpha + r\alpha) & (\beta + r\beta) & 1 \\ (\alpha + s\alpha) & (\beta + s\beta) & 1 \end{vmatrix} = \begin{vmatrix} \alpha & \beta & 1 \\ [\alpha]-[\alpha + r\alpha] & [\beta]-[\beta + r\beta] & -r \\ [\alpha]-[\alpha + s\alpha] & [\beta]-[\beta + s\beta] & -s \end{vmatrix}$$

in the expansion on the right the coefficient of $\alpha$ must be 0, i.e.

$$\frac{[\beta]-[\beta+r\beta]}{-r} = \frac{[\beta]-[\beta+s\beta]}{-s};$$

if this happened for infinitely many s, then since as $s \to \infty$ the right side tends to $\beta$,

$$\frac{[\beta]-[\beta+r\beta]}{-r} = \beta$$

which contradicts (i);

vi) by (iv), (v) and #2 (iv) there is an $\epsilon$-dense set of $P_n$ for each $\epsilon > 0$; therefore the $P_n$ are dense.

## VI Beatty, Skolem Theorems ~ Solutions

1. If $0<\alpha<1$ then given $n$ there is an $m$ such that $m\alpha - 1 < n \le m\alpha$; hence $[m\alpha] = n$.

2. Each of $A(\alpha)$ and $A(\beta)$ contains infinitely many positive integers; if $0<\alpha\le 1$ or $0<\beta\le 1$ then $A(\alpha)\cap A(\beta)$ is $A(\alpha)$ or $A(\beta)$ and, therefore, is not empty; thus each of $\alpha$, $\beta$ is $>1$.

3. $[n\alpha] = [nm\frac{\alpha}{m}]$.

4. 1 is in $A(\sqrt{2})$ but 1 is not in $A(1+\sqrt{2})$; that $A(1+\sqrt{2}) \subset A(\sqrt{2})$ is the content of IV #22.

5. If $\alpha = \frac{a}{b}$ and $\beta = \frac{c}{d}$ then $[nbc\alpha] = [nda\beta]$ for all $n$.

6. Since $\alpha > 1$ and $\beta > 1$ each of $S(\alpha)$, $S(\beta)$ is a sequence of distinct terms; the total number of terms not exceeding $n$ taken together is $\left[\frac{n}{\alpha}\right] + \left[\frac{n}{\beta}\right]$; from $0 < \frac{n}{\alpha} - \left[\frac{n}{\alpha}\right] < 1$ and $0 < \frac{n}{\beta} - \left[\frac{n}{\beta}\right] < 1$ we have $0 < n - \left[\frac{n}{\alpha}\right] - \left[\frac{n}{\beta}\right] < 2$ and, therefore, $\left[\frac{n}{\alpha}\right] + \left[\frac{n}{\beta}\right] = n - 1$; hence the total number of terms not exceeding $n$ in $S(\alpha)$ and $S(\beta)$ is $n - 1$; since this is true for each $n \geq 1$ each interval $n$ to $n+1$ contains exactly one such term; this implies

$$A(\alpha) \cup A(\beta) = Z \quad \text{and} \quad A(\alpha) \cap A(\beta) = \phi$$

which yields the desired conclusion.

7. This follows immediately from #6 since

$$\frac{1}{\sqrt{2}} + \frac{1}{2 + \sqrt{2}} = 1 .$$

8. This follows immediately from #6 since

$$\frac{1}{\tau} + \frac{1}{\tau^2} = 1 .$$

9. As $n$ runs over the positive integers the sequences $\{\tau n\}$ and $\{\tau^2 n\}$ run disjointly over these integers; thus the sequences $\{\tau[\tau n]\}$, $\{\tau[\tau^2 n]\}$ run disjointly over all positive integral multiples of $\tau$; therefore the 1st two sequences yield $\{[\tau n]\}$ and this with $\{[\tau^2 n]\}$ disjointly exhausts the positive integers.

10. Put $B_0 = \{[\tau^2 n] \mid n \in \mathbb{Z}\}$, $B_{m+1} = \{[\tau^2 n] \mid n \in B_m\}$ for $m \geq 0$; then $B_{m+1} \subset B_m$ for $m \geq 0$ and by #8, $A_0 \cup B_0 = \mathbb{Z}$ and $A_{m+1} \cup B_{m+1} = B_m$; thus each $A_j \subset B_m$ for $m < j$; thus $A_j \cap A_m = \phi$ for $m < j$ since for such $m$, $A_m \cap B_m = \phi$; now $A_{m+1} = B_m \setminus B_{m+1}$ so $\bigcup_{j \geq 1} A_j = \bigcup_{j \geq 1} (B_{j-1} \setminus B_j) = B_0$ and $\bigcup_{j \geq 0} A_j = A_0 \cup B_0 = \mathbb{Z}$; a picture to go with the argument is:

$$\{[\tau n]\}, \{[\tau^2[\tau n]]\}, \{[\tau^2[\tau^2[\tau n]]]\}, \cdots$$

$$\{[\tau^2 n]\}, \{[\tau^2[\tau^2 n]]\}, \{[\tau^2[\tau^2[\tau^2 n]]]\}, \cdots.$$

11. The proof is the same as that given for #10.

12. Since $A(\alpha) \cap A(\beta)$ is finite $\frac{[\frac{n}{\alpha}] + [\frac{n}{\beta}]}{n} \to 1$; but, by IV #15, this quantity tends to $\frac{1}{\alpha} + \frac{1}{\beta}$.

13. Immediate from #5, 6 and 12.

14. From #6 and 12.

15. i) Since 1 is in exactly one of the sequences, it must be in $S(\alpha_1)$; thus since $\alpha_1 > 1$ (not all positive integers are in $S(\alpha_1)$) we must have

$$\alpha_1 = 1 + \delta, \quad 0 < \delta < 1;$$

ii) this follows from

$[(k+1)\alpha_1] - [k\alpha_1] \leq (k+1)\alpha_1 - (k\alpha_1 - 1) = \alpha_1 + 1 = 2 + \delta < 3$;

iii) suppose $(j-1)\delta < 1 \leq j\delta$; then for $i \leq j-1$ we have $[i\alpha_1] = i + [i\delta] = i$ so $i$ is in $S(\alpha_1)$ and

$[j\alpha_1] = j + [j\delta] \geq j+1$ so $j$ is not in $S(\alpha_1)$;

iv) since $\alpha_2 < \cdots < \alpha_n$ it is clear that if $m \neq [\alpha_2]$ then $[\alpha_2]$ is an integer smaller than $m$ and must then be in $S(\alpha_1)$ in violation of the disjointness of $S(\alpha_1)$ and $S(\alpha_2)$; thus

$$\alpha_2 = m + \epsilon \ , \quad 0 \leq \epsilon < 1 \ ;$$

v) since $x$ is not in $S(\alpha_1)$ there is a $k$ such that $x = [k\alpha_1] + 1$ and $[(k+1)\alpha_1] - [k\alpha_1] > 1$; this implies $[(k+1)\delta] > [k\delta]$ and the existence of an integer $t$ such that $k\delta < t \leq (k+1)\delta$; since $\delta < 1$, $[k\delta] = t-1$ and $x = k+1+[k\delta] = k+t$; further, $k\delta < t \leq (k+1)\delta < \cdots < (k+m-1)\delta$

$$< k\delta + 1 < (k+m)\delta < (k+m+1)\delta$$

and $t+1$ either lies, Case 1, between the 3rd last and 2nd last terms or, Case 2, between the last two terms;

Case 1: $t+1 \le (k+m)\delta$ : then

$$[(k+m)\alpha_1] = k+m+t+1 \text{ and}$$

$$[(k+m-1)\alpha_1] = k+m+t-1;$$

thus $k+m+t$ is not in $S(\alpha_1)$ ;

Case 2: $t+1 > (k+m)\delta$ : then

$$[(k+m+1)\alpha_1] = k+m+t+2$$

$$\text{and } [(k+m)\alpha_1] = k+m+t;$$

thus $k+m+t+1$ is not in $S(\alpha_1)$;

since $[(k+j)\alpha_1] = k+j+[(k+j)\delta] = k+j+t$ for $1 \le j \le m-1$ we conclude that the next term after $x$ missing from $S(\alpha_1)$ is either

$$x+m \text{ or } x+m+1 ;$$

vi) $\alpha_2 = m+\epsilon$ so $[n\alpha_2] = nm+[n\epsilon]$ and

$$[(n+1)\alpha_2] = (n+1)m+[(n+1)\epsilon]$$

$$= [n\alpha_2]+m+[(n+1)\epsilon]-[n\epsilon]$$

and this last quantity is either $[n\alpha_2]+m$

$$\text{or } [n\alpha_2]+m+1 ;$$

vii) the 1st integer missing from $S(\alpha_1)$ is $m$ which is also the 1st element of $S(\alpha_2)$; suppose now that the proposition is true up to $k$; thus if $x_k$ is the $k^{th}$ positive integer missing from $S(\alpha_1)$ then $x_k$ is also the $k^{th}$ element of $S(\alpha_2)$; by earlier results

$$x_{k+1} = x_k + m \quad \text{or} \quad x_k + m + 1$$

and $[(k+1)\alpha_2] = [k\alpha_2] + m$ or $[k\alpha_2] + m + 1$; since $[k\alpha_2] = x_k$ this means

$$[(k+1)\alpha_2] = x_k + m \quad \text{or} \quad x_k + m + 1 \quad ;$$

whichever value $x_{k+1}$ assumes, the other, by (ii), must be in $S(\alpha_1)$ and thus not be in $S(\alpha_2)$; consequently $x_{k+1} = [(k+1)\alpha_2]$ and the induction is complete;

viii) by (vii) all integers are in either $S(\alpha_1)$ or $S(\alpha_2)$ so $S(\alpha_3) = \phi$ contrary to assumption.

16. i) $[a n \beta] = [-b n \alpha]$;

ii) the point $(1,1)$ is on the line $ax + by = a + b$ and the slope of this line is $-\frac{a}{b}$ and this quantity is $> 0$; every line of positive slope passing through $(1,1)$ goes through $S$;

iii) since $a(1-\frac{1}{\alpha}) + b(1-\frac{1}{\beta}) < a + b - 1 < a + b$ the points $(1-\frac{1}{\alpha}, 1-\frac{1}{\beta})$ and $(1,1)$ are on opposite sides of $ax + by = a + b - 1$ and, therefore, this line must pass through $S$.

17. i) By I#5 there are positive integers $x, y$ such that $ax - by = d$; consequently there are integers $u, v$ such that $au + bv = -cd$; this yields $au + bv = -a\frac{d}{\alpha} - b\frac{d}{\beta}$;

ii) $\frac{d}{b}(u + \frac{d}{\alpha})$ is irrational so the $w_n$ are dense in the unit interval (see V#1(v)); thus the $x_n$ are dense in the interval $[0, \frac{b}{d}]$, the $y_n$ are dense in the interval $[0, -\frac{a}{d}]$, and the $(x_n, y_n)$ all lie on the line with slope $-\frac{a}{b}$ passing through the origin; the conclusion follows;

iii) this follows immediately from the facts that $(c,d) = (a,b,c) = 1 = (\frac{a}{d}, \frac{b}{d})$;

iv) immediate upon substitution and the use of $\frac{a}{\alpha} + \frac{b}{\beta} = c$ along with the equalities in (iii) and the values of $x_n$, $y_n$ given in (ii);

v) by (ii), for each fixed $m$ the points $(x_{nm}, y_{nm})$ are dense on the line segment joining $(\frac{mb}{d} + u_1 + \frac{s}{\alpha}, -\frac{ma}{d} + v_1 + \frac{s}{\beta})$ and $(\frac{(m+1)b}{d} + u_1 + \frac{s}{\alpha}, -\frac{(m+1)a}{d} + v_1 + \frac{s}{\beta})$; varying $m$ over all the integers yields the desired result;

vi) taking $q = a+b$ and $q = a+b-1$, respectively, this follows from (v) and #16 (ii) in the 1st case and from (v) and #16 (iii) in the 2nd case;

vii) if $b<0$, $c=0$ this was proved in #16 (i); otherwise, by (vi), infinitely many $(x_{nm}, y_{nm})$ lie in S; if $(x_{nm}, y_{nm})$ is in the interior of S then

$$1-\frac{1}{\alpha} < x_n + \frac{mb}{d} + u_1 + \frac{s}{\alpha} = nu + \frac{nd}{\alpha} - \frac{b}{d}\left[\frac{nd}{b}\left(u+\frac{d}{\alpha}\right)\right] + u_1 + \frac{s}{\alpha}$$

$$= \frac{nd+s}{\alpha} + u_2 < 1 ,$$

$$1-\frac{1}{\beta} < y_n - \frac{ma}{d} + v_1 + \frac{s}{\beta} = -\frac{na}{b}\left(u+\frac{d}{\alpha}\right) + \frac{a}{d}\left[\frac{nd}{b}\left(u+\frac{d}{\alpha}\right)\right] + v_1 + \frac{s}{\beta}$$

$$= \frac{nd+s}{\beta} + v_2 < 1 ,$$

where $u_2, v_2$ are integers and we have used (i) at the last equality in the 2nd line; these inequalities yield

$$nd+s < (1-u_2)\alpha < nd+s+1 ,$$

$$nd+s < (1-v_2)\beta < nd+s+1$$

from which we conclude

$$[(1-u_2)\alpha] = nd+s = [(1-v_2)\beta]$$

and, therefore, that $A(\alpha) \cap A(\beta) \neq \phi$.

18. By Kronecker's theorem in 2 dimensions, see V #3 (vi), there are integers $n$ such that

$$0 < \frac{n}{\alpha} - \left[\frac{n}{\alpha}\right] < \frac{1}{\alpha}, \quad 0 < \frac{n}{\beta} - \left[\frac{n}{\beta}\right] < \frac{1}{\beta};$$

thus $0 < n - \left[\frac{n}{\alpha}\right]\alpha < 1$ and $0 < n - \left[\frac{n}{\beta}\right]\beta < 1$ so

$$\left[\left[\frac{n}{\alpha}\right]\alpha\right] = \left[\left[\frac{n}{\beta}\right]\beta\right].$$

19. If $\frac{a}{\alpha} + \frac{b}{\beta} = 1$ then by Beatty's theorem, see #6, $A\left(\frac{\alpha}{a}\right) \cap A\left(\frac{\beta}{b}\right) = \phi$; since, by #3,

$$A(\alpha) \subseteq A\left(\frac{\alpha}{a}\right) \text{ and } A(\beta) \subseteq A\left(\frac{\beta}{b}\right)$$

we conclude $A(\alpha) \cap A(\beta) = \phi$ and the sufficiency of the stated conditions is proved;
on the other hand if $A(\alpha) \cap A(\beta) = \phi$ then by #18 it is not true that $1, \frac{1}{\alpha}, \frac{1}{\beta}$ are rationally independent while from #17 it is clear that they cannot be dependent unless

$$a > 0, b > 0, c = 1.$$

20. If the theorem were false then by #19 there are positive integers $a, b, c, d, e, f$ such that

$$\frac{a}{\alpha}+\frac{b}{\beta}=1, \quad \frac{c}{\alpha}+\frac{d}{\gamma}=1, \quad \frac{e}{\beta}+\frac{f}{\gamma}=1;$$

thus

$$\frac{ae}{\alpha}+\frac{be}{\beta}=e \qquad \frac{ace}{\alpha}-\frac{bcf}{\gamma}=ce-cb$$

$$\frac{be}{\beta}+\frac{bf}{\gamma}=b \qquad \frac{ace}{\alpha}+\frac{ade}{\gamma}=ae$$

$$\frac{ae}{\alpha}-\frac{bf}{\gamma}=e-b \qquad \frac{bcf+ade}{\gamma}=ae-ce+cb;$$

this means that $\gamma$, and, therefore, also $\alpha$ are rational; but then, by #5,

$$A(\alpha)\cap A(\beta)\neq\phi.$$

21. The line passing through the 2 given points is $\frac{x}{\alpha}+\frac{y}{\beta}=1$; the conclusion now follows from #19.

## VII The Game of Wythoff – Solutions

1. Direct calculation shows this for the 1st pair and also shows that any B move from any other pair enables A to follow with a move to an earlier pair or to an immediate win.

2. Assume $\{a, b\}$ is not in the list and $a < b$, $a < 12$; now $a$ occurs in some pair in #1 and either $\{a, s\}$, $a < s$, is there or $\{s, a\}$, $s < a$, is there; when $\{a, s\}$, $a < s$, is there then

if $b > s$ the desired move is $\{a, b\} \to \{a, s\}$;

if $b < s$ the desired move is $\{a, b\} \to \{c, d\}$, where $b - a = d - c$;

when $\{s, a\}$, $s < a$, is there the desired move is $\{a, b\} \to \{a, s\}$.

3. By #1 it is clear that $\{m, n\}$ must not be in the list if A can force a win for himself; note now that there is always a move A can make which leaves to B one of the sets on the list.

4. The $n^{th}$ element of the desired list is $\{a, b\}$, $a < b$, where $b - a = n$ and a is the smallest positive integer not appearing in the $1^{st}$ $n-1$ sets; now follow the proofs given in #2, 3.

5. Since $\frac{1}{\tau} + \frac{1}{\tau^2} = 1$ each positive integer occurs precisely once among the numbers $[n\tau]$, $[n\tau^2]$, $n \geq 1$, as was proved in VI #8 as a consequence of Beatty's theorem (see VI #6); further,

$$[n\tau^2] - [n\tau] = [n(\tau+1)] - [n\tau] = n.$$

## VIII $\tau, \sigma, \varphi$ ~ Solutions

1\. i) Each divisor of $ab$ is of the form $d\delta$, where $d$ divides $a$ and $\delta$ divides $b$; further if either $d \neq d_1$ or $\delta \neq \delta_1$ then $d\delta \neq d_1\delta_1$ since $(a, b) = 1$; thus since the number of possible $d$'s is $\tau(a)$ and the number of possible $\delta$'s is $\tau(b)$ the number of divisors of $ab$ is just $\tau(a)\tau(b)$; one may write the above argument as follows: $\tau(ab) = \sum_{d|ab} 1 =$

$$\sum_{d|a}\sum_{\delta|b} 1 = \Big(\sum_{d|a} 1\Big)\Big(\sum_{\delta|b} 1\Big) = \tau(a)\tau(b);$$

ii) if $\sigma(a) = a_1 + \cdots + a_s$, $\sigma(b) = b_1 + \cdots + b_t$ then $\sigma(ab) = \sum_{i,j} a_i b_j = (a_1 + \cdots + a_s)(b_1 + \cdots + b_t)$ $= \sigma(a)\sigma(b)$, where no divisor of $ab$ is counted twice because $(a, b) = 1$;

iii) a number $m$ is prime to $ab$ if and only if it is prime to each of $a$ and $b$ (recall

$(a,b)=1$); thus $st$, $1 \le s < a$, $1 \le t < b$, is prime to $ab$ only if $(s,a)=1$ and $(t,b)=1$; thus the number of pairs is just $\varphi(a)\varphi(b)$.

2\. i) The divisors of $p^\alpha$ are exactly $1, p, p^2, \cdots, p^\alpha$ so $\tau(p^\alpha)=\alpha+1$;

ii) by (i) we see that

$$\sigma(p^\alpha) = 1 + p + \cdots + p^\alpha = \frac{p^{\alpha+1}-1}{p-1};$$

iii) all numbers except the multiples of $p$ are prime to $p^\alpha$ so

$$\varphi(p^\alpha) = p^\alpha - \frac{p^\alpha}{p} = p^\alpha\left(1-\frac{1}{p}\right).$$

3\. By an induction argument each of the results of #1 may be extended to a product of $k$ pairwise relatively prime factors. Thus:

i) $$\tau(n) = \tau(p_1^{\alpha_1}) \cdots \tau(p_k^{\alpha_k}) = (\alpha_1+1)\cdots(\alpha_k+1);$$

ii) $\sigma(n) = \sigma(p_1^{\alpha_1}) \cdots \sigma(p_k^{\alpha_k}) = \prod_{j=1}^{k} \frac{p_j^{\alpha_j}-1}{p_j-1}$ ;

iii) $\varphi(n) = \varphi(p_1^{\alpha_1}) \cdots \varphi(p_k^{\alpha_k}) = n \prod_{p|n} (1 - \frac{1}{p})$ .

4. If $\tau(n)$ is odd and $n$ is as in #3 then all $\alpha$ are even; consequently $n$ is a square.

5. As $d$ runs over the divisors of $n$ so also does $\frac{n}{d}$; thus $\prod_{d|n} d = \prod_{d|n} \frac{n}{d} = n^{\tau(n)} \prod_{d|n} \frac{1}{d}$ and, therefore, $(\prod_{d|n} d)^2 = n^{\tau(n)}$ from which the desired result is immediate.

6. If $\delta | n$ then $2^\delta - 1 \mid 2^n - 1$ and the result follows.

7. If $\delta$ is an odd divisor of $n$ then $2^{\frac{n}{\delta}} + 1 \mid 2^n + 1$; however the divisor 1 of $2^n+1$ does not correspond to any divisor of $n$.

8. Let $n = n_1 \cdots n_k$, where $n_j = p_j^{\alpha_j}$, $1 \le j \le k$; then $\sum_{d|n} \tau^3(d) = \sum_{a_1|n_1,\cdots,a_k|n_k} \tau^3(a_1)\cdots\tau^3(a_k)$

$$= \prod_{j=1}^{k} \sum_{a_j|n_j} \tau^3(a_j) = \prod_{j=1}^{k} (1^3 + \cdots + (\alpha_j+1)^3)$$

$$= (\prod_{j=1}^{k} (1 + \cdots + (\alpha_j+1)))^2 = (\sum_{d|n} \tau(d))^2.$$

9. For each $j$ there are $\tau(j)$ pairs $x, y$ such that $xy = j$. Hence there are $\tau(1) + \cdots + \tau(n)$ pairs $x, y$ such that $xy \le n$. For each $j$ there are $[\frac{n}{j}]$ values of $x$ such that $xj \le n$. Hence there are $[\frac{n}{1}] + \cdots + [\frac{n}{n}]$ pairs $x, y$ such that $xy \le n$. Now this last sum is just the number of lattice points in the first quadrant under the curve $xy = n$. That number is twice the number in the shaded region diminished by the number in the doubly shaded region.

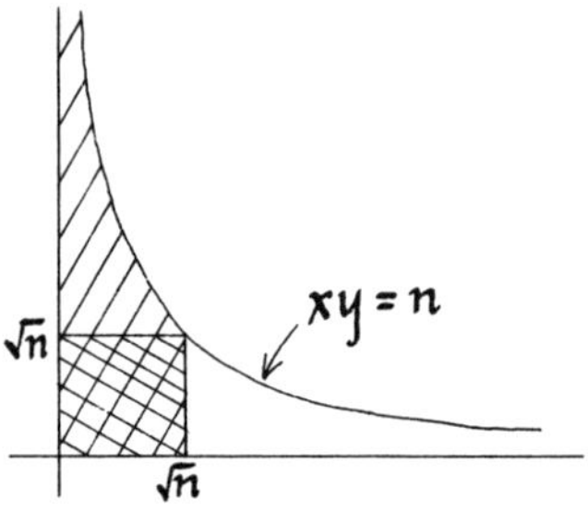

10. The argument is the same as that given for #3 (ii).

11. $\sigma(a)b < \sigma(ab)$ since 1 contributes to the right but not to the left; hence $\frac{\sigma(a)}{a} < \frac{\sigma(ab)}{ab}$; clearly $\sigma(ab) \le \sigma(a)\sigma(b)$ since every divisor of $ab$ will appear on the right and some of them may appear more than once; consequently the right inequality follows.

12. If $a = p^{\alpha}$, $b = p^{\alpha+\beta}$ then $(a,b) = p^{\alpha}$ and $\sigma(a)\sigma(b) = \frac{p^{\alpha+1}-1}{p-1} \cdot \frac{p^{\alpha+\beta+1}-1}{p-1}$ and $\sum_{d|(a,b)} d\,\sigma\left(\frac{ab}{d^2}\right) =$

$$\sum_{j=0}^{\alpha} p^j \sigma(p^{2\alpha+\beta-2j}) = \sum_{j=0}^{\alpha} p^j \frac{p^{2\alpha+\beta-2j+1}-1}{p-1} =$$

$$\frac{1}{p-1}\left\{p^{\alpha+\beta+1}\sum_{j=0}^{\alpha} p^{\alpha-j} - \sum_{j=0}^{\alpha} p^j\right\} = \frac{p^{\alpha+1}-1}{p-1} \cdot \frac{p^{\alpha+\beta+1}-1}{p-1} =$$

$$\sigma(a)\sigma(b);$$

if $a = p_1^{\alpha_1}\cdots p_k^{\alpha_k}$, $b = p_1^{\beta_1}\cdots p_k^{\beta_k}$ then $\sigma(a)\sigma(b) =$

$$\prod_{j=1}^{k} \sigma(p_j^{\alpha_j})\sigma(p_j^{\beta_j}) = \prod_{j=1}^{k} \sum_{\delta_j|(p_j^{\alpha_j}, p_j^{\beta_j})} \delta_j \cdot \sigma\left(\frac{p_j^{\alpha_j} p_j^{\beta_j}}{\delta_j}\right)$$

$$= \sum_{d|(a,b)} d\,\sigma\left(\frac{ab}{d^2}\right).$$

13. $\sigma(1)+\cdots+\sigma(n) = \sum_{d=1}^{n} d \sum_{\substack{k \\ kd \le n}} 1 = \sum_{d=1}^{n} d \left[\frac{n}{d}\right]$.

14. Direct verification.

15. i) $\varphi(n^2) = n^2 \prod_{p|n^2}\left(1-\frac{1}{p}\right) = n^2 \prod_{p|n}\left(1-\frac{1}{p}\right)$
$= n\,\varphi(n)$;

ii) since $n \ge 2$, $n$ does not contribute to $\varphi(n)$ so $\varphi(n) < n$;

iii) $\varphi(n^2)+\varphi((n+1)^2) = n\,\varphi(n)+(n+1)\varphi(n+1)$ $\le n(n-1)+(n+1)n = 2n^2$, with equality only if $n$ and $n+1$ are primes, which cannot happen with $n \ge 3$. (For generalizations see the solution to Luthar [1972].)

16. This follows from the fact that

$$(m,n)=(n-m,n).$$

17. Let $n = p_1^{\alpha_1} \cdots p_k^{\alpha_k}$ and suppose $\varphi(n) \mid n$; then $(p_1 - 1) \cdots (p_k - 1) \mid p_1 \cdots p_k$; since $p_1 \cdots p_k$ has at most one 2 there cannot be as many as 2 odd primes in $n$; thus $n = 2^{\alpha_1} p^{\alpha_2}$ and $\varphi(n) = 2^{\alpha_1 - 1} p^{\alpha_2 - 1}(p-1)$; if $p \neq 3$ then $p-1$ could not divide $2p$; hence $p = 3$; if $\alpha_1 = 0$ then $p$ cannot be 3 since otherwise again $p-1$ could not divide $n$; the only cases left are those enumerated in the problem and they all work.

18. $\varphi(ab) = ab \prod_{p \mid ab} \left(1 - \frac{1}{p}\right) =$

$$a \prod_{p \mid a} \left(1 - \frac{1}{p}\right) b \prod_{p \mid b} \left(1 - \frac{1}{p}\right) \frac{1}{\prod_{p \mid c} \left(1 - \frac{1}{p}\right)} = \varphi(a)\varphi(b)\frac{c}{\varphi(c)} .$$

19. Let $c_d$ be the number of numbers among $1, 2, \cdots, n$ having a gcd of $d$ with $n$; then

$$n = \sum_{d \mid n} c_d = \sum_{d \mid n} \varphi\left(\frac{n}{d}\right) = \sum_{d \mid n} \varphi(d).$$

20. $\sum_{d=1}^{n} \varphi(d)\left[\frac{n}{d}\right] = \sum_{j=1}^{n} \sum_{d|j} \varphi(d) = \sum_{j=1}^{n} j = \frac{n(n+1)}{2}$;

alternatively this may be done by induction with the induction step: $\sum_{d=1}^{n+1} \varphi(d)\left[\frac{n+1}{d}\right] =$

$$\sum_{d=1}^{n} \varphi(d)\left[\frac{n+1}{d}\right] + \varphi(n+1)$$

$$= \sum_{d=1}^{n} \varphi(d)\left[\frac{n}{d}\right] + \sum_{\substack{d|n+1 \\ d\le n}} \varphi(d) + \varphi(n+1) = \frac{n(n+1)}{2} + n+1$$

$$= \frac{(n+1)(n+2)}{2}.$$

21. ( See Pólya, Szegö II #69 p. 130 )

$$\sum_{n=1}^{\infty} \frac{\varphi(n)x^n}{1-x^n} = \sum_{n=1}^{\infty} \sum_{j=1}^{\infty} \varphi(n)\, x^{jn} = \sum_{k=1}^{\infty} \Big( \sum_{d|k} \varphi(d) \Big) x^k$$

$$= \sum_{k=1}^{\infty} k x^k = \frac{x}{(1-x)^2}.$$

22. If there were only finitely many primes and their product was $P$ then, for all $k$,

$$1 = \varphi(kP) = kP \prod_p \left(1 - \frac{1}{p}\right),$$

where $\prod_p \left(1 - \frac{1}{p}\right)$ is a fixed positive constant.

23. If $n = p_1 \cdots p_k - \frac{1}{p_j} p_1 \cdots p_k$ then

$$\varphi(n) = n \prod_{\substack{i=1 \\ i\ne j}}^{k} \left(1 - \frac{1}{p_i}\right) = p_1 \cdots p_k \prod_{i=1}^{k} \left(1 - \frac{1}{p_i}\right).$$

24.i) If $m \le x$, $(m,n)=1$ then $m$ contributes 1 to the 1st term and 0 to all other terms; on the other hand if $m \le x$ and $m$ is divisible by exactly $j$ of the prime factors of $n$ then $m$ contributes 1 to the 1st term, $j$ to the 2nd term, $\binom{j}{2}$ to the 3rd term, $\cdots$ ; in all, $m$ contributes $1 - j + \binom{j}{2} - \binom{j}{3} + \cdots \pm \binom{j}{j} = (1-1)^j = 0$; hence the result;

ii) taking $x = n$ we find $\varphi(n) = \varphi(n,n) =$

$$n - \sum_i \frac{n}{p_i} + \sum_{\substack{i,j \\ i \ne j}} \frac{n}{p_i p_j} - \cdots + (-1)^k \frac{n}{p_1 \cdots p_k} = n \prod_{p \mid n} \left(1 - \frac{1}{p}\right);$$

iii) if an integer falls between $\sqrt{x}$ and $x$ and is not divisible by any prime smaller than $\sqrt{x}$ then it is a prime; since $\varphi(x, p_1 \cdots p_t)$ enumerates these as well as 1 the result follows.

25. i) By direct calculation;

ii) $\sigma(2^{n-1}(2^n-1)) = \sigma(2^{n-1})\sigma(2^n-1)$
$= \frac{2^n-1}{2-1}\cdot 2^n = 2\{2^{n-1}(2^n-1)\}$ ;

iii) let $n = 2^{k-1}m$, where $m$ is odd and $k \geq 2$; then $\sigma(n) = \sigma(2^{k-1})\sigma(m) = (2^k-1)\sigma(m) = 2^k m$ ; since $(2^k-1, 2^k) = 1$ we have $\sigma(m) = 2^k t$, $m = (2^k-1)t$ ; hence $\sigma(m) = 2^k t = (2^k-1)t + t$ $= m + t$ ; therefore $t = 1$ and $m$ is prime; if $k$ were not prime then $2^k - 1 = m$ would not be prime (see the proof of #6) ;

iv) since $2^4$ ends in 6 we see that $2^{4a}$, $2^{4a+1}-1$, $2^{4a+2}$, $2^{4a+3}-1$ end, respectively, in 6, 1, 4, 7 ; thus $2^{4a}(2^{4a+1}-1)$, $2^{4a+2}(2^{4a+3}-1)$ end, respectively, in 6, 8 ; this proves the first assertion ; noting that for $n = 13, 17$ we have consecutive even perfect numbers (see e.g. the remarks at the end of the chapter) and observing that each of 13, 17 is of the form $4a+1$ we see that the $5^{th}$ and $6^{th}$ consecutive even perfect numbers end in 6 ;

v) since $\sigma(n) = \sum_{d|n} d = \sum_{d|n} \frac{n}{d} = n \sum_{d|n} \frac{1}{d}$ the conclusion is obvious ;

vi) if $n = p^\alpha q^\beta$ then $\sigma(n) = \frac{p^{\alpha+1}-1}{p-1} \cdot \frac{q^{\beta+1}-1}{q-1}$
$< \frac{p^{\alpha+1} q^{\beta+1}}{(p-1)(q-1)} = \frac{npq}{(p-1)(q-1)} = \frac{n}{(1-\frac{1}{p})(1-\frac{1}{q})} \leq \frac{15}{8} n.$

26. i) $H(n) = \frac{\tau(n)}{\sum_{d|n} \frac{1}{d}} = \frac{\tau(n)}{\sum_{d|n} \frac{d}{n}} = \frac{n\tau(n)}{\sigma(n)}$ ; the multiplicativity of H follows from that of all of $n, \tau(n), \sigma(n)$ ;

ii) when $n > 1$, $\sigma(n)$ is smaller than $n\tau(n)$ since $n\tau(n)$ may be obtained from $\sigma(n)$ by replacing each contributor to $\sigma(n)$ by $n$ ;

now ( $p$ and $q$ are distinct primes ) :

$$H(p^\alpha) = \frac{p^\alpha(\alpha+1)(p-1)}{p^{\alpha+1}-1} = \frac{(\alpha+1)(1-\frac{1}{p})}{1-\frac{1}{p^{\alpha+1}}}$$

$$> (\alpha+1)(1-\tfrac{1}{p}) \begin{cases} \geq 3(1-\frac{1}{p}) \geq 3(1-\frac{1}{3}) = 2 \\ \qquad \text{for } p \text{ odd and } \alpha \geq 2\,; \\ \geq 4(1-\frac{1}{2}) = 2 \text{ for } p=2 \text{ and } \alpha \geq 3\,; \end{cases}$$

$H(pq) = \frac{4pq}{(p+1)(q+1)} \geq \frac{4\cdot 2\cdot 3}{(2+1)(3+1)} = 2$ since $\frac{x}{1+x}$ is strictly monotone increasing ; further, in this last expression we have strict inequality except for $p=2$, $q=3$ ; these results and multiplicativity of $H$ guarantee $H(n)>2$ except for primes and 1, 2, 4, 6 ; checking these we find $H(n) \leq 2$ for all of them ;

iii) $H(m) = \frac{m\tau(m)}{\sigma(m)} = \frac{m\cdot n\cdot 2}{2m} = n$ ;

iv) $H(n) = H(2^{H(n)-1})H(2^{H(n)}-1) =$
$\frac{2^{H(n)-1}H(n)}{2^{H(n)}-1}H(2^{H(n)}-1) > \frac{1}{2}H(n)H(2^{H(n)}-1)$,
from which the conclusion follows ;

v) by (ii) every odd composite number $m$ has $H(m) > 2$ ; since $2^{H(n)}-1$ is odd and $H(2^{H(n)}-1) < 2$ we conclude $2^{H(n)}-1$ is prime ; thus $n$ is perfect.

27. i) $g(ab) = \sum_{d|ab} f(d) = \sum_{d|a} \sum_{\delta|b} f(d\delta)$

$$= \sum_{d|a} f(d) \sum_{\delta|b} f(\delta) = g(a)g(b) ;$$

ii) these are all clear except possibly for $\sigma^\circ$ ; let $a = 2^s a'$ , $b = 2^t b'$ where $a', b'$ are odd ; then $\sigma^\circ(ab) = \sigma^\circ(a'b') = \sigma(a'b') = \sigma(a')\sigma(b')$

$$= \sigma^\circ(a')\sigma^\circ(b') = \sigma^\circ(a)\sigma^\circ(b) .$$

# IX Fermat, Wilson, Chevalley – Solutions

1\. i) Since $\binom{p}{j} = \frac{p!}{(p-j)!\,j!}$ it is clear that $p \mid \binom{p}{j}$ when $1 \le j < p$; consequently

$$(m+n)^p - (m^p + n^p) = \sum_{j=1}^{p-1} \binom{p}{j} m^j n^{p-j}$$

is divisible by $p$;

ii) by (i), $p$ divides $(m+1)^p - (m^p+1)$, hence the conclusion follows from

$$(m+1)^p - (m+1) = ((m+1)^p - (m^p+1)) + (m^p - m);$$

iii) by induction on $m$; clearly this is true for $m=1$ and (ii) is just the induction step.

2\. Use the multinomial theorem in exactly the same way the binomial theorem was used in #1(i); to deduce Fermat's theorem put $k=m$ and all $m_j = 1$.

3. By #1 (i) and the hypothesis we see that $p \mid (m+n)^p$ so $m = -n + pt$ for some $t$ ; thus $m^p + n^p = (-n+pt)^p + n^p = \sum_{j=0}^{p-1} \binom{p}{j} (-n)^j (pt)^{p-j}$ and this sum is divisible by $p^2$ since each summand is divisible by $p^2$.

4. i) Clearly there are $n^p$ strings of length $p$ altogether and of these exactly $n$ of them are monocolored ;

ii) each distinguishable necklace has exactly $p$ rotations ;

iii) the number of necklaces in (ii) is clearly an integer ;

iv) since $p$ divides the even number $n^p - n$ and since $p$ and 2 are relatively prime, the result follows.

5. i) All primes $\leq n$ divide $N$ and hence do not divide $N+1$; also 2 cannot divide $N+1$ since $N+1$ is odd;

ii) $N^m+1=(N+1)(N^{m-1}-N^{m-2}+\cdots-N+1)$, when $m$ is odd;

iii) if $p=4k+3$ is a prime factor of $N+1$ then, since $N+1 \mid N^{2k+1}+1$, $p$ also divides

$$N^{2k+1}+1=N^{\frac{p-1}{2}}+1=(n!)^{p-1}+1;$$

but by Fermat's theorem $p$ divides

$$(n!)^p-n! \quad (=n!\{(n!)^{p-1}-1\});$$

since $p$ does not divide $N$ (it does divide $N+1$) it cannot divide $n!$, thus $p \mid (n!)^{p-1}-1$ and $p \mid (n!)^{p-1}+1$ which implies $p \mid 2$; but $p$ is odd so we have a contradiction;

iv) by (iii) all prime factors of $(n!)^2+1$ are of the form $4k+1$; further if $p \mid (n!)^2+1$ then

$p$ does not divide $(m!)^2+1$ for $m > p$ so we can generate infinitely many $4k+1$ primes.

6. i) $na \equiv nb \pmod p$ implies $p \mid n(a-b)$; since $(n,p)=1$ the result follows from III #3;

ii) by (i), $n, 2n, \cdots, (p-1)n$ are congruent, in some order, to $1, 2, \cdots, p-1$; multiplying these congruences yields the stated congruence;

iii) using (i) after noting that $((p-1)!, p) = 1$ we have the desired conclusion.

7. i-iii) Same argument as given in #6(i)-(iii);

iv) when $m$ is a prime, $\varphi(m) = m-1$.

8. i) Let $a = 2k+1$; then $a^2-1 = 4k(k+1)$ and since one of $k, k+1$ is even, the conclusion follows;

ii) by induction; the case $\alpha = 3$ is (i); suppose true for $\alpha$; then

$$a^{2^{(\alpha+1)-2}} - 1 = (a^{2^{\alpha-2}})^2 - 1 = (a^{2^{\alpha-2}} - 1)(a^{2^{\alpha-2}} + 1);$$

now $2^\alpha$ divides the 1st factor on the right and 2 divides the 2nd factor on the right; hence $2^{\alpha+1}$ divides the product and the induction is complete.

9. i) For $p$ an odd prime or $p=2$ and $0 \leq \alpha \leq 2$, $a^{\chi(p^\alpha)} = a^{\varphi(p^\alpha)} \equiv 1 \pmod{p^\alpha}$, by #7, and for $p=2$, $\alpha > 2$, $a^{\chi(2^\alpha)} = a^{2^{\alpha-2}} \equiv 1 \pmod{2^\alpha}$, by #8; thus for all $p$ and $\alpha$, $p^\alpha$ divides $a^{\chi(p^\alpha)} - 1$ which, in turn, divides $a^{\chi(m)} - 1$, when $p^\alpha | m$; this yields the desired result;

ii) for $m$ prime $\varphi(m) = m - 1$;

iii) this is clear because $\chi(m)$ is a divisor of $\varphi(m)$;

iv) since $m$ is odd, $(2,m)=1$, so $2^{\chi(m)} \equiv 1 \pmod{m}$; but $2^{\chi(m)}-1 \mid 2^{m-1}-1$ when $\chi(m) \mid m-1$;

v) $561 = 3 \cdot 11 \cdot 17$, $\varphi(561) = 2 \cdot 10 \cdot 16 = 320$, $\chi(561) = \operatorname{lcm}\{2, 10, 16\} = 80$; further $80 \mid 560$ but 320 does not divide 560;

$341 = 11 \cdot 31$, $\chi(341) = \operatorname{lcm}\{10, 30\} = 30$; $\varphi(341) = 300$; $2^{300} \equiv 1 \pmod{341}$, $2^{40} \equiv 2^{10} \equiv 1 \pmod{341}$ so $2^{340} \equiv 1 \pmod{341}$; but $\chi(341) = 30$ does not divide 340.

10. This is clear since $p$ divides $2^{p-1}-1 =$

$$(2^{\frac{p-1}{2}}-1)(2^{\frac{p-1}{2}}+1).$$

11. i) a) $b \geq n-3$ implies $n = ab \geq 2(n-3) =$

$$2n-6 = n+(n-6) > n;$$

b) if $2a \geq n-3$ then $n = a^2 \geq \frac{n^2-6n+9}{4}$ so $(n-1)^2+8 \leq 0$, contrary to fact;

ii) true for $n = 6$ by direct checking; otherwise, if $a < b$, $ab(n-3)(n-2) \mid (n-2)!$, while if $a = b$, $2ab \mid (n-2)!$; in any event the conclusion follows.

12. i) This follows from #6(i);

ii) if $x \equiv a \pmod{m}$ then $ax \equiv x^2 \equiv 1 \pmod{m}$ so $x \equiv \pm 1 \pmod{m}$;

iii) by (i) and (ii) the numbers $2, \cdots, p-2$ split into disjoint pairs with the product of the numbers in each pair $\equiv 1 \pmod{p}$; multiplying these congruences yields the result;

iv) multiply the congruence in (iii) by the congruence $p-1 \equiv -1 \pmod{p}$;

v) if $n$ is prime then $n \mid (n-1)! + 1$ by (iv); if $n \mid (n-1)! + 1$ then, by #11, $n$ may not be a composite integer $> 4$; but 4 does not divide $3!+1$ so the conclusion follows.

13. If $p = 4k+1$ then
$-1 \equiv p-1,\ -2 \equiv p-2, \cdots, -2k \equiv 2k+1 \pmod{p}$;
multiplying these congruences we find

$$(2k)! \equiv (2k+1)\cdots(p-1) \pmod{p};$$

multiplying both sides by $(2k)!$ and using Wilson's theorem we obtain

$$(2k)!^2 \equiv (p-1)! \equiv -1 \pmod{p},$$

from which the result follows.

14. i) The number of different paths, starting from a given vertex, traversing the p-gons is clearly $(p-1)!$; but each p-gon corresponds to two paths from the starting vertex, since each vertex is on 2 sides; consequently $T = \frac{(p-1)!}{2}$;

ii) a regular $p$-gon is determined once two consecutive vertices are specified; from any vertex there are $p-1$ other possible "next" vertices; but again the two edges on a given vertex causes $p-1$ to be twice the number of regular $p$-gons;

iii) let $\alpha = \langle 0, 1, \cdots, p-1 \rangle$ be a polygon in $T$; if the view of the polygon is the same from each vertex of the pair $0, j$ then the same will be true of the pair $a, a+j$ for all integers $a$ (modulo $p$); taking $a$ successively equal to $j, 2j, \cdots, (p-1)j$ we see the view from vertices $0, j, 2j, \cdots, (p-1)j$ is always the same; i.e. $\alpha$ is regular; consequently each element of $T - R$ gives rise to exactly $p$ such elements by rotation;

iv) this follows immediately from (iii).

15. i) $(m+n-1)! = (m+n-1)\cdots(m+n-n)(m-1)!$
$$\equiv (-1)^n n!\,(m-1)! \pmod{m+n};$$

ii) noting that $m \equiv -n \pmod{m+n}$ reduce the left side to the right side by substituting on the left in accordance with (i);

iii) that $(p,k)=1$ and $k$ is even is clear; now in (ii) put $n=k$, $m=p$ and use Wilson's theorem to obtain the result;

iv) with $p=13$, $k=22$ we have $(p,k)=1$, $k$ even, and the congruence in (iii) holds; however $p+k=35$ is not prime;

v) mod $p$, the congruence of (iii) implies $(p+1)!+1 \equiv 0 \pmod p$ so the primality of $p$ follows from # 12(v); mod $p+k$, since $p \equiv -k$ the congruence of (iii) yields

$$k!(k!(p-1)!+1) \equiv 0 \pmod{p+k};$$

but by (i),

$$k!(p-1)! \equiv (-1)^k (k+p-1)! \pmod{p+k};$$

thus (recall $k$ is even)

$$k!(k!(p-1)!+1) \equiv k!((k+p-1)!+1) \equiv 0 \pmod{p+k}$$

and $p+k$ is prime, again using #12 (v);

vi) taking $p=n$, $k=2$ this result follows from (iii), (v).

16. Write $n = qs + r$, $0 \leq r < s$; then

$$a^n = (a^s)^q a^r \equiv a^r \equiv 1 \pmod{m};$$

hence if $r \neq 0$ we would have a contradiction with the definition of $s$.

17. i) This follows from the proof of #9(v);

ii) if $a$ is as described and $s$ is the smallest power of $a$ which is congruent to 1 modulo $m$

then by #16 $s \mid m-1$ and $s \mid \varphi(m)$; by hypothesis this means $s = m-1$ so $m-1 \mid \varphi(m)$; but

$$\varphi(m) = m \prod_{p|m} \left(1 - \frac{1}{p}\right) < m-1$$

when $m$ is composite; hence $m$ is prime under the conditions stated.

18. (See Guy [1967] for the following solutions.) For $p = 1093$ we have, where all the congruences are modulo $p^2$: $2^{10} = 1024 = p - 69$ so

$2^{14} = 16p - 1104 = 15p - 11 \equiv -1078p - 11 = -11(1 + 98p)$;

thus $$11^3 = 1331 = p + 238,$$

$11^4 = 11p + 2618 = 13p + 432 \equiv 432 - 1080p = 2^3 \cdot 3^3 (2 - 5p)$;

$$3^7 = 2187 = 2p + 1;$$

using these we find

$$\begin{aligned} 2^{392} &\equiv 11^{28}(1+98p)^{28} \equiv \{2^3 \cdot 3^3(2-5p)\}^7 (1 + 2744p) \\ &\equiv 2^{21} \cdot 3^{21} (2^7 - 2^6 \cdot 5 \cdot 7p)(1 + 558p) \\ &\equiv 2^{27}(2p+1)^3 (2 - 35p)(1 + 558p) \\ &\equiv 2^{27}(1+6p)(2-12p) \equiv 2^{28} \end{aligned}$$

so $2^{1092} = 2^{3(392-28)} \equiv 1$;

For $p = 3511$ we have, where again all congruences are modulo $p^2$:

$$2^6 \cdot 5 \cdot 11 = 3520 = 3^2 + p \equiv 3^2 - (p-1)p = 3^2(1 - 390p)$$

$$3^8 = 81^2 = 6561 = 2p - 461 ,$$

$$3^{10} = 18p - 4149 = 17p - 638 ;$$

$$3^{10} \cdot 11 = 187p - 7018 = 185p + 4 \equiv 4 + 3696p = 2^2(1 + 924p);$$

$$3^{12} \cdot 11 (1 - 390p) \equiv 2^8 \cdot 5 \cdot 11 (1 + 924p)$$

$$2^8 \cdot 5 \equiv 3^{12}(1 - 1314p), \qquad 5^5 = 3125 = p - 386 ,$$

$$5^7 = 25p - 9650 = 22p + 883 ;$$

$$5^9 = 550p + 22075 = 556p + 1009 ;$$

$$5^{11} = 13900p + 25225 = 13907p + 648 \equiv 648 - 3648p$$

$$= 2^3 \cdot 3 (3^3 - 152p) \equiv 2^3 \cdot 3 (3^3 - 3663p) = 2^3 \cdot 3^3 (3 - 407p)$$

$$\equiv 2^3 \cdot 3^3 (3 - 3918p) = 2^3 \cdot 3^4 (1 - 1306p),$$

$$2^3 \cdot 3^4 \equiv 5^{11}(1 + 1306p) \text{ and } 2^{12} = 4096 = p + 585$$

$$\text{so } 2^{13} \cdot 3 = 6p + 3510 = 7p - 1 ; \text{ now}$$

$$2^{1755} \cdot 3^{132} \cdot 5^{11} = (2^8 \cdot 5)^{11} \, 2^3 \cdot 3^4 (2^{13} \cdot 3)^{128}$$

$$\text{so } 2^{1755} \cdot 3^{132} \cdot 5^{11} \equiv \{3^{12}(1 - 1314p)\}^{11} \, 5^{11} (1 + 1306p)(7p - 1)^{128};$$

$$2^{1755} \equiv (1 - 1314p)^{11} (1 + 1306p)(1 - 7p)^{128} \equiv 1 - 13140p - 8p - 896p$$

$$= 1 - 14044p = 1 - 4p^2 \equiv 1 \text{ and } 2^{3510} \equiv 1 .$$

19. i) This was shown for 341 in #9(v) and for 561 in #17(i); for $n = 161\,038 = 2 \cdot 73 \cdot 1103$, $n-1 = 3^2 \cdot 29 \cdot 617$, $2^9 - 1 = 7 \cdot 73$, $2^{29} - 1 = 233 \cdot 1103 \cdot 2089$; since 9 and 29 divide $n-1$ we see that $2^9 - 1$ and $2^{29} - 1$ divide $2^{n-1} - 1$; thus, using $73 \mid 2^9 - 1$ and $1103 \mid 2^{29} - 1$, we find $2^{n-1} - 1$ is divisible by 73 and by 1103; since $2^n - 2$ is even and divisible by 73 and by 1103 we are done.
(This even pseudoprime was found by Lehmer and the verification is due to Sierpinski.)

ii) we need to show $F_n \mid 2^{F_n} - 2$; this would follow from $F_n \mid 2^{F_n - 1} - 1$; this, in turn, follows from
if $m < n$ then $2^{2^n} - 1 = (2^{2^m} - 1) \prod_{j=m}^{n-1} (2^{2^j} + 1)$;

iii) since $n$ is odd and divides the even $2^n - 2$ there is a $k$ such that $2^n - 2 = 2kn$; hence $2^{2^n - 2} - 1 = (2^n)^{2k} - 1$ and, since $2^n - 1 \mid (2^n)^{2k} - 1$, we conclude $2^n - 1 \mid 2^{2^n - 2} - 1$; thus $2^n - 1 \mid 2^{2^n - 1} - 2$, and, since $n$ is odd, $2^n - 1$ is composite; therefore $2^n - 1$ is a pseudoprime;

iv) $n - 1 = 4(2^{p-1} + 1)(2^{p-1} - 1)/3$ and $2^{p-1} - 1$ is divisible by each of $p$ and 3 and hence by $3p$; thus $2p \mid n - 1$; since $n \mid 2^{2p} - 1$, which, in turn, divides $2^{n-1} - 1$ we see that $n \mid 2^{n-1} - 1$;

v) immediate from either (iii) or (iv).

20. i-a) Because one can add, subtract, and multiply congruences term by term, if respective coefficients in $F$ and $G$ are congruent modulo $p$, then certainly $F(c_1, \cdots, c_n) \equiv G(c_1, \cdots, c_n) \pmod{p}$;

b) by Fermat's theorem $c^p \equiv c \pmod{p}$ for all $c$ so if $F(x) = x^p$, $G(x) = x$ then $F \sim G \pmod{p}$;

c) if $x^n$, $n \geq p$, appears write $n = qp + r$, $0 \leq r < p$, and then replace $x^n (= x^{qp+r})$ by $x^r$; this is permitted since, by Fermat's theorem,

$$x^{qp+r} = (x^p)^q x^r \equiv x^r \pmod{p};$$

d) if $P(x) = c_0 + c_1 x + \cdots + c_n x^n$ and $P(c) \equiv 0 \pmod{p}$ then

$$P(x) - P(c) = (x-c)(d_0 + d_1 x + \cdots + d_{n-1} x^{n-1})$$

so $P(x) = P(c) + (x-c)(d_0 + \cdots + d_{n-1} x^{n-1})$; thus if $d_0 + \cdots + d_{n-1} x^{n-1}$ has no more than $n-1$ zeros (mod $p$) then $P$ has no more than $n$; since a linear polynomial has exactly 1 zero mod $p$, we know a polynomial of degree $n$ has no more than $n$ zeros mod $p$; thus if $F \sim G$ (mod $p$) then, since $F - G$, of degree $p-1$, has $p$ zeros $F - G$ must be identically zero; i.e.

$$F \equiv G \pmod{p};$$

e) by induction; let

$$F(x_1, \cdots, x_{n+1}) = \sum_{j=0}^{p-1} F_j(x_1, \cdots, x_n) x_{n+1}^j,$$

$$G(x_1, \cdots, x_{n+1}) = \sum_{j=0}^{p-1} G(x_1, \cdots, x_n) x_{n+1}^j;$$

given $c_1, \cdots, c_n$ let $F_{(c_1, \cdots, c_n)}(x) = F(c_1, \cdots, c_n, x)$, $G_{(c_1, \cdots, c_n)}(x) = G(c_1, \cdots, c_n, x)$, and note that $F_{(c_1, \cdots, c_n)}(c) \equiv G_{(c_1, \cdots, c_n)}(c) \pmod{p}$ for all $c$; thus

$F_{(c_1,\cdots,c_n)} \sim G_{(c_1,\cdots,c_n)}$ (mod $p$) and, therefore, by (d), $F_{(c_1,\cdots,c_n)} \equiv G_{(c_1,\cdots,c_n)}$ (mod $p$); this means $F_j(c_1,\cdots,c_n) \equiv G_j(c_1,\cdots,c_n)$ (mod $p$), $0 \le j \le n$; since this is true for all $n$-tuples $c_1,\cdots,c_n$ the induction hypothesis guarantees $F \equiv G$ (mod $p$), $0 \le j \le p-1$, and the result follows;

u-a(1)) clear;

(2) in this case one of the terms in the product is congruent to 0 modulo $p$ so the entire product is congruent to 0;

b) $G(a_1,\cdots,a_n) \equiv 1$ (mod $p$) and if $(x_1,\cdots,x_n) \ne (a_1,\cdots,a_n)$ then $F^{p-1}(x_1,\cdots,x_n) \equiv 1$ (mod $p$) so $G(x_1,\cdots,x_n) \equiv 0$ (mod $p$); thus $F \sim G$ (mod $p$);

c) $H$ is reduced so this follows from (b) and (i-e);

d) clear;

iii) if the assertion were not true the hypothesis of (ii) would be satisfied and this would lead to the contradiction $n \le \deg F$ (see (ii-d));

iv) since a form always has the trivial (all $x_j = 0$) solution this result follows from Chevalley's theorem (iii);

v-a) equivalent reduced forms are identical;

b) this is clear from the expression in (a);

c) if $p$ does not divide $s$ then

$$n(p-1) = \deg H^* \le \deg H = r(p-1)$$

and $n \le r$, contrary to fact;

d) immediate from (c);

vi) $\deg F \le n-1$ so $\deg H \le (n-1)(p-1)$; now

$$\prod_{j=1}^{n} (1-(x_j - a_j^{(i)})^{p-1}) = Q_i(x) + (-1)^n \prod_{j=1}^{n} (x_j - a_j^{(i)})^{p-1},$$

where $\deg Q_i(x) < n(p-1)-(p-1) = (n-1)(p-1)$,

so that

$$\mathcal{H}^*(x_1,\cdots,x_n) = \sum_{i=1}^{s} Q_i(x) + (-1)^n \sum_{i=1}^{s} \prod_{j=1}^{n} (x_j - a_j^{(i)})^{p-1} \quad ;$$

since $\deg \mathcal{H}^* \le \deg \mathcal{H} \le (n-1)(p-1)$ all terms in the 2nd term on the right of degree $> (n-1)(p-1)$ must have coefficients divisible by $p$; this 2nd term equals

$$\sum_{i=1}^{s} \sum_{\substack{0 \le t_j \le p-1 \\ 1 \le j \le n}} \binom{p-1}{t_1} \cdots \binom{p-1}{t_n} (-a_1^{(i)})^{p-1-t_1} \cdots (-a_n^{(i)})^{p-1-t_n} x_1^{t_1} \cdots x_n^{t_n}$$

$$= \sum_{\substack{0 \le t_j \le p-1 \\ 1 \le j \le n}} (-1)^{n(p-1)-(t_1+\cdots+t_n)} \binom{p-1}{t_1} \cdots \binom{p-1}{t_n} \sum_{i=1}^{s} (a_1^{(i)})^{p-1-t_1} \cdots (a_n^{(i)})^{p-1-t_n} x_1^{t_1} \cdots x_n^{t_n} \quad ;$$

thus all coefficients of terms for which $t_1 + \cdots + t_n > (n-1)(p-1)$ are divisible by $p$; this implies $\sum_{i=1}^{s} (a_1^{(i)})^{p-1-t_1} \cdots (a_n^{(i)})^{p-1-t_n} \equiv 0 \pmod p$ for $t_1 + \cdots + t_n > (n-1)(p-1)$; taking all $t_i = p-1$ except for $t_j$, which is taken $\ge 1$, yields $t_1 + \cdots + t_n > (n-1)(p-1)$ and $\sum_{i=1}^{s} (a_j^{(i)})^k \equiv 0 \pmod p$ for $0 \le k \le p-2$;

vii) put $\mathcal{H}(x_1,\cdots,x_n) = \prod_{i=1}^{n} (1 - F_i^{p-1}(x_1,\cdots,x_n))$ and show $\mathcal{H}^*(x_1,\cdots,x_n) = \sum_{i=1}^{s} \prod_{j=1}^{n} (1 - (x_j - a_j^{(i)})^{p-1})$, where $(a_1^{(i)},\cdots,a_n^{(i)})$, $1 \le i \le s$, are all the solutions of the system; now

$\deg \mathcal{H}^* \le \deg \mathcal{H} = (p-1)(r_1+\cdots+r_m) \le (n-1)(p-1)$ so $\deg \mathcal{H}^* < n(p-1)$ so the highest degree term in $\mathcal{H}^*$ must have a coefficient divisible by $p$; this coefficient is $s$ so $p|s$.

## X Divisibility Criteria - Solutions

1. i & ii) Let $n = a_m 10^m + \cdots + a_0$, $0 \le a_j < 10$ for all $j$; then $n - S_{10}(n) = a_m(10^m - 1) + \cdots + a_1(10-1)$ and, since both 3 and 9 divide $10^j - 1$ for $j = 1, \cdots, m$, the conclusions follow;

iii) let $n = a_m k^m + \cdots + a_0$, $0 \le a_j < k$ for all $j$; then $n - S_k(n) = a_m(k^m - 1) + \cdots + a_1(k-1)$; since $d$ divides $k-1$, which in turn divides $k^j - 1$, $j = 1, \cdots, m$, the conclusion follows as before.

2. i) By #1 (iii), any common divisor of 6 and $S_7(p)$ would have to divide $p$ since (by #1 (iii)) 6 divides $p - S_7(p)$; hence $(6, S_7(p)) = 1$;

ii) the smallest $S_7(p)$ which is composite must be 25 since by (i) we know 5 is the smallest possible prime divisor of $S_7(p)$; now the smallest $n$ with $S_7(n) = 25$ is $(16666)_7$ and this number in base 10 is

$$6(1+7+7^2+7^3)+7^4 = 4801,$$

which is a prime ;

iii) the next larger composite value for $S_7(p)$ is $5 \cdot 7 = 35$ and the smallest integer $n$ with $S_7(n) = 35$ is $(566666)_7$ , which is

$$6(1+7+7^2+7^3+7^4)+5 \cdot 7^5 > 100\,000.$$

3. i) This follows from the fact that 11 divides $10^{2s}-1$ and $10^{2s+1}+1$ for all non-negative integral $s$ ;

ii) as in (i), $k+1$ divides $k^{2s}-1$ and $k^{2s+1}+1$ for all non-negative integral $s$; hence $d$ also divides these quantities.

4. i) clear;

ii) $7 \cdot 11 \cdot 13 = 1001$ and $Q(n) - R(n) = 1001\,Q(n) - n$; thus $c \mid Q(n) - R(n)$ if and only if $c \mid n$;

iii) we illustrate by means of an example; it shows 14 824 017 659 to be divisible by 11 but not by either 7 or 13; since −451 is divisible by 11 but not by 7 or 13 the above assertion is correct.

```
14 824 017 659
          -659
--------------
14 823 358
      -358
----------
14 465
  -465
------
  -451
```

5. i) This is a consequence of #1 (iii);

ii) if $n = a_0 + a_1 k + \cdots + a_s k^s$ then the highest power of $k$ in $n!$ is $\left[\frac{n}{k}\right] + \left[\frac{n}{k^2}\right] + \cdots + \left[\frac{n}{k^s}\right] =$

$$(a_1 + \cdots + a_s k^{s-1}) + (a_2 + \cdots + a_s k^{s-2}) + \cdots + a_s =$$

$$a_1 + a_2(k+1) + a_3(k^2 + k + 1) + \cdots + a_s(k^{s-1} + \cdots + 1) =$$

$$a_1 \frac{k-1}{k-1} + a_2 \frac{k^2-1}{k-1} + a_3 \frac{k^3-1}{k-1} + \cdots + a_s \frac{k^s-1}{k-1} =$$

$$\frac{1}{k-1}\left\{(a_1 k + a_2 k^2 + \cdots + a_s k^s) - (a_1 + \cdots + a_s)\right\} =$$

$$\frac{1}{k-1}\left\{n - S_k(n)\right\} = T_k(n) \quad ;$$

iii) by (ii) the highest power of 2 in $n!$ is $T_2(n) = n - S_2(n)$ and $S_2(n)$ is just the quantity $\nu$ described ;

iv) this is proved by induction ; it is clear for $n=1$ so we will assume it to be true for $N$ ;

case 1 : $k$ does not divide $N+1$ ; then $N+1 = b_0 + b_1 k + \cdots + b_t k^t$ , $0 < b_0 \leq k-1$ , and $N = (b_0 - 1) + b_1 k + \cdots + b_t k^t$ ; hence $S_k(N+1) = S_k(N) + 1$ , $T_k(N+1) = T_k(N)$ , and $k$ divides $N+1-b_0$ ; therefore,

$$\left\{ \frac{(N+1)!}{(-k)^{T_k(N+1)}} - b_0! \cdots b_t! \right\} \equiv$$

$$b_0 \left\{ \frac{N!}{(-k)^{T_k(N)}} - (b_0 - 1)!\, b_1! \cdots b_t! \right\} \pmod{k},$$

and , since by the induction hypothesis the 2nd bracketed expression is divisible by $k$ , the induction is complete ;

Case 2: $k$ divides $N+1$; then

$N+1 = b_s k^s + \cdots + b_{s+t} k^{s+t}$, $0 < b_s$, $1 \le s$, and

$N = (k-1) + \cdots + (k-1)k^{s-1} + (b_s - 1)k^s + \cdots + b_{s+t} k^{s+t}$;

hence $S_k(N+1) = S_k(N) - s(k-1) + 1$,

$T_k(N+1) = T_k(N) + s$

and $k$ divides $\frac{N+1}{k^s} - b_s$; therefore, since by Wilson's theorem, $k$ also divides $(k-1)!^s - (-1)^s$, we have

$$\frac{(N+1)!}{(-k)^{T_k(N+1)}} - b_s! \cdots b_{s+t}! \equiv$$

$$\frac{N+1}{(-k)^s} \frac{N!}{(-k)^{T_k(N)}} - (-1)^s (k-1)!^s b_s! \cdots b_{s+t}! \equiv$$

$$(-1)^s b_s \left\{ \frac{N!}{(-k)^{T_k(N)}} - (k-1)!^s (b_s - 1)! \cdots b_{s+t}! \right\} \pmod{k};$$

since, by the induction hypothesis $k$ divides the 3rd expression, the conclusion follows.

## XI Squares ~ Solutions

1. Put $N = n(2n+1)$; then

$$N^2 + (N+1)^2 + \cdots + (N+n)^2 = (N+n+1)^2 + \cdots + (N+2n)^2$$

as is easily verified; the given equalities are all special cases of this identity.

2. i) Direct verification yields the result;

ii) suitable additions of one or more of $1^2, 2^2, 3^2$ to the following equalities suffice to prove this:

$129 = 10^2 + 5^2 + 2^2$

$130 = 9^2 + 7^2$

$132 = 9^2 + 5^2 + 4^2 + 3^2 + 1^2$

$133 = 9^2 + 6^2 + 4^2$

$138 = 8^2 + 7^2 + 5^2$

$141 = 10^2 + 5^2 + 4^2$

$149 = 10^2 + 7^2$

$152 = 10^2 + 6^2 + 4^2$

$155 = 9^2 + 7^2 + 5^2$

$166 = 9^2 + 7^2 + 6^2$

$171 = 9^2 + 7^2 + 5^2 + 4^2$

$173 = 10^2 + 8^2 + 3^2$

$174 = 10^2 + 7^2 + 5^2$

$182 = 9^2 + 7^2 + 6^2 + 4^2$

$189 = 10^2 + 8^2 + 5^2$;

iii) $1^2+\cdots+10^2=\frac{10\cdot 11\cdot 21}{6}=385$, so the numbers from 193 to 256 are the "complementary" sums to those from 129 to 192 ;

iv) $256-129+1=128>11^2$ so adding $11^2$ to the sums yielding 129 to 256 extends the range to $256+121=377$ ;

v) same argument as that given to (iv) ;

vi) this follows from the fact that $256+11^2+\cdots+k^2-129+1>(k+1)^2$ for $k\geq 11$, which is easy to prove by induction.

3. i) Solving $x^2+y^2=1$, $y=\lambda(x+1)$ simultaneously for $x$ yields $(1+\lambda^2)x^2+2\lambda^2x+(\lambda^2-1)=0$ so $x=\frac{-\lambda^2\pm 1}{1+\lambda^2}$ and, since $x\neq -1$, we find $x=\frac{1-\lambda^2}{1+\lambda^2}$, $y=\lambda\left(\frac{1-\lambda^2}{1+\lambda^2}+1\right)=\frac{2\lambda}{1+\lambda^2}$; clearly every rational point on $C'$ corresponds to a line $L_\lambda$ with rational $\lambda$ ;

ii) from $x^2+y^2=z^2$ we have $(\frac{x}{z})^2+(\frac{y}{z})^2=1$ so $(\frac{x}{z}, \frac{y}{z})$ is a rational point on C′ and corresponds to a rational λ, say $\lambda=\frac{r}{s}$, $(r,s)=1$; then

$$\frac{x}{z}=\frac{1-\lambda^2}{1+\lambda^2}=\frac{s^2-r^2}{s^2+r^2}, \quad \frac{y}{z}=\frac{2\lambda}{1+\lambda^2}=\frac{2rs}{s^2+r^2};$$

if r and s are of opposite parity then

$$(s^2-r^2, s^2+r^2)=(2rs, s^2+r^2)=1$$

and we may put $u=r, v=s$ giving the indicated expressions for x, y, z; if r and s are of the same parity they must be odd and in this case we put $u=\frac{s-r}{2}$, $v=\frac{s+r}{2}$ giving the indicated expressions for x, y, z with x and y interchanged.

4. Every odd number leaves a remainder of 1 or 3 when divided by 4 and, therefore, the square of an odd number leaves a remainder of 1 when divided by 4; consequently the sum of two odd numbers squared always leaves a remainder of 2 when divided by 4; but a square never leaves a remainder other than 0 or 1.

5. i) The number of $am+n$ with $(m,n) \in A$ is $([\sqrt{p}]+1)^2 > p$ so there must be two of them which are congruent modulo $p$;

ii) without loss of generality assume in (i) that $m > m'$ ($m$ may not be equal to $m'$ since then $n = n'$ and $(m,n) = (m',n')$ contrary to fact) and note $a(m-m') \equiv n'-n \pmod{p}$;

put $y = m-m'$ and $x = |n'-n|$.

6. i) There are $ef$ pairs $(s,t)$ with $0 \le s < e$, $0 \le t < f$; thus there are $ef$ numbers $at+s$; since $m < ef$ there must be two of these $ef$ numbers which are congruent modulo $m$;

proceed as in #5(ii);

ii) take $e = f = [\sqrt{p}]+1$, $m = p$.

7. We know $ay \equiv \pm x \pmod{p}$ so $a^2y^2 \equiv x^2 \pmod{p}$; thus $(a^2+1)y^2 \equiv x^2+y^2 \equiv 0 \pmod{p}$; since $0 < x < \sqrt{p}$, $0 < y < \sqrt{p}$ we find $0 < x^2+y^2 < 2p$; since $p \mid x^2+y^2$ we must have $p = x^2+y^2$.

8. i) See the proof of #4;
ii) by (i) and #7.

9. i) $(au+bv)^2+(av-bu)^2 = (a^2+b^2)(u^2+v^2)$
$\equiv 0 \pmod{p}$;

ii) since $(a,b)=1$ we know there are $u, v$ such that $av-bu=1$; choosing such $u, v$ in (i) yields the desired result;

iii) suppose an odd prime $p$ divides $a^2+b^2$ where $(a,b)=1$; then by (ii) $p$ divides a number of the form $x^2+1$ so, by #7, $p$ is a sum of two squares, which, by #8, means $p$ is of the form $4k+1$ ; thus all odd divisors of a sum of relatively prime squares, being products of $4k+1$ primes, are themselves of the form $4k+1$.

10. i) Let $p_n$ be a prime factor of $(n!)^2+1$, $n \geq 1$ ; then, since $p_n > n$, the sequence $p_1, p_2, \cdots$ contains infinitely many different primes and, by #9 (iii), they are all of the form $4k+1$ ;

ii) since $F_n = 2^{2^n}+1 = (2^{2^{n-1}})^2+1$, the prime factors of $F_n$, all odd, are of the form $4k+1$ ; since the Fermat numbers are relatively prime in pairs - see III #9 (iv-b) - a sequence of primes whose $n^{th}$ term is a factor of $F_n$ is a sequence of pairwise distinct $4k+1$ primes.

11. i) See the proof of #9 (i);

ii) put $a = -u$ and $b = v$ in the identity given in the proof of #9 (i).

12. i) By #9 (iii) no $4k+3$ prime may divide a sum of 2 relatively prime squares and since only even powers of such primes may divide a sum of 2 non-relatively prime squares the result follows;

ii) $2 = 1^2 + 1^2$, all $p_i$ are sums of 2 squares, $q_i^2 = q_i^2 + 0^2$ and the result now follows from #11(i);

iii) immediate from (i) and (ii).

13. This is merely an algebraic verification.

14. i) The total number of such numbers in $A \cup B$ is $p+1$ so the result is immediate;

ii) $n^2 \equiv -1 - m^2 \pmod{p}$ leads to

$$0 < sp = n^2 + m^2 + 1^2 + 0^2 \le \left(\frac{p-1}{2}\right)^2 + \left(\frac{p-1}{2}\right)^2 + 1 < p^2;$$

iii) the integers $t$, $-\frac{1}{2}s < t \le \frac{1}{s}s$ constitute a complete system of residues modulo $s$, so we may choose the $A_j$ as indicated; from $\Sigma A_j^2 \equiv \Sigma a_j^2 \equiv 0 \pmod{s}$ we know there is an $r$ such that $\Sigma A_j^2 = sr$ and $0 \le r = \frac{1}{s}\Sigma A_j^2 \le s$; if $r = 0$ or $r = s$ then all $A_j = 0$ or all $A_j = \frac{s}{2}$ so $sp = \Sigma a_j^2 \equiv 0 \pmod{s^2}$ which implies

$$p \equiv 0 \pmod{s}, \text{ contrary to } 1 < s < p;$$

iv) $rs^2p = \boxed{4}$ follows from #13; further, each of the right hand parenthetical expressions (see #13) is clearly congruent to 0 modulo $s^2$; hence $s^2$ divides both sides and

$$rp = \boxed{4};$$

v) by (iv) each $s > 1$ for which $sp = \boxed{4}$ leads to an $r$, $0 < r < s$, $rp = \boxed{4}$ ; repetition leads to $p = \boxed{4}$ ;

vi) immediate from (v) and # 13.

15. If we regard $x, y, z$ and $y, x, z$ as the same triple then by #3 (ii) all Pythagorean triples are obtained from the equations

$$x = v^2 - u^2, \quad y = 2vu, \quad z = v^2 + u^2,$$

where $u$ and $v$ are relatively prime integers of opposite parity ; further, all such triples are indeed Pythagorean triples as one may easily check ; we first show that any single application of one of the matrices U, A, D to a primitive Pythagorean triple leads again to a primitive Pythagorean triple ; this follows from

$$(v^2-u^2,\, 2vu,\, v^2+u^2)\begin{pmatrix} \epsilon & 2\epsilon & 2\epsilon \\ 2\nu & \nu & 2\nu \\ 2 & 2 & 3 \end{pmatrix}$$

$$= (v'^2-u'^2,\, 2v'u',\, v'^2+u'^2),$$

where $v' = \frac{1}{2}((1-\epsilon+2\nu)u + (3+\epsilon)v)$,

$u' = \frac{1}{2}((1-\epsilon)u + (1+\epsilon)v)$

in each of the three cases

$$\epsilon = -\nu = 1, \quad \epsilon = \nu = 1, \quad \epsilon = -\nu = -1 ;$$

also in each of these cases $u', v'$ are relatively prime integers of opposite parity; we now need to show that every such triple is of the form $(3,4,5)\mathcal{A}$; in the three cases above we have

$$(v', u') = \begin{cases} (2v-u,\, v) \\ (2v+u,\, v) \\ (2u+v,\, u) \end{cases} ;$$

let $v', u'$ be given relatively prime integers of opposite parity and suppose $v' > u'$; there are three cases

(a) $u' < v' < 2u'$, (b) $2u' < v' < 3u'$, (c) $3u' < v'$;

in these three cases the triple corresponding to $u', v'$ comes by applying, respectively, U, A, D to the triple corresponding to $u, v$ where

(a) $v = u'$, $u = 2u' - v'$ ;

(b) $v = u'$, $u = v' - 2u'$ ;

(c) $v = v' - 2u'$, $u = u'$ ;

in each case $u, v$ are relatively prime, of opposite parity, and $v > u$ ; further, we note that $v < v'$ and $u \leq u'$ ; the process must stop when $v = 2$, in which case $u = 1$ and the triple $(3, 4, 5) = (2^2 - 1^2, 2 \cdot 2 \cdot 1, 2^2 + 1^2)$ is reached.

# XII Sums of Powers - Solutions

1. *i)* Using the binomial theorem we find

$$b_1^t+\cdots+b_n^t+(c_1+h)^t+\cdots+(c_n+h)^t=$$
$$b_1^t+\cdots+b_n^t+\sum_{j=0}^{t}\binom{t}{j}h^{t-j}\sum_{i=1}^{n}c_i^j,$$
$$c_1^t+\cdots+c_n^t+(b_1+h)^t+\cdots+(b_n+h)^t=$$
$$c_1^t+\cdots+c_n^t+\sum_{j=0}^{t}\binom{t}{j}h^{t-j}\sum_{i=1}^{n}b_i^j;$$

for $0\le t\le m$, these are equal since

$$b_1,\dots,b_n\overset{m}{=}c_1,\dots,c_n;$$

for $t=m+1$ the two expressions become

$$b_1^{m+1}+\cdots+b_n^{m+1}+\sum_{j=0}^{m}\binom{m+1}{j}h^{m+1-j}\sum_{i=1}^{n}c_i^j+\sum_{i=1}^{n}c_i^{m+1}$$
$$c_1^{m+1}+\cdots+c_n^{m+1}+\sum_{j=0}^{m}\binom{m+1}{j}h^{m+1-j}\sum_{i=1}^{n}b_i^j+\sum_{i=1}^{n}b_i^{m+1};$$

again, for the same reason, these are equal;

*ii)* $(b_1+x)^m+\cdots+(b_n+x)^m=$

$$\sum_{j=0}^{m}\left\{\binom{m}{j}\sum_{i=1}^{n}b_i^{m-j}\right\}x^j$$

$(c_1+x)^m+\cdots+(c_n+x)^m=$

$$\sum_{j=0}^{m}\left\{\binom{m}{j}\sum_{i=1}^{n}c_i^{m-j}\right\}x^j;$$

if $b_1,\dots,b_n\overset{m}{=}c_1,\dots,c_n$ it is clear that these two expressions are equal for all x ; on the other hand, if they are equal for all x they must have identical coefficients which proves the other direction ;

iii) this follows immediately from (i) starting with 1 & 2.

2. The four instances are clear ; suppose that

$$\sum_{n=1}^{2^{k+1}} (1-a_{n-1})n^t = \sum_{n-1}^{2^{k+1}} a_{n-1}n^t \quad \text{for } 0 \le t \le k ;$$

then, by #1 (i),

$$\sum_{n=1}^{2^{k+1}} (1-a_{n-1})n^t + \sum_{n=1}^{2^{k+1}} a_{n-1}(n+2^{k+1})^t = \sum_{n=1}^{2^{k+1}} a_{n-1}n^t + \sum_{n=1}^{2^{k+1}} (1-a_{n-1})(n+2^{k+1})^t$$

for $0 \le t \le k+1$ ; noting that

$$\sum_{n=1}^{2^{k+1}} a_{n-1}(n+2^{k+1})^t = \sum_{n=2^{k+1}+1}^{2^{k+2}} a_{n-1-2^{k+1}}n^t ,$$

$$\sum_{n=1}^{2^{k+1}} (1-a_{n-1})(n+2^{k+1})^t = \sum_{n=2^{k+1}+1}^{2^{k+2}} (1-a_{n-1-2^{k+1}})n^t$$

and $a_{n-1-2^{k+1}} = \begin{cases} 0 & \text{if } a_{n-1} = 1 \\ 1 & \text{if } a_{n-1} = 0 \end{cases}$ we see that the above is equivalent to $\sum_{n=1}^{2^{k+2}} (1-a_{n-1})n^t = \sum_{n=1}^{2^{k+2}} a_{n-1}n^t$ for $0 \le t \le k+2$ , and the induction step is complete.

3. i) Expanding $(rn+s)^t$ using the binomial theorem and interchanging sums leads to the result by coefficient by coefficient application of #2;

ii) follows as does (i) through using the multinomial theorem.

4. i) Put $r=2$, $s=-1$ in #3(i) and use #1(ii) with $m=k$, $n=2^k$;

ii) since the $d_i$ may be chosen as large as we please this is clear;

iii) by (i), with $x$ in (i) replaced by $x+d_i$, we see that $L_i=R_i$ for each $i$; thus since for each two products $U_1 \cdots U_k$ the factors are respectively equal so also are the products;

iv) the 1st part of the assertion is clear; since the $d_i + b_j$ are all distinct, if we choose $x$ larger than the largest of the $d_i + b_j$ then the $x + d_i + b_j$ will necessarily be distinct; that they are odd follows from the fact that $x$ and the $d_i$ are even while the $b_j$ are odd;

v) this is immediate from (iii) & (iv) since all of the products $U_1 \cdots U_k$ equal $s$.

5. i) $ts = S_1 + \cdots + S_t$ and each $S_j$ is such a sum;

ii) by (i) each $t_i s$ is a sum of $k^{th}$ powers; consequently so also is $t_i s 2^{jk}$ and, therefore, $ms$;

iii) since all the summands in the $t_i s$ terms are odd and distinct and since for different $i$ these are multiplied by distinct powers of 2 none of the terms arising can equal any other;

iv) this follows from (i)–(iii) and the possibility, guaranteed by #4(v), that the hypothesis of (i) are realizable.

6. Each $(is+1) \le (s-1)s+1 \le s^2$ ; hence $S_r < s^{2k+1}$ ; the $S_r$ exhaust the residue classes modulo s so certainly all integers $\ge s^{2k+1}$ are writeable in the form $ms + S_r$ ;
the conclusion now follows from the fact that ms is the sum of distinct $k^{th}$ powers all smaller than the smallest $k^{th}$ power making up $S_r$ .

## XIII Continued Fractions ~ Solutions

1. i) This follows immediately from the definition of $E(x_0, \cdots, x_n)$;

ii) the summands of $E(x_0, \cdots, x_{n+1})$ either contain $x_{n+1}$ or they do not; those containing $x_{n+1}$ are precisely the summands of $E(x_0, \cdots, x_n)x_{n+1}$ while those not containing $x_{n+1}$ are those of $E(x_0, \cdots, x_{n-1})$;

iii) for $n=2$, $E(x_0,x_1,x_2)E(x_1) - E(x_0,x_1)E(x_1,x_2)$

$= (x_0x_1x_2 + x_0 + x_2)x_1 - (x_0x_1+1)(x_1x_2+1) = -1$

$= (-1)^{2-1}$ ; assume true for $n$, then

$$E(x_0, \cdots, x_{n+1})E(x_1, \cdots, x_n) - E(x_0, \cdots, x_n)E(x_1, \cdots, x_{n+1})$$

$$= (x_{n+1}E(x_0, \cdots, x_n) + E(x_0, \cdots, x_{n-1}))E(x_1, \cdots, x_n)$$

$$- E(x_0, \cdots, x_n)(x_{n+1}E(x_1, \cdots, x_n) + E(x_1, \cdots, x_{n-1}))$$

$$= -(E(x_0, \cdots, x_n)E(x_1, \cdots, x_{n-1}) - E(x_0, \cdots, x_{n-1})E(x_1, \cdots, x_n))$$

$$= -(-1)^{n-1} = (-1)^{(n+1)-1};$$

iv) for $n=3$,

$$E(x_0,x_1,x_2,x_3)E(x_1)-E(x_0,x_1)E(x_1,x_2,x_3) =$$

$$(x_0x_1x_2x_3+x_0x_1+x_0x_3+x_2x_3+1)x_1-(x_0x_1+1)(x_1x_2x_3+x_1+x_3)$$

$= -x_3 = (-1)^3 x_3$ ; assume true for $n$, then

$$E(x_0,\cdots,x_{n+1})E(x_1,\cdots,x_{n-1})-E(x_0,\cdots,x_{n-1})E(x_1,\cdots,x_{n+1})$$

$$=(x_{n+1}E(x_0,\cdots,x_n)+E(x_0,\cdots,x_{n-1}))E(x_1,\cdots,x_{n-1})$$

$$-E(x_0,\cdots,x_{n-1})(x_{n+1}E(x_1,\cdots,x_n)+E(x_1,\cdots,x_{n-1}))$$

$$=x_{n+1}(E(x_0,\cdots,x_n)E(x_1,\cdots,x_{n-1})-E(x_0,\cdots,x_{n-1})E(x_1,\cdots,x_n))$$

$$=x_{n+1}(-1)^{n-1}=(-1)^{n+1}x_{n+1} \quad ;$$

v) for $s=1$, $n=3$ we must have $t=2$ and, in this case,

$$E(x_0,x_1,x_2,x_3)E(x_1,x_2)-E(x_0,x_1,x_2)E(x_1,x_2,x_3)=$$

$$(x_0x_1x_2x_3+x_0x_1+x_0x_3+x_2x_3+1)(x_1x_2+1)$$

$$-(x_0x_1x_2+x_0+x_2)(x_1x_2x_3+x_1+x_3) = 1$$

$= (-1)^{2-1+1}E(x_{-1})E(x_4)$ ; assume ok for $s=1$, $n=n$; then

$$E(x_0,\cdots,x_{n+1})E(x_1,\cdots,x_t)-E(x_0,\cdots,x_t)E(x_1,\cdots,x_{n+1})$$

$$= (x_{n+1} E(x_0, \cdots, x_n) + E(x_0, \cdots, x_{n-1})) E(x_1, \cdots, x_t)$$
$$- E(x_0, \cdots, x_t)(x_{n+1} E(x_1, \cdots, x_n) + E(x_1, \cdots, x_{n-1}))$$
$$= x_{n+1}(-1)^t E(x_{t+2}, \cdots, x_n) + (-1)^t E(x_{t+2}, \cdots, x_{n-1})$$
$$= (-1)^t E(x_{t+2}, \cdots, x_{n+1}) = (-1)^{t-s+1} E(x_{-1}) E(x_{t+2}, \cdots, x_{n+1});$$

this proves the result for $s = 1$, $n \geq 3$; similarly one shows ok for $s = 2$, $n = 4$ and that if true for $s = 2$, $n = n$ then true for $s = 2$, $n \to n+1$; thus true for $s = 2$, $n \geq 4$; assume now that the formula is correct for $s \leq q$, $n \geq q+2$; then

$$E(x_0, \cdots, x_n) E(x_{q+1}, \cdots, x_t) - E(x_0, \cdots, x_t) E(x_{q+1}, \cdots, x_n)$$
$$= E(x_0, \cdots, x_n)(E(x_{q-1}, \cdots, x_t) - x_{q-1} E(x_q, \cdots, x_t))$$
$$- E(x_0, \cdots, x_t)(E(x_{q-1}, \cdots, x_n) - x_{q-1} E(x_q, \cdots, x_n))$$
$$= E(x_0, \cdots, x_n) E(x_{q-1}, \cdots, x_t) - E(x_0, \cdots, x_t) E(x_{q-1}, \cdots, x_n)$$
$$- x_{q-1}(E(x_0, \cdots, x_n) E(x_q, \cdots, x_t) - E(x_0, \cdots, x_t) E(x_q, \cdots, x_n))$$
$$= (-1)^{t-(q-1)+1} E(x_0, \cdots, x_{q-3}) E(x_{t+2}, \cdots, x_n)$$
$$- x_{q-1}(-1)^{t-q+1} E(x_0, \cdots, x_{q-2}) E(x_{t+2}, \cdots, x_n)$$
$$= (-1)^{t-q}(E(x_0, \cdots, x_{q-3}) + x_{q-1} E(x_0, \cdots, x_{q-2})) E(x_{t+2}, \cdots, x_n)$$
$$= (-1)^{t-(q+1)+1} E(x_0, \cdots, x_{(q+1)-2}) E(x_{t+2}, \cdots, x_n);$$

The above argument is sketched as follows: verify the assertion for $s=1, n=3$ and $s=1, n=4$; then show if the assertion is true for $s=1, n\le k$ then it is also true for $s=1, n=k+1$; repeat for $s=2, n=4$; $s=2, n=5$; $s=2, n\le k$ implies $s=2, n=k+1$; the above shows for $s=1$ or $2$ and $n\ge 3$ or $4$ that the assertion is correct; suppose now true for $s\le q$ and $n\ge q+2$; then this leads to the truth for $s\le q+1, n\ge q+3$; this implies the truth for $s>0, n\ge s+2$.

vi) Putting $s=m, t=m+1, n=2m$ in (v) yields, after using (ii), and noting $x_{m+1}=x_m, \cdots, x_{2m}=x_0$,

$$E(x_0,\cdots,x_{2m})E(x_m,x_{m+1}) = E^2(x_0,\cdots,x_{m+1})+E^2(x_0,\cdots,x_{m-2})$$
$$=\{x_{m+1}E(x_0,\cdots,x_m)+E(x_0,\cdots,x_{m-1})\}^2+E^2(x_0,\cdots,x_{m-2})$$
$$=(x_m^2+1)\{E^2(x_0,\cdots,x_m)+E^2(x_0,\cdots,x_{m-1})\}+E^2(x_0,\cdots,x_{m-2})$$
$$-\{E^2(x_0,\cdots,x_m)-2x_mE(x_0,\cdots,x_m)E(x_0,\cdots,x_{m-1})$$
$$+x_m^2E^2(x_0,\cdots,x_{m-1})\}$$
$$=(x_m^2+1)\{E^2(x_0,\cdots,x_m)+E^2(x_0,\cdots,x_{m-1})\};$$

since $E(x_m, x_{m+1}) = x_m x_{m+1} + 1 = x_m^2 + 1$ , cancellation of this factor yields the desired result.

vii) Put $s = m+1$, $t = m+2$, $n = 2m+1$ in (v) to obtain

$$E(x_0, \cdots, x_m, x_{m+1}, x_m, \cdots, x_0) E(x_{m+1}, x_{m+2})$$

$$= E(x_0, \cdots, x_m, x_{m+1}, x_{m+2}) E(x_{m+1}, x_m, \cdots, x_0)$$

$$+ E(x_0, \cdots, x_{m-1}) E(x_{m+4}, \cdots, x_0)$$

$$= x_{m+2} E^2(x_0, \cdots, x_m, x_{m+1}) + E(x_0, \cdots, x_m) E(x_0, \cdots, x_{m+1})$$

$$+ E(x_0, \cdots, x_{m-1}) E(x_0, \cdots, x_{m-2})$$

$$= x_{m+2}\big(x_{m+1} E(x_0, \cdots, x_m) + E(x_0, \cdots, x_{m-1})\big) E(x_0, \cdots, x_m, x_{m+1})$$

$$+ E(x_0, \cdots, x_m) E(x_0, \cdots, x_{m+1}) + E(x_0, \cdots, x_{m-1}) E(x_0, \cdots, x_{m-2})$$

$$= (x_{m+1} x_{m+2} + 1) E(x_0, \cdots, x_m) E(x_0, \cdots, x_m, x_{m+1})$$

$$+ x_{m+2} E(x_0, \cdots, x_{m-1})\big(x_{m+1} E(x_0, \cdots, x_m) + E(x_0, \cdots, x_{m-1})\big)$$

$$+ E(x_0, \cdots, x_{m-1}) E(x_0, \cdots, x_{m-2})$$

$$= (x_{m+1} x_{m+2} + 1) E(x_0, \cdots, x_m) \big(E(x_0, \cdots, x_m, x_{m+1})$$

$$+ E(x_0, \cdots, x_{m-1})\big) \; ;$$

cancelling the factor $E(x_{m+1}, x_{m+2}) = x_{m+1} x_{m+2} + 1$ yields the desired result.

2. $E_n = E(1, \cdots, 1)$, where there are $n$ places;

i) clearly $E_1 = E(1) = 1$ and, from #1(ii),

$$E_{n+2} = E_{n+1} + E_n \; ;$$

ii) from (i) and the definition of $u_n$ ;

iii) the 1st equality is by induction on $n$ and the 2nd follows from the 1st using the binomial theorem ;

iv-a) #1 (iii) reads, when one puts all $x = 1$,

$$E_{n+1} E_{n-1} - E_n^2 = (-1)^{n-1}$$

so the result follows from (ii) ;

b) this follows from #1 (iv) almost identically to the argument given in (a), except that the index needs to be changed by unity ;

c) follows from #1(v) as in (a) & (b) above;

d) follows from #1(vi) as in (a) & (b) above;

e) follows from #1 (vii) as in (a) & (b) above;

f) $u_{n-3}u_{n-1}+u_{n-1}u_{n-2}+u_{n-1}u_{n-2}+u_{n-1}u_{n-2}+u_{n-2}^2$

$= u_{n-1}^2+u_{n-1}u_{n-2}+u_{n-2}u_n = u_{n-1}u_n+u_{n-2}u_n = u_n^2$;

g) $u_{n-3}^2+u_{n-2}u_{n-3}+u_{n-2}u_{n-3}+u_{n-2}u_{n-3}+u_{n-2}u_{n-4}$

$= u_{n-3}u_{n-1}+u_{n-2}u_{n-3}+u_{n-2}^2 = u_{n-3}u_{n-1}+u_{n-2}u_{n-1}=u_{n-1}^2$;

v) for $n=2,3$ the assertion is correct; suppose it to be correct for $n, n+1$; then

$$a_{n+2}= a_{n+1}+a_n = (u_{n-1}a+u_nb)+(u_{n-2}a+u_{n-1}b)$$
$$= u_na+u_{n+1}b$$

and the assertion is also true for $n+2$.

3-i) a) $[a_0,\cdots,a_n]= a_0+\dfrac{1}{[a_1,\cdots,a_n]}$

$$= a_0+\frac{1}{\dfrac{E(a_1,\cdots,a_n)}{E(a_2,\cdots,a_n)}}$$

$$= \frac{a_0E(a_1,\cdots,a_n)+E(a_2,\cdots,a_n)}{E(a_1,\cdots,a_n)} = \frac{E(a_0,\cdots,a_n)}{E(a_1,\cdots,a_n)} = \frac{p_n}{q_n};$$

b) $[a_n,\cdots,a_1] = \dfrac{E(a_n,\cdots,a_1)}{E(a_{n-1},\cdots,a_1)}$

$$= \frac{E(a_1,\cdots,a_n)}{E(a_1,\cdots,a_{n-1})} = \frac{q_n}{q_{n-1}};$$

ii) $p_k = E(a_0, \cdots, a_k) = a_k E(a_0, \cdots, a_{k-1}) + E(a_0, \cdots, a_{k-2})$

$= a_k p_{k-1} + p_{k-2}$;

a similar argument works for $q_k$;

iii) $[a_0, \cdots, a_{n-1}, a_n + \frac{1}{b}] = [a_0, \cdots, a_n, b]$

$$= \frac{E(a_0, \cdots, a_n, b)}{E(a_1, \cdots, a_n, b)} = \frac{b E(a_0, \cdots, a_n) + E(a_0, \cdots, a_{n-1})}{b E(a_1, \cdots, a_n) + E(a_1, \cdots, a_{n-1})}$$

$$= \frac{b p_n + p_{n-1}}{b q_n + q_{n-1}} \quad ;$$

iv-a) this follows from #1 (iii) by taking $x_j = a_j$ for all $j$;

b) this follows from #1 (iv) by taking $x_j = a_j$ for all $j$;

v) from (iv) we have $\frac{p_{n-2}}{q_{n-2}} + \frac{(-1)^n a_n}{q_{n-2} q_n} = \frac{p_n}{q_n} = \frac{p_{n-1}}{q_{n-1}} + \frac{(-1)^{n-1}}{q_{n-1} q_n}$ and the inequalities not involving $\alpha$ follow by examining the cases with $n$ even (odd); for the inequalities involving $\alpha$ note that if $\alpha = [a_1, \cdots, a_n + \frac{1}{b}]$ then

$$\alpha - \frac{p_n}{q_n} = \frac{b p_n + p_{n-1}}{b q_n + q_{n-1}} - \frac{p_n}{q_n} = \frac{p_{n-1} q_n - p_n q_{n-1}}{q_n (b q_n + q_{n-1})} = \frac{(-1)^n}{q_n (b q_n + q_{n-1})};$$

vi) this follows from the fact that $\frac{a+c}{b+d}$ always lies between $\frac{a}{b}$ and $\frac{c}{d}$;

vii) in the argument above for (v) we found

$$\left|\alpha - \frac{p_n}{q_n}\right| = \frac{1}{q_n(bq_n + q_{n-1})},$$

where $a_{n+1} < b < a_{n+1} + 1$; thus

$$q_{n+1} = a_{n+1}q_n + q_{n-1} < bq_n + q_{n-1} < (a_{n+1}+1)q_n + q_{n-1} = q_n + q_{n+1}$$

and the result follows; (note that $a_{n+1} < b$ since it is assumed that $q_{n+1}$ exists);

alternative: the last four terms in the sequence given in (vi) are $\frac{p_{n+1}}{q_{n+1}}, \alpha, \frac{p_{n+1}+p_n}{q_{n+1}+q_n}, \frac{p_n}{q_n}$; thus

$$\frac{1}{q_n(q_{n+1}+q_n)} = \left|\frac{p_n}{q_n} - \frac{p_{n+1}+p_n}{q_{n+1}+q_n}\right| < \left|\frac{p_n}{q_n} - \alpha\right|$$
$$< \left|\frac{p_n}{q_n} - \frac{p_{n+1}}{q_{n+1}}\right| = \frac{1}{q_n q_{n+1}};$$

viii-a) $\left|\alpha - \frac{p_n}{q_n}\right| < \frac{1}{q_n q_{n+1}} = \frac{1}{q_n(a_{n+1}q_n + q_{n-1})}$

$$\le \frac{1}{q_{n-1}(q_{n-1}+q_n)} < \left|\alpha - \frac{p_{n-1}}{q_{n-1}}\right|;$$

b) $|\alpha q_n - p_n| < \frac{1}{q_{n+1}} = \frac{1}{a_{n+1}q_n + q_{n-1}}$

$$\le \frac{1}{q_n + q_{n-1}} < |\alpha q_{n-1} - p_{n-1}|;$$

ix) the limit exists since bounded monotone sequences always converge ; the limits are equal if and only if

$$\left|\frac{p_{n+1}}{q_{n+1}}-\frac{p_n}{q_n}\right| = \frac{1}{q_n q_{n+1}} \to 0 \text{ as } n \to \infty ;$$

x) immediate from (ix) ;

xi) for $n=1$, $\frac{p_1}{q_1}=[a_0, a_1]=a_0+\frac{1}{a_1}=a_0+\frac{1}{q_0 q_1}$ ; suppose ok for $n$, then using (iv-a) with $n$ replaced by $n+1$ and having divided by $q_n q_{n+1}$ we have

$$\frac{p_{n+1}}{q_{n+1}}=\frac{p_n}{q_n}+\frac{(-1)^n}{q_n q_{n+1}}=a_0+\frac{1}{q_0 q_1}-\frac{1}{q_1 q_2}+\cdots+\frac{(-1)^{n-1}}{q_{n-1}q_n}+\frac{(-1)^n}{q_n q_{n+1}} ;$$

xii-a) this follows immediately from (iv-a) ;

b) $q_n = E(a_1, \cdots, a_n) \geq E_n \to \infty$ ;

since $\frac{p_n}{q_n}$ has a limit and $q_n \to \infty$ it must be true that $|p_n| \to \infty$ ;

c) for $n = 2, 3$ we have

$$q_2 = E(a_1, a_2) = a_1 a_2 + 1 \geq 2 \geq 2^{\frac{2-1}{2}};$$

$$q_3 = E(a_1, a_2, a_3) = a_1 a_2 a_3 + a_1 + a_3 \geq 3 \geq 2^{\frac{3-1}{2}};$$

assume ok up to and including $n$, then

$$q_{n+1} = a_{n+1} q_n + q_{n-1} \geq q_n + q_{n-1} \geq 2^{\frac{n}{2} - \frac{1}{2}} + 2^{\frac{n}{2} - 1}$$

$$= 2^{\frac{n}{2}}\left(\frac{1}{\sqrt{2}} + \frac{1}{2}\right) > 2^{\frac{(n+1)-1}{2}};$$

d) by (ix) since $q_n q_{n+1} \to \infty$ as $n \to \infty$;

xiii-a) suppose $\alpha = [a_0, \cdots, a_{n-1} + \frac{1}{\alpha'}] = \frac{\alpha' p_{n-1} + p_{n-2}}{\alpha' q_{n-1} + q_{n-2}}$,

$$\beta = [a_0, \cdots, a_{n-1} + \frac{1}{\beta'}] = \frac{\beta' p_{n-1} + p_{n-2}}{\beta' q_{n-1} + q_{n-2}}$$

let $\alpha' = a_n + \alpha''$, $\beta' = b_n + \beta''$, where $0 < \alpha'' < 1$, $0 < \beta'' < 1$; then $\beta' - \alpha' = (b_n - a_n) + \beta'' - \alpha''$

$$\geq 1 + \beta'' - \alpha'' > 0; \text{ now}$$

$$\alpha - \beta = \frac{\alpha' p_{n-1} + p_{n-2}}{\alpha' q_{n-1} + q_{n-2}} - \frac{\beta' p_{n-1} + p_{n-2}}{\beta' q_{n-1} + q_{n-2}} = \frac{(\beta' - \alpha')(p_{n-2} q_{n-1} - p_{n-1} q_{n-2})}{(\alpha' q_{n-1} + q_{n-2})(\beta' q_{n-1} + q_{n-2})}$$

$$= \frac{(\beta' - \alpha')(-1)^{n-1}}{(\alpha' q_{n-1} + q_{n-2})(\beta' q_{n-1} + q_{n-2})} \begin{cases} < 0 \text{ if } n \text{ is even;} \\ > 0 \text{ if } n \text{ is odd;} \end{cases}$$

b) put $\delta = [d_0, d_1, d_2, \cdots]$, where $d_{2k} = c_{2k}$, $d_{2k-1} = b_{2k-1}$; if $\alpha < \delta$ then, using (a), we see that if $j$ is the first place where $\delta$ and $\alpha$ differ then

$$d_j > a_j \text{ for } j \text{ even}, \quad d_j < a_j \text{ for } j \text{ odd};$$

but then if $j = 2k$, $c_{2k} > a_{2k}$, and if $j = 2k-1$, $b_{2k-1} < a_{2k-1}$; since these violate the hypothesis, we must have $\delta \le \alpha$; the right inequality is proved in the same way;

c) take all $c_j = 1$ and all $b_j = 2$; then

$$\frac{1+\sqrt{3}}{2} = [1,2,1,2,\cdots] \le \alpha \le [2,1,2,1,\cdots] = 1+\sqrt{3}.$$

4. Let $\alpha = \frac{a}{b}$ and suppose

$$\begin{aligned}
a &= a_0 b + r_0 & 0 &\le r_0 < b \\
b &= a_1 r_0 + r_1 & 0 &\le r_1 < r_0 \\
r_0 &= a_2 r_1 + r_2 & 0 &\le r_2 < r_1 \\
r_1 &= a_3 r_2 + r_3 & 0 &\le r_3 < r_2 \\
&\vdots \\
r_{k-3} &= a_{k-1} r_{k-2} + r_{k-1} & 0 &\le r_{k-1} < r_{k-2} \\
r_{k-2} &= a_k r_{k-1} + 0 \ ;
\end{aligned}$$

then

$$\frac{a}{b} = a_0 + \frac{r_0}{b} = a_0 + \frac{1}{\frac{b}{r_0}} = a_0 + \frac{1}{a_1 + \frac{r_1}{r_0}}$$

$$= a_0 + \frac{1}{a_1 + \frac{1}{\frac{r_0}{r_1}}}$$

$$= a_0 + \frac{1}{a_1 + \frac{1}{a_2 + \cdots + \frac{1}{a_k}}} = [a_0, \cdots, a_k] \;;$$

if $a_k = 1$, then this equals $[a_0, \cdots, a_{k-1} + 1]$; from $\alpha = [a_0, a_1, a_2, \cdots, a_n] = [b_0, b_1, \cdots, b_n] = \beta$ we conclude, unless $n = 1$ and one of $a_1, b_1$ is 1, that $a_0 = [\alpha] = [\beta] = b_0$; continue by induction to get uniqueness.

5. Put $a_0 = [\alpha]$, $a_1 = \left[\frac{1}{\alpha - a_0}\right]$, $a_2 = \left[\frac{1}{\frac{1}{\alpha - a_0} - a_1}\right]$, $a_3 = \left[\frac{1}{\frac{1}{\frac{1}{\alpha - a_0} - a_1} - a_2}\right]$, $\cdots$ ; i.e. define $\alpha, \alpha_1, \alpha_2, \cdots$ by $\alpha_1 = \frac{1}{\alpha - a_0}$, $\alpha_{j+1} = \frac{1}{\alpha_j - a_j}$; then $a_j = [\alpha_j]$; uniqueness follows from the fact that if $\alpha = [a_0, a_1, \cdots]$ then $a_0 = [\alpha]$, etc.

6 - i) Such a scheme is illustrated by the diagram:

| $k$ | $-2$ | $-1$ | $0$ | $1$ | $2$ | $\cdots$ | $s-1$ | $s$ | $s+1$ | $\cdots$ |
|---|---|---|---|---|---|---|---|---|---|---|
| $a_k$ | | | $a_0$ | $a_1$ | $a_2$ | | | | | |
| $p_k$ | $0$ | $1$ | $a_0$ | $a_0a_1+1$ | $a_2(a_0a_1+1)+a_0$ | $\cdots$ | $p_{s-1}$ | $p_s$ | $a_{s+1}p_s+p_{s-1}$ | $\cdots$ |
| $q_k$ | $1$ | $0$ | $1$ | $a_1$ | $a_2a_1+1$ | | $q_{s-1}$ | $q_s$ | $a_{s+1}q_s+q_{s-1}$ | |

ii)

| -2 | -1 | 0 | 1 | 2 | 3 | 4 | 5 | 6 | 7 |
|---|---|---|---|---|---|---|---|---|---|
| | | 2 | 2 | 1 | 3 | 1 | 1 | 4 | 3 |
| 0 | 1 | 2 | 5 | 7 | 26 | 33 | 59 | 269 | 866 |
| 1 | 0 | 1 | 2 | 3 | 11 | 14 | 25 | 114 | 367 |

so the convergents are

$$\frac{2}{1}, \frac{5}{2}, \frac{7}{3}, \frac{26}{11}, \frac{33}{14}, \frac{59}{25}, \frac{269}{114}, \frac{866}{367};$$

iii & iv)

$$\frac{2227}{9911} = 0 + \cfrac{1}{4 + \cfrac{1}{2 + \cfrac{1}{4 + \cfrac{1}{1 + \cfrac{1}{1 + \cfrac{1}{6}}}}}}$$

since

$$2227 = 0 \cdot 9911 + 2227$$

$$9911 = 4 \cdot 2227 + 1003$$

$$2227 = 2 \cdot 1003 + 221$$

$$1003 = 4 \cdot 221 + 119$$

$$221 = 1 \cdot 119 + 102$$

$$119 = 1 \cdot 102 + 17$$

$$102 = 6 \cdot 17 + 0 \quad ;$$

| | | | | | | | | |
|---|---|---|---|---|---|---|---|---|
| | | 0 | 4 | 2 | 4 | 1 | 1 | 6 |
| 0 | 1 | 0 | 1 | 2 | 9 | 11 | 20 | 131 |
| 1 | 0 | 1 | 4 | 9 | 40 | 49 | 89 | 583 |

thus $\frac{2227}{9911} = \frac{131}{583}$ (cancel 17) ;

$$\frac{34453}{10349} = 3 + \cfrac{1}{3 + \cfrac{1}{26}}$$

since

$$34453 = 3 \cdot 10349 + 3406$$

$$10349 = 3 \cdot 3406 + 131$$

$$3406 = 26 \cdot 131 + 0 \quad ;$$

| | | | | |
|---|---|---|---|---|
| | | 3 | 3 | 26 |
| 0 | 1 | 3 | 10 | 263 |
| 1 | 0 | 1 | 3 | 79 |

thus $\frac{34453}{10349} = \frac{263}{79}$ (cancel 131) ;

v) $\pi = [3, 7, 15, 1, 292, 1, \cdots]$ so

| | | | | | | | |
|---|---|---|---|---|---|---|---|
| | | 3 | 7 | 15 | 1 | 292 | 1 |
| 0 | 1 | 3 | 22 | 333 | 355 | 103993 | 104348 |
| 1 | 0 | 1 | 7 | 106 | 113 | 33102 | 33215 |

$$\left|\pi - \frac{355}{113}\right| < \left|\frac{355}{113} - \frac{103993}{33102}\right| = \frac{1}{113 \cdot 33102}$$

$$= \frac{10^{-7}}{.3740526} < 3 \cdot 10^{-7} ;$$

vi) one obtains the sequence

$\frac{3}{1}, \frac{25}{8}, \frac{47}{15}, \frac{69}{22}, \frac{91}{29}, \frac{113}{36}, \frac{135}{43}, \frac{157}{50}, \frac{179}{57}, \frac{201}{64}, \frac{223}{71}, \frac{245}{78}, \frac{267}{85}, \frac{289}{92}, \frac{311}{99}, \frac{333}{106}$ ; one knows to stop since the next term is $\frac{355}{113}$ which is on the other side of $\pi$ and is, incidentally, the next convergent after $\frac{333}{106}$ ;

vii) $\frac{1+\sqrt{5}}{2} = 1 + \frac{\sqrt{5}-1}{2} = 1 + \frac{1}{\frac{2}{\sqrt{5}-1}} = 1 + \frac{1}{\frac{1+\sqrt{5}}{2}}$

$= [1, 1, 1, 1, \cdots]$ ;

viii) $\alpha = a + \frac{1}{a + \frac{1}{a+\alpha}} = \frac{a^3 + \alpha a^2 + 2a + \alpha}{a^2 + \alpha a + 1}$ ;

hence

$\alpha^2 = a^2 + 2$ and $\alpha = \sqrt{a^2+2}$ ;

ix) $\sqrt{a^2-2} = a - 1 + \left(\sqrt{a^2-2} - (a-1)\right) = a - 1 + \frac{1}{\frac{\sqrt{a^2-2}+a-1}{2a-3}}$

$\frac{\sqrt{a^2-2}+a-1}{2a-3} = 1 + \frac{\sqrt{a^2-2}-(a-2)}{2a-3} = 1 + \frac{1}{\frac{\sqrt{a^2-2}+a-2}{2}}$

$\frac{\sqrt{a^2-2}+a-2}{2} = a - 2 + \frac{\sqrt{a^2-2}-(a-2)}{2} = a - 2 + \frac{1}{\frac{\sqrt{a^2-2}+a-2}{2a-3}}$

$\frac{\sqrt{a^2-2}+a-2}{2a-3} = 1 + \frac{\sqrt{a^2-2}-(a-1)}{2a-3} = 1 + \frac{1}{\sqrt{a^2-2}+a-1}$

$\sqrt{a^2-2} + a - 1 = 2(a-1) + \left(\sqrt{a^2-2} - (a-1)\right)$ ;

thus

$\sqrt{a^2-2} = [a-1, \overline{1, a-2, 1, 2(a-1)}]$ ;

putting $a = 5$ we have $\sqrt{23} = [4, \overline{1, 3, 1, 8}]$;
the expansion for $\sqrt{a^2-2}$ was originally given by Euler, see Perron I, p 99 [1954].

x) (see Brousseau [1971])

using the schema of (i) we find

| | | | | | | | | | |
|---|---|---|---|---|---|---|---|---|---|
| | | | $\overbrace{\phantom{x}}^{n-3\ 1\text{'s}}$ | | | | | $\overbrace{\phantom{x}}^{n-2\ 1\text{'s}}$ | |
| | | 2 | 1 | $\cdots$ | 1 | 3 | 1 | $\cdots$ | 1 |
| 0 | 1 | 2 | 3 | $\cdots$ | $u_{n-1}$ | $3u_{n-1}+u_{n-2}$ | | $\cdots$ | $\alpha$ |
| 1 | 0 | 1 | 1 | | $u_{n-3}$ | $3u_{n-3}+u_{n-4}$ | | | $\beta$ |

where $\alpha, \beta$ are the respective $n^{th}$ terms of sequences constructed as in #2 (v) where the $a, b$ are as follows:

for $\alpha$, $a = u_{n-1}$, $b = 3u_{n-1}+u_{n-2}$

for $\beta$, $a = u_{n-3}$, $b = 3u_{n-3}+u_{n-4}$,

using the results of #2(v) and #2(iv-f, g) we have the desired conclusion.

7. i-a) $q_k = a_k q_{k-1} + q_{k-2} = q_{k-1}(1+a_k) + q_{k-2} - q_{k-1}$
$< q_{k-1}(1 + a_k + a_k^2 + \cdots) = \frac{q_{k-1}}{1-a_k}$, when $q_{k-2} < q_{k-1}$;

and $q_k = a_k q_{k-1} + q_{k-2} \le (a_k+1)\, q_{k-2} < \frac{q_{k-2}}{1-a_k}$ ,

when $q_{k-2} \ge q_{k-1}$ ;

b) for suitably large $k_0$ the convergence of $\sum_{n=1}^{\infty} a_n$ guarantees that $a_k < 1$ for $k \ge k_0$ ; now iteratively using (a) we obtain

$$q_k < q_s (1-a_{i_1})^{-1} \cdots (1-a_{i_r})^{-1} ,$$

where $s \le k_0$ and $k = i_1 > \cdots > i_r > k_0$ ;

c) from (b) we see that $q_k$ is bounded since $\prod_{i=k_0}^{\infty} (1-a_i)^{-1}$ converges ; consequently $q_k q_{k+1}$ does not tend to $\infty$ and #3(x) guarantees the divergence of $[a_0, a_1, \cdots]$ ;

ii-a) $q_0 \ge \min\{q_0, q_1\} = c$ , and, similarly, $q_1 \ge c$ ; if $q_k \ge c$ for $k \le n$ then $q_{n+1} = a_{n+1} q_n + q_{n-1} \ge c$ ;

b) this follows from $q_k = a_k q_{k-1} + q_{k-2}$, the positivity of $a_k$ and $q_{k-1} \ge c$ ;

c) $q_k + q_{k-1} \ge q_{k-1} + q_{k-2} + c\, a_k$ for $k \ge 2$ ; iterating this we find

$$q_k + q_{k-1} \geq q_0 + q_{-1} + c\sum_{n=1}^{k} a_n > c\sum_{n=1}^{k} a_n \ ;$$

d) $q_{k-1} \geq c$, $q_k \geq c$, so

$$q_k q_{k-1} \geq c\ \frac{q_k + q_{k-1}}{2} > \frac{c^2}{2}\sum_{n=1}^{k} a_n \ ;$$

e) by #3(x), $q_k q_{k+1} \to \infty$ as $k \to \infty$ ;

iii) this merely combines (i) & (ii) ;

iv) the series $\sum_{n=1}^{\infty} a_n$ always diverges under the given conditions.

8. i-a) There are positive integers $\epsilon, \nu$ such that $bx - ay = \epsilon$, $dx - cy = -\nu$ ;

solving these equations yields

$$x = \frac{\nu a + \epsilon c}{bc - ad} \geq \frac{a + c}{bc - ad},\ y = \frac{\nu b + \epsilon d}{bc - ad} \geq \frac{b + d}{bc - ad};$$

$bc - ad \neq 0$ since $\frac{a}{b} < \frac{c}{d}$ ;

b) every fraction lying between $\frac{a}{b}$ and $\frac{c}{d}$ has a denominator which is $\geq b + d$ ; if one of $\frac{a}{b}, \frac{c}{d}$ is closer to $\alpha$ then that fraction is a BA1 to $\alpha$ (the other fraction may or may not be a BA1 in this case);

ii) let $\frac{p_j}{q_j}$ be a convergent to $\alpha$; then either

$$\frac{p_{j-1}}{q_{j-1}} < \alpha < \frac{p_j}{q_j} \quad \text{or} \quad \frac{p_j}{q_j} < \alpha < \frac{p_{j-1}}{q_{j-1}}$$

and, since $|p_j q_{j-1} - p_{j-1} q_j| = 1$, we conclude from (i-b) that $\frac{p_j}{q_j}$ is a BA1 to $\alpha$;

iii) by problem #6(v):

$$\frac{3}{1}, \frac{22}{7}, \frac{333}{106}, \frac{355}{113}, \frac{103993}{33102} \quad ;$$

iv) by problem #6(vii)

$$\frac{1+\sqrt{5}}{2} = [1, 1, 1, \cdots] = \lim \frac{E_{n+1}}{E_n} = \lim \frac{u_{n+1}}{u_n};$$

v) if $\frac{p_{j-1}}{q_{j-1}} < \frac{a}{b} < \frac{p_{j+1}}{q_{j+1}}$ then, using (ii-a) & #3(iv-b), noting that j must be odd,

$$b \geq \frac{q_{j-1} + q_{j+1}}{q_{j-1}p_{j+1} - q_{j+1}p_{j-1}} = \frac{a_{j+1}q_j + 2q_{j-1}}{a_{j+1}} > q_j \;;$$

if $\frac{p_{j+1}}{q_{j+1}} < \frac{a}{b} < \frac{p_{j-1}}{q_{j-1}}$ then, using (ii-a) & #3(iv-b), noting that j must be even,

$$b \geq \frac{q_{j+1} + q_{j-1}}{q_{j+1}p_{j-1} - p_{j+1}q_{j-1}} = \frac{a_{j+1}q_j + 2q_{j-1}}{a_{j+1}} > q_j \;.$$

9.i) $\Phi_1 = \left\{\frac{0}{1}, \frac{1}{1}\right\}$ ;

$\Phi_2 = \left\{\frac{0}{1}, \frac{1}{2}, \frac{1}{1}\right\}$ ;

$\Phi_3 = \left\{\frac{0}{1}, \frac{1}{3}, \frac{1}{2}, \frac{2}{3}, \frac{1}{1}\right\}$ ;

$\Phi_4 = \left\{\frac{0}{1}, \frac{1}{4}, \frac{1}{3}, \frac{1}{2}, \frac{2}{3}, \frac{3}{4}, \frac{1}{1}\right\}$ ;

$\Phi_5 = \left\{\frac{0}{1}, \frac{1}{5}, \frac{1}{4}, \frac{1}{3}, \frac{2}{5}, \frac{1}{2}, \frac{3}{5}, \frac{2}{3}, \frac{3}{4}, \frac{4}{5}, \frac{1}{1}\right\}$ ;

$\Phi_6 = \left\{\frac{0}{1}, \frac{1}{6}, \frac{1}{5}, \frac{1}{4}, \frac{1}{3}, \frac{2}{5}, \frac{1}{2}, \frac{3}{5}, \frac{2}{3}, \frac{3}{4}, \frac{4}{5}, \frac{5}{6}, \frac{1}{1}\right\}$ ;

ii) clear ;

iii) a) $(a, b) = 1$ so the congruence $ay \equiv -1 \pmod{b}$ is solvable ; choose that residue class for $y$ so that $n - b < y \leq n$ ;

b) with $x_0$ and $y_0$ as specified suppose $(c, d) = 1$ and $\frac{a}{b} < \frac{c}{d} < \frac{x_0}{y_0}$ ; then $ad + 1 \leq bc$ and $cy_0 + 1 \leq dx_0 = d \cdot \frac{ay_0 + 1}{b}$ ; this implies $bcy_0 + b \leq ady_0 + d$ and, therefore, $n < y_0 + b \leq (bc - ad)y_0 + b \leq d$ ; this means $\frac{c}{d}$ is not in $\Phi_n$ while $\frac{x_0}{y_0}$ is in $\Phi_n$ ;

c) put $a = 79$, $b = 101$, $n = 101$; then $79 y_0 \equiv -1 \pmod{101}$ and $0 < y_0 \leq 101$ are satisfied by $y_0 = 23$; thus $x_0 = 18$ and $\frac{18}{23}$ is the desired fraction in $\Phi_{101}$; since only the inequality changes to work in $\Phi_{200}$, and then because $99 < y_0 \leq 200$ we see $y_0 = 23 + 101 = 124$ in the second case; thus $x_0 = 97$ and $\frac{97}{124}$ is the next fraction after $\frac{79}{101}$ in $\Phi_{200}$;

d) let $m = \max\{b, d\}$; then since $ad \equiv -1 \pmod{b}$ and $m - b < d \leq m$ we conclude

$$\frac{\frac{ad+1}{b}}{d} = \frac{ad+1}{bd} = \frac{bc}{bd} = \frac{c}{d}$$

is the element next after $\frac{a}{b}$ in $\Phi_m$;

iv-a) if $b + d \leq n$ then the fraction $\frac{a+c}{b+d}$ would be in $\Phi_n$ and then $\frac{a}{b}, \frac{c}{d}$ would not be consecutive in $\Phi_n$;

b) by (iii), $\frac{c}{d} = \frac{x_0}{y_0}$ and, therefore,
$bc - ad = bx_0 - ay_0 = 1$;

c) immediate from (b) ;

d) from (b), if $b = d$ then $b = d = 1$ and, since $\frac{a}{1}, \frac{b}{1}$ are consecutive elements only in $\Phi_1$, $n = 1$ ; but this violates $n > 1$ ;

v) this follows from (iv-b) and #8 (i-b) ;

vi-a) by (v-b), $bx - ay = 1$ and $cy - dx = 1$ ; thus $x = \frac{a+c}{bc-ad}$ and $y = \frac{b+d}{bc-ad}$ ; since $(x,y) = 1$ this implies $(a+c, b+d) = bc - ad$ ;

b) from (a) $\frac{x}{y} = \frac{a+c}{b+d}$ ;

vii) clear from (vi) ;

viii) $\Phi_7 = \left\{ \frac{0}{1}, \frac{1}{7}, \frac{1}{6}, \frac{1}{5}, \frac{1}{4}, \frac{2}{7}, \frac{1}{3}, \frac{2}{5}, \frac{3}{7}, \frac{1}{2}, \frac{4}{7}, \frac{3}{5}, \frac{2}{3}, \frac{5}{7}, \frac{3}{4}, \frac{4}{5}, \frac{5}{6}, \frac{6}{7}, \frac{1}{1} \right\}$ ;

ix-a) all fractions in $\Phi_{n+1} \setminus \Phi_n$ are Farey mediants of fractions in $\Phi_n$ by (vii) ;

consequently the only fraction in $\Phi_{n+1} \setminus \Phi_n$ lying between $\frac{a}{b}$ and $\frac{c}{d}$ is $\frac{a+c}{b+d}$ ; since no fraction of $\Phi_{n+1} \setminus \Phi_n$ not lying between $\frac{a}{b}$ and $\frac{c}{d}$ can be as close to $\alpha$ as each of $\frac{a}{b}$ and $\frac{c}{d}$ the conclusion follows ;

6) by (v) at least one of $\frac{a}{b}$ and $\frac{c}{d}$ is a BA1 to $\alpha$ ; consequently all fractions with denominators $\leq n$ are further from $\alpha$ than the closer of $\frac{a}{b}$, $\frac{c}{d}$ ; if $\frac{a+c}{b+d}$ is in $\Phi_{n+1} \setminus \Phi_n$ and is closer to $\alpha$ than the nearer of $\frac{a}{b}$, $\frac{c}{d}$ then clearly it is nearer to $\alpha$ than all other elements of $\Phi_{n+1} \setminus \Phi_n$ and, therefore, is a BA1 to $\alpha$ ;

x) the double underlined terms are the requisite fractions for $\pi - 3$ : $\frac{0}{\underline{\underline{1}}}$ $\frac{1}{8}$ $\frac{2}{15}$ $\frac{3}{22}$ $\frac{4}{29}$ $\frac{5}{36}$ $\frac{6}{43}$ $\frac{7}{50}$ $\frac{8}{\underline{\underline{57}}}$ $\frac{9}{\underline{\underline{64}}}$ $\frac{10}{\underline{\underline{71}}}$ $\frac{11}{\underline{\underline{78}}}$ $\frac{12}{\underline{\underline{85}}}$ $\frac{13}{\underline{\underline{92}}}$ $\frac{14}{\underline{\underline{99}}}$ $\frac{15}{\underline{\underline{106}}}$ $\Big\}$ $\frac{16}{\underline{\underline{113}}}$ $\frac{1}{\underline{\underline{7}}}$ $\frac{1}{\underline{\underline{6}}}$ $\frac{1}{\underline{\underline{5}}}$ $\frac{1}{\underline{\underline{4}}}$ $\frac{1}{3}$ $\frac{1}{2}$ $\frac{1}{1}$

hence the desired fractions for $\pi$ are :

$$3, \frac{7}{4}, \frac{16}{5}, \frac{19}{6}, \frac{22}{7}, \frac{179}{57}, \frac{201}{64}, \frac{223}{71}, \frac{245}{78}, \frac{257}{85}, \frac{279}{92}, \frac{301}{99}, \frac{323}{106}, \frac{355}{113}.$$

10. i) In the contrary case

$$\frac{1}{bd} = \left|\frac{c}{d} - \frac{a}{b}\right| = \left|\alpha - \frac{a}{b}\right| + \left|\alpha - \frac{c}{d}\right| \geq \frac{1}{2b^2} + \frac{1}{2d^2}$$

so $(b-d)^2 = b^2 - 2bd + d^2 \leq 0$, which is impossible unless $b = d$, which does not happen for $n > 1$;

ii) let $\frac{a}{b} \leq \alpha < \frac{c}{d}$, where $\frac{a}{b}$ and $\frac{c}{d}$ are neighboring elements of $\Phi_m$;

if $\left|\alpha - \frac{a}{b}\right| > \frac{1}{b(m+1)}$ and $\left|\alpha - \frac{c}{d}\right| > \frac{1}{d(m+1)}$ we also have $\frac{1}{bd} = \frac{c}{d} - \frac{a}{b} = \left|\alpha - \frac{a}{b}\right| + \left|\alpha - \frac{c}{d}\right|$

$$\geq \frac{b+d}{bd(m+1)} \geq \frac{1}{bd};$$

since, by #9 (iv-a), $b+d > m$; thus, either

$$\left|\alpha - \frac{a}{b}\right| \leq \frac{1}{b(m+1)} \text{ or } \left|\alpha - \frac{c}{d}\right| \leq \frac{1}{d(m+1)};$$

since $b \leq m$ and $d \leq m$ we may choose $\frac{s}{t}$ to be $\frac{a}{b}$ in the 1st case and $\frac{c}{d}$ in the 2nd case.

11. i) See #4;

ii) by (i) we may expand each of the numbers $\frac{p}{t}$, $1 \leq t \leq s$, into a scf so that $a_0 > 1$, $a_n > 1$;

doing so yields $\frac{p}{t} = \frac{E(a_0,\cdots,a_n)}{E(a_1,\cdots,a_n)}$ ;

iii) by #3 (xii-a), $p = E(a_0,\cdots,a_n)$ for each of the numbers t in (ii); if, on the other hand, $p = E(a_0,\cdots,a_n)$ and $a_0 > 1$ , $a_n > 1$ , then $\frac{p}{t} = \frac{E(a_0,\cdots,a_n)}{E(a_1,\cdots,a_n)}$ is one of the above expansions with $1 \le t \le s$; the last conclusion follows from #1 (i) ;

iv) by (iii) if $p = E(a_0,\cdots,a_n)$ also $p = E(a_n,\cdots,a_0)$; thus these sequences may be paired and, since $p = E(p)$, the sequence p is paired with itself; now the evenness of the number of sequences means that one of the other sequences must be the same forwards and backwards; i.e. $a_j = a_{n-j}$ ;

v) by #1 (vii), $E(a_0,\cdots,a_m, a_{m+1}, a_m,\cdots,a_0)$ is not a prime when $a_0 > 1$ ;

thus $p = E(a_0, \cdots, a_m, a_m, \cdots, a_0)$, as in (iv), and by #1 (vi), $p = E^2(a_0, \cdots, a_m) + E^2(a_0, \cdots, a_{m-1})$ ;

vi) $s = \left[\frac{13}{2}\right] = 6$, $13 = E(13)$, $\frac{13}{2} = 6 + \frac{1}{2} = \frac{E(6,2)}{E(2)}$,

$$\frac{13}{3} = 4 + \frac{1}{3} = \frac{E(4,3)}{E(3)}, \quad \frac{13}{4} = 3 + \frac{1}{4} = \frac{E(3,4)}{E(4)},$$

$$\frac{13}{5} = 2 + \cfrac{1}{1 + \cfrac{1}{1 + \cfrac{1}{2}}} = \frac{E(2,1,1,2)}{E(1,1,2)}, \quad \frac{13}{6} = 2 + \frac{1}{6} = \frac{E(2,6)}{E(6)};$$

the pairs are : $\frac{13}{2} \longleftrightarrow \frac{13}{6}$, $\frac{13}{3} \longleftrightarrow \frac{13}{4}$,

$13 \longleftrightarrow 13$, $\frac{13}{5} \longleftrightarrow \frac{13}{5}$, thus

$$13 = E(2,1,1,2) = E^2(2,1) + E^2(2) = 3^2 + 2^2 .$$

12. i-a) $\left| \frac{a}{b} - \frac{s}{t} \right| \leq \frac{1}{t([\sqrt{b}]+1)} < \frac{1}{t\sqrt{b}}$

and $0 < t \leq [\sqrt{b}] \leq \sqrt{b}$ ; thus

$$0 < (at - bs)^2 + t^2 < \frac{b^2}{b} + b = 2b ;$$

b) from (a),

$$0 < (at - bs)^2 + t^2 = (a^2+1)t^2 - 2abst + b^2s^2 < 2b$$

and, since $b$ divides $a^2+1$, it also divides the middle expression which implies

$$(a^2+1)t^2 - 2abst + b^2s^2 = b$$

or, what is the same, $\frac{a^2+1}{b}t^2 - 2ast - bs^2 = 1$ ; if $(at-bs,t) = \delta$ then $\delta$ divides $t$ and $at-bs$ so $\delta | bs$ ; since $(s,t) = 1$ either $\delta = 1$ or $\delta | b$; in the latter case $\delta$ divides the left side of the above equation and therefore divides the right side of the equation and this implies $\delta = 1$ ;

c) this was shown in the proof of (a) ;

ii) let $b$ divide $a^2+1$ ; then $(a,b) = 1$ and we may take $\alpha = \frac{a}{b}$ in (i) ; the conclusion $(at-bs)^2 + t^2 = b$ derived in the proof of (i-a) yields the desired result since by (b)

$$(at-bs,t) = 1 ;$$

iii) Wilson's theorem tells us

$$(p-1)! = (4k)! \equiv -1 \pmod{4k+1} ;$$

now

$$\begin{aligned}(4k)! &= (1\cdot 2\cdots(2k))((2k+1)\cdots(4k)) \\ &= (1\cdot 2\cdots(2k))(p-2k)\cdots(p-1) \\ &\equiv (1\cdot 2\cdots(2k))^2(-1)^{(2k)} = a^2 \pmod{p} ;\end{aligned}$$

i.e. taking $a = (2k)!$, $p$ divides $a^2+1$;
the conclusion follows from (ii);

iv) if $(b,c) = \delta$ then $\delta | a^2$ so $\delta = 1$; thus there is a $u$ such that $cu \equiv 1 \pmod b$; thus $(a^2+c^2)u^2 = (au)^2 + 1 \equiv 0 \pmod b$ and we are done;

v) this follows immediately from (ii) & (iv).

13. i) $\left|\alpha - \frac{a}{b}\right| \geq \left|\alpha - \frac{c}{d}\right|$ for $d \leq b$ implies
$$d\,|\alpha b - a| \geq b\,|\alpha d - c|$$
which in turn implies
$$|\alpha b - a| \geq \frac{b}{d}|\alpha d - c| \geq |\alpha d - c|;$$

ii) $\frac{1}{3}$ is a BA1 but not a BA2 to $\frac{1}{5}$;

iii-a) if $\frac{a}{b} < \frac{p_0}{q_0} = a_0$ then $\left|\frac{a}{b} - \alpha\right| > \left|\frac{a_0}{1} - \alpha\right|$ so $\frac{a}{b}$ would not be a BA1 to $\alpha$;

b) if $\frac{a}{b} > \frac{p_1}{q_1}$ then $\alpha < \frac{p_1}{q_1} < \frac{a}{b}$ so

$$|1 \cdot \alpha - a_0| \leq \frac{1}{a_1} = \frac{1}{q_1} = b \cdot \frac{1}{bq_1} \leq b\left|\frac{a}{b} - \frac{p_1}{q_1}\right|$$
$$< b\left|\alpha - \frac{a}{b}\right| = |b\alpha - a|$$

so $\frac{a}{b}$ would not be a BA2 to $\alpha$;

c) $b > q_k$ follows from #8(v); clearly

$$\left|\alpha - \frac{a}{b}\right| \geq \left|\frac{a}{b} - \frac{p_{k+1}}{q_{k+1}}\right| = \frac{|aq_{k+1} - bp_{k+1}|}{bq_{k+1}} \geq \frac{1}{bq_{k+1}};$$

$$\left|\alpha - \frac{p_k}{q_k}\right| < \left|\frac{p_{k+1}}{q_{k+1}} - \frac{p_k}{q_k}\right| = \frac{1}{q_k q_{k+1}};$$

d) from (c) we conclude

$$|b\alpha - a| \geq \frac{1}{q_{k+1}} > |q_k\alpha - p_k|$$

and this violates our supposition that $\frac{a}{b}$ is a BA2 to $\alpha$;

e) we have shown in (a)~(d) that $\frac{a}{b}$ may not lie between consecutive convergents and also may not lie to the left or right of all convergents; consequently $\frac{a}{b}$ must be a convergent to $\alpha$;

iv-a) we use #3(vii) to derive

$$|q_n\alpha - p_n| < \frac{1}{q_{n+1}} = \frac{1}{a_{n+1}q_n + q_{n-1}} \leq \frac{1}{q_n + q_{n-1}} < |q_{n-1}\alpha - p_{n-1}|;$$

b) since $\alpha$ is between $\frac{p_{n-1}}{q_{n-1}}$ and $\frac{p_n}{q_n}$ we know $\left|\alpha - \frac{p_{n-1}}{q_{n-1}}\right| + \left|\alpha - \frac{p_n}{q_n}\right| = \frac{1}{q_n q_{n-1}}$ and multiplying by $q_n q_{n-1}$ yields the result;

c) no matter how $\frac{a}{b}$, $\alpha$, $\frac{p_{n-1}}{q_{n-1}}$ are arranged on the line

$$\frac{1}{bq_{n-1}} \leq \left|\frac{p_{n-1}}{q_{n-1}} - \frac{a}{b}\right| \leq \left|\frac{p_{n-1}}{q_{n-1}} - \alpha\right| + \left|\alpha - \frac{a}{b}\right|;$$

now multiply by $bq_{n-1}$;

d) immediate since the left side is obviously less than or equal to the left side of the inequality in (b);

e) follows by a comparison of the inequalities in (c) and (d);

f) immediate from (e);

v) this is merely a restatement of (iii-e) and (iv-f).

14. Assume the contrary for $\frac{p_k}{q_k}$, $\frac{p_{k+1}}{q_{k+1}}$ and deduce $(q_k - q_{k+1})^2 \leq 0$.

15. i) If $\frac{a}{b} = \alpha$ then $\alpha' = 0$ ; otherwise

$$\left|\alpha - \frac{p_s}{q_s}\right| < \frac{1}{q_s^2} < \frac{1}{q_s q_{s-1}} = \left|\frac{p_{s-1}}{q_{s-1}} - \frac{p_s}{q_s}\right|$$

so either

(n even)

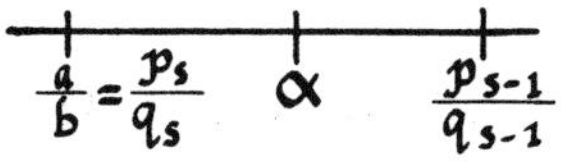


or (n odd)

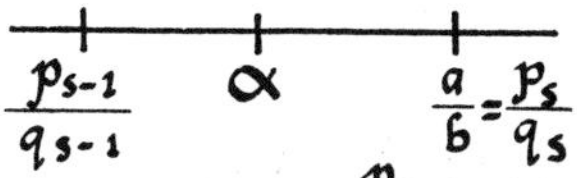


in either event $\alpha' = \dfrac{\alpha - \frac{p_s}{q_s}}{\frac{p_{s-1}}{q_{s-1}} - \alpha} \cdot \dfrac{q_s}{q_{s-1}} > 0$ ;

now, by #3 (iii),

$$[a_0, \cdots, a_s + \alpha'] = \frac{\frac{1}{\alpha'} p_s + p_{s-1}}{\frac{1}{\alpha'} q_s + q_{s-1}} = \alpha \ ;$$

ii) $$\frac{1}{\alpha'} + \frac{q_{s-1}}{q_s} = \frac{p_{s-1} - \alpha q_{s-1}}{\alpha q_s - p_s} + \frac{q_{s-1}}{q_s}$$

$$= \frac{(-1)^s}{q_s^2(\alpha - \frac{a}{b})} = \frac{1}{b^2 |\alpha - \frac{a}{b}|} > 2 \ ;$$

thus $\frac{1}{\alpha'} > 2 - \frac{q_{s-1}}{q_s} > 1$ so $\alpha' < 1$ ; also we know $\alpha' \geq 0$ by (i) ; since $\alpha' < 1$ and $\alpha = [a_0, \cdots, a_s + \alpha']$, by (i), the final conclusion follows from (i) ;

iii) this is an immediate consequence of (ii);

iv) this follows from the fact that the interval $\left(\frac{a}{b} - \frac{1}{2b^2}, \frac{a}{b} + \frac{1}{2b^2}\right)$ contains uncountably many real numbers for each of which $\frac{a}{b}$ is a convergent ;

v) (see Gessel [1972]) suppose $n$ is a Fibonacci number ; then for some positive integer $k$, $n = \frac{1}{\sqrt{5}}\left\{\left(\frac{1+\sqrt{5}}{2}\right)^k - \left(\frac{1-\sqrt{5}}{2}\right)^k\right\}$ ; thus $5n^2 = \left(\frac{1+\sqrt{5}}{2}\right)^{2k} + 2(-1)^{k+1} + \left(\frac{1-\sqrt{5}}{2}\right)^{2k}$

and, therefore,

$$\left\{\left(\frac{1+\sqrt{5}}{2}\right)^k + \left(\frac{1-\sqrt{5}}{2}\right)^k\right\}^2 = \begin{cases} 5n^2+4 \text{ when } k \text{ is even;} \\ 5n^2-4 \text{ when } k \text{ is odd;} \end{cases}$$

the converse is true for $n = 1$ so let $n \geq 2$ and suppose $5n^2 \pm 4 = m^2$ ; then $k = \frac{m+n}{2}$ is an integer since $m$ and $n$ have the same parity ; further $m^2 = 4n^2 + n^2 \pm 4 \geq 4n^2$ so $m \geq 2n$ ; substituting $2k - n$ for $m$ in our equation yields $5n^2 \pm 4 = 4k^2 - 4kn + n^2$ or

$$\frac{\pm 1}{n^2} = \left(\frac{k}{n}\right)^2 - \left(\frac{k}{n}\right) - 1 = \left(\frac{k}{n} - \tau\right)\left(\frac{k}{n} - \tau'\right),$$

where $\tau = \frac{1+\sqrt{5}}{2}$, $\tau' = \frac{1-\sqrt{5}}{2}$; hence

$\left|\frac{k}{n} - \tau\right| = \frac{1}{n^2\left|\frac{k}{n} - \tau'\right|} < \frac{1}{2n^2}$ since

$\frac{k}{n} - \tau' = \frac{1}{2}\left(\frac{m}{n}+1\right) - \tau' \geq \frac{3}{2} - \frac{1-\sqrt{5}}{2} = 1 + \frac{\sqrt{5}}{2} > 2$;

therefore, by (iii), $\frac{k}{n}$ is a convergent to $\tau$, say $\frac{k}{n} = \frac{u_{s+1}}{u_s}$, and $n = u_s$ as desired.

16. i-a) $\frac{1}{q_s q_{s-1}} = \left|\frac{p_s}{q_s} - \frac{p_{s-1}}{q_{s-1}}\right| = \left|\alpha - \frac{p_s}{q_s}\right| + \left|\alpha - \frac{p_{s-1}}{q_{s-1}}\right|$

$\geq \frac{1}{\sqrt{5}}\left(\frac{1}{q_s^2} + \frac{1}{q_{s-1}^2}\right)$;

b) each of $\frac{q_s}{q_{s-1}}$ and $\frac{q_{s-1}}{q_s}$ when substituted for $x$ in $x^2 - \sqrt{5}x + 1$ makes this quantity $\leq 0$; thus these rational numbers must lie strictly between the irrational zeros $\frac{\sqrt{5} \pm 1}{2}$ of this quadratic;

c) each of $\frac{q_{s+1}}{q_s}$ and $\frac{q_{s-1}}{q_s}$ is between $\frac{\sqrt{5}-1}{2}$ and $\frac{\sqrt{5}+1}{2}$ so their distance apart is less than $\frac{\sqrt{5}+1}{2} - \frac{\sqrt{5}-1}{2} = 1$;

d) by (c), in the contrary case, $a_{s+1}$ would be a positive integer smaller than 1;

e) this follows immediately from (d) since there exist infinitely many disjoint triplets of consecutive convergents to $\alpha$ ;

ii) by #8(iv) & #15(iii) if $\left|\frac{1+\sqrt{5}}{2} - \frac{a}{b}\right| < \frac{\beta}{\sqrt{5}\,b^2}$ then $\frac{a}{b} = \frac{u_n}{u_{n-1}}$ for some $n$ ; hence $\left|\frac{1+\sqrt{5}}{2} u_{n-1} - u_n\right| < \frac{\beta}{\sqrt{5}\,u_{n-1}}$ and, therefore, using #2(iii), $\frac{1}{\sqrt{5}}\left|\left(\frac{1+\sqrt{5}}{2}\right)^{n+1} + \left(\frac{1-\sqrt{5}}{2}\right)^{n-1} - \left(\frac{1+\sqrt{5}}{2}\right)^{n+1} + \left(\frac{1-\sqrt{5}}{2}\right)^{n+1}\right|$

$$< \frac{\beta}{\left(\frac{1+\sqrt{5}}{2}\right)^n - \left(\frac{1-\sqrt{5}}{2}\right)^n}$$

or

$$\left|\left(\frac{\sqrt{5}-1}{2}\right)^{2n-1} + \left(\frac{\sqrt{5}-1}{2}\right)^{2n+1} \pm \sqrt{5}\right| < \beta\sqrt{5} \ ;$$

since $\beta < 1$ and the left side tends to $\sqrt{5}$ this happens only for finitely many values of $n$ ;

(alternate argument)

suppose $\lambda < \frac{1}{\sqrt{5}}$ and $\left|\frac{1+\sqrt{5}}{2} - \frac{a}{b}\right| < \frac{\lambda}{b^2}$ ; then $\frac{1+\sqrt{5}}{2} - \frac{\lambda}{b^2} < \frac{a}{b} < \frac{1+\sqrt{5}}{2} + \frac{\lambda}{b^2}$ so

$$\frac{\sqrt{5}\,b}{2} - \frac{\lambda}{b} < a - \frac{b}{2} < \frac{\sqrt{5}\,b}{2} + \frac{\lambda}{b} \ ;$$

this implies $\frac{\lambda^2}{b^2} - \sqrt{5}\lambda < a^2 - ab - b^2 < \frac{\lambda^2}{b^2} + \sqrt{5}\lambda$ and, therefore, for large enough $b$,

$-1 < a^2 - ab - b^2 < 1$ ; hence $a^2 - ab - b^2 = 0$ and $\frac{a}{b} = \frac{1 \pm \sqrt{5}}{2}$ which is impossible ;

iii-a) as in (i-b), each of $\frac{q_{s+1}}{q_s}$ and $\frac{q_{s-1}}{q_s}$ lies in $\left(\frac{\sqrt{m^2+4}-m}{2}, \frac{\sqrt{m^2+4}+m}{2}\right)$ ; hence $a_{s+1} = \frac{q_{s+1}}{q_s} - \frac{q_{s-1}}{q_s} < m$ ;

b) if no such $s_0$ exists then there are infinitely many disjoint triples $s-1, s, s+1$ for which $a_{s+1} \geq m$ ; consequently by (a) there are infinitely many convergents satisfying $\left|\alpha - \frac{p_n}{q_n}\right| < \frac{1}{q_n^2 \sqrt{m^2+4}}$ ;

c) put $m = 2$ in (b) ;

iv-a) from #3 (vii) we find $q_n|q_n\alpha - p_n| < \frac{q_n}{q_{n+1}} < 1$ and, therefore, $\nu(\alpha) \leq 1$ ;

b) by Hurwitz' theorem $q_n|q_n\alpha - p_n| < \frac{1}{\sqrt{5}}$ infinitely often ; consequently $\nu(\alpha) \leq \frac{1}{\sqrt{5}}$ ;

c, d) same argument as in (b) ;

e) by (d) ;

v-a) $q_n|q_n\alpha - p_n| = q_n^2\left|\alpha - \frac{p_n}{q_n}\right|$

$= q_n^2\left|\frac{\alpha_n p_n + p_{n-1}}{\alpha_n q_n + q_{n-1}} - \frac{p_n}{q_n}\right| = \frac{1}{\alpha_n + \frac{q_{n-1}}{q_n}}$ ;

b) #3(i-b) gives the equality and the 1st inequality is a consequence of #3(xiii-c); finally, the result $q_n|q_n\alpha - p_n| > \frac{1}{2\sqrt{3}} = \frac{1}{\sqrt{12}}$ shows $\nu(\alpha) \geq \frac{1}{\sqrt{12}}$ ;

c) either infinitely many $a_j$ are $\geq 3$ or all but a finite number are $\leq 2$ ; in the 1st case $\nu(\alpha) \leq \frac{1}{\sqrt{13}}$ by (iv-e) and in the 2nd case $\nu(\alpha) \geq \frac{1}{\sqrt{12}}$ by (b).

17. i) Let $\alpha = [a_0, \cdots, a_s, \overline{a_{s+1}, \cdots, a_{s+t}}]$,

$\alpha_s = [\overline{a_{s+1}, \cdots, a_{s+t}}]$ ; then

$$\alpha = [a_0, \cdots, a_s, \alpha_s] = [a_0, \cdots, a_{s+t}, \alpha_s]$$

$$= \frac{\alpha_s p_s + p_{s-1}}{\alpha_s q_s + q_{s-1}} = \frac{\alpha_s p_{s+t} + p_{s+t-1}}{\alpha_s q_{s+t} + q_{s+t-1}} \;;$$

from the 4th equality we see $\alpha_s$ satisfies a quadratic equation and, since its scf expansion is infinite, must be irrational;

hence $\alpha_s$ and, therefore also, $\alpha$ is a quadratic irrational;

ii-a) by (i):

b, c) $\alpha = [a_0, \cdots, a_n, \alpha_n] = \frac{\alpha_n p_n + p_{n-1}}{\alpha_n q_n + q_{n-1}}$ so

$0 = A\alpha^2 + B\alpha + C = A(\alpha_n p_n + p_{n-1})^2 + B(\alpha_n p_n + p_{n-1})(\alpha_n q_n + q_{n-1}) + C(\alpha_n q_n + q_{n-1})^2$
$= q_n^2 f\left(\frac{p_n}{q_n}\right)\alpha_n^2 + (2A p_{n-1} p_n + B p_n q_{n-1} + B p_{n-1} q_n + 2C q_{n-1} q_n)\alpha_n + q_{n-1}^2 f\left(\frac{p_{n-1}}{q_{n-1}}\right)$;

direct calculation shows

$$B_n^2 - 4A_n C_n = B^2 - 4AC;$$

the expressions for $A_n$, $C_n$ are clear; since $f(\alpha) = 0$ and $\alpha$ is between $\frac{p_{n-1}}{q_{n-1}}$ and $\frac{p_n}{q_n}$ then the values of $f$ at these points are of opposite sign since the other zero of the quadratic does not lie between $\frac{p_{n-1}}{q_{n-1}}$ and $\frac{p_n}{q_n}$; this shows $A_n C_n < 0$;

d) since $B_n^2 - 4A_nC_n = B_n^2 + |4A_nC_n|$ is a constant it is clear that only finitely many such triples exist ;

e) by the Dirichlet box principle such $k$ and $n$ exist for which $A_{n+k} = A_k$, $B_{n+k} = B_k$, $C_{n+k} = C_k$ ; thus $\alpha_k = \alpha_{n+k}$, and this means $\alpha$ is periodic ;

iii-a) if $\alpha$ is purely periodic then $a_0 > 0$ so $\alpha > 1$ ; further, $\alpha = [a_0, \cdots, a_s + \frac{1}{\alpha}] = \frac{\alpha p_s + p_{s-1}}{\alpha q_s + q_{s-1}}$ so $q_s\alpha^2 + (q_{s-1} - p_s)\alpha - p_{s-1} = 0$; the roots are $\frac{P \pm \sqrt{D}}{Q}$, where $P = p_s - q_{s-1}$, $D = (q_{s-1} - p_s)^2 + 4q_s p_{s-1}$, $Q = 2q_s$ ; clearly $\sqrt{D} > P$ so $\alpha = \frac{P+\sqrt{D}}{Q}$ and $\alpha' = \frac{P-\sqrt{D}}{Q} < 0$ is the other zero ; thus

$$\alpha\alpha' = -\frac{p_{s-1}}{q_s} = -\frac{p_{s-1}}{q_{s-1}}\,\frac{q_{s-1}}{q_s} > -\alpha\,\frac{q_{s-1}}{q_s} > -\alpha$$

for s odd ; hence $\alpha' > -1$ ( s even is similar ) ;

(alternative) at 0 and $-1$ the quadratic equals $-p_{s-1} < 0$ and $q_s - q_{s-1} + p_s - p_{s-1} > 0$ ; hence $\alpha'$, the other zero, lies between 0 and $-1$ ;

b) from $-1 < \alpha' < 0 < 1 < \alpha$ we deduce

$$0 > -\frac{1}{1+a_0} > \frac{1}{\alpha' - a_0} = \alpha'_0 > -\frac{1}{a_0} \geq -1 \; ;$$

c) from (b) by iteration ;

d) $\alpha_{s-1} - \alpha_{s+t-1} =$

$$[a_s, \overline{a_{s+1}, \cdots, a_{s+t}}] - [a_{s+t}, \overline{a_{s+1}, \cdots, a_{s+t}}] =$$

$$a_s + \frac{1}{[\overline{a_{s+1}, \cdots, a_{s+t}}]} - a_{s+t} - \frac{1}{[\overline{a_{s+1}, \cdots, a_{s+t}}]} = a_s - a_{s+t} \; ;$$

the second equation follows from the first since the conjugate operator is additive ;

e) since each of $\alpha'_{s-1}$ and $\alpha'_{s+t-1}$ are strictly between $-1$ and 0 their difference must lie strictly between $-1$ and 1 ; but then $a_s = a_{s+t}$ contrary to our assumption ;

f) this is a direct consequence of (a) & (e) ;

iv-a) since $\frac{1+\sqrt{13}}{3} > 1$ and $-1 < \frac{1-\sqrt{13}}{3} < 0$ the scf expansion, by (iii-f), is purely periodic; since $\frac{1-\sqrt{13}}{3} < 1$ and $\frac{2+\sqrt{3}}{4} < 1$ neither of these is reduced, hence neither has a purely periodic scf expansion;

b) $$\frac{1+\sqrt{13}}{3} = 1 + \frac{\sqrt{13}-2}{3} = 1 + \cfrac{1}{1 + \cfrac{1}{\cfrac{\sqrt{13}+1}{4}}}$$

$$= 1 + \cfrac{1}{1 + \cfrac{1}{1 + \cfrac{1}{\sqrt{13}+3}}} = 1 + \cfrac{1}{1 + \cfrac{1}{1 + \cfrac{1}{6 + \cfrac{1}{\cfrac{\sqrt{13}+3}{4}}}}}$$

$$= 1 + \cfrac{1}{1 + \cfrac{1}{1 + \cfrac{1}{6 + \cfrac{1}{1 + \cfrac{1}{\cfrac{1+\sqrt{13}}{3}}}}}} = [\overline{1,1,1,6,1}];$$

noting that $-\left(\frac{1-\sqrt{13}}{3}\right)^{-1} = \frac{\sqrt{13}+1}{4}$ the above calculation tells us

$$-\left(\frac{1-\sqrt{13}}{3}\right)^{-1} = [\overline{1,6,1,1,1}];$$

c) let $\alpha_{-1} = \alpha = [\overline{a_0, \cdots, a_s}]$ ; then for $0 \le j \le s$ we have $\alpha_{j-1} = a_j + \frac{1}{\alpha_j}$ ; in the particular case $j = s$ we have

$$\alpha_{s-1} = a_s + \frac{1}{\alpha_s} = a_s + \frac{1}{\alpha} ;$$

consequently $-(\alpha_j')^{-1} = a_j + \frac{1}{-(\alpha_{j-1}')^{-1}}$ for $0 \le j \le s$, and putting $\beta_j = -(\alpha_j')^{-1}$ yields

$$\beta_j = a_j + \frac{1}{\beta_{j-1}} \quad \text{for } 0 \le j \le s ;$$

hence

$$-(\alpha')^{-1} = \beta_s = a_s + \frac{1}{\beta_{s-1}} = a_s + \cfrac{1}{a_{s-1} + \cfrac{1}{\beta_{s-2}}}$$

$$= a_s + \cfrac{1}{a_{s-1} + \cfrac{1}{a_{s-2} + \cfrac{1}{a_{s-3} + \cdots + \cfrac{1}{a_0 + \cfrac{1}{\beta_{-1}}}}}}$$

$$= [\overline{a_s, \cdots, a_0}] ;$$

v) put $a_0 = [\alpha]$ and let $\beta = \frac{1}{\alpha - a_0}$ ; since $0 < \alpha - a_0 < 1$ , $\beta > 1$ ; further,

$$-1 < \beta' = \frac{1}{\alpha' - a_0} = \frac{1}{-\frac{\sqrt{D}}{Q} - a_0} < 0$$

so $\beta$ is reduced and, therefore, has a purely periodic scf expansion, say $\beta = [\overline{a_1, \cdots, a_k, b}]$;

now $-(\beta')^{-1} = [\overline{b, a_k, \cdots, a_1}]$ while

$$-(\beta')^{-1} = a_0 + (-\alpha') = a_0 + \alpha \text{; thus}$$

$b = [-(\beta')^{-1}] = [a_0 + \alpha] = a_0 + [\alpha] = 2a_0$ ;

therefore $\alpha = a_0 + \frac{1}{\beta} = [a_0, \overline{a_1, \cdots, a_k, 2a_0}]$ ;

vi-a) clearly $\frac{1}{\sqrt{D} - a_0} > 1$ ; further, since $-1 < \frac{1}{-\sqrt{D} - a_0} < 0$, the number $\frac{1}{\sqrt{D} - a_0}$ is reduced ;

b) $\sqrt{D} + a_0 = -\left(\frac{1}{-\sqrt{D} - a_0}\right)^{-1}$ so its period is the reverse of that of $\frac{1}{\sqrt{D} - a_0}$ ; also $\sqrt{D} + a_0 > 1$ and $-1 < -\sqrt{D} + a_0 < 0$ so $\sqrt{D} + a_0$ is reduced ;

c) by (v), $\sqrt{D} = [a_0, \overline{a_1, \cdots, a_k, 2a_0}]$ so $\frac{1}{\sqrt{D} - a_0} = [\overline{a_1, \cdots, a_k, 2a_0}]$ and, therefore, by (iii-f) $\frac{1}{\sqrt{D} - a_0}$ is reduced ; hence by (b)

$$\sqrt{D} + a_0 = [\overline{2a_0, a_k, \cdots, a_1}]$$

but $\sqrt{D} + a_0 = [2a_0, \overline{a_1, \cdots, a_k, 2a_0}] = [\overline{2a_0, a_1, \cdots, a_k}]$ ;

consequently, by uniqueness,

$$a_1 = a_k \ , a_2 = a_{k-1} \ , \cdots$$

and the desired result follows ;

vii-a) $x_0^2 - Dy_0^2 = 1$ implies $\left(\frac{x_0}{y_0}\right)^2 = D + \frac{1}{y_0^2}$;
thus $\frac{x_0}{y_0} = \sqrt{D + \frac{1}{y_0^2}} > \sqrt{D}$ and $\frac{x_0}{y_0} + \sqrt{D} > 2\sqrt{D} > 2$;
this means, since $(x_0 - \sqrt{D}y_0)(x_0 + \sqrt{D}y_0) = 1$, that

$$\left|\frac{x_0}{y_0} - \sqrt{D}\right| = \frac{1}{y_0^2\left|\frac{x_0}{y_0} + \sqrt{D}\right|} < \frac{1}{2y_0^2}$$

and, by #15 (iii), $\frac{x_0}{y_0}$ is a convergent to $\sqrt{D}$;

6-A) $\sqrt{D} = [a_0, \cdots, a_s + \frac{1}{\alpha_s}] = \frac{\alpha_s p_s + p_{s-1}}{\alpha_s q_s + q_{s-1}}$ so

$$\alpha_s^2(p_s^2 - Dq_s^2) + 2\alpha_s(p_{s-1}p_s - Dq_{s-1}q_s) + (p_{s-1}^2 - Dq_{s-1}^2) = 0$$ and

$A_s = p_s^2 - Dq_s^2 = x_0^2 - Dy_0^2 = 1$;
$C_s = p_{s-1}^2 - Dq_{s-1}^2$;
$B_s^2 = 4(p_{s-1}^2 - Dq_{s-1}^2 + D(p_{s-1}q_s - p_s q_{s-1})^2) = 4(C_s + D)$;

B) $\alpha_s = \frac{-B_s \pm \sqrt{B_s^2 - 4C_s}}{2} = \frac{-B_s \pm \sqrt{4D}}{2} = -\frac{1}{2}B_s \pm \sqrt{D}$;

note now that since $p_s^2 - Dq_s^2 = 1 > 0$ we must have $C_s = p_{s-1}^2 - Dq_{s-1}^2 < 0$ so $-\frac{1}{2}B_s - \sqrt{D} = \pm\sqrt{C_s + D} - \sqrt{D} < 0$; since $\alpha_s > 0$ we then have $\alpha_s = -\frac{1}{2}B_s + \sqrt{D}$;

C) from (B), $\sqrt{D} = \frac{1}{2}B_s + \alpha_s$

$= [\frac{1}{2}B_s + a_{s+1}, \overline{a_{s+2}, \cdots, a_{s+k+1}}]$ and since

$\sqrt{D} = [a_0, \overline{a_1, \cdots, a_{k-1}, 2a_0}]$ we must have

$$a_0 = \frac{1}{2}B_s + a_{s+1},\ 2a_0 = a_{s+k+1};$$

since $a_{s+1} = a_{s+k+1}$ we conclude $-\frac{1}{2}B_s = a_0$ and $\alpha_s = a_0 + \sqrt{D}$;

D) $\alpha_s = [a_{s+1}, a_{s+2}, \cdots] = [2a_0, a_1, a_2, \cdots]$ and the uniqueness guarantees $a_j = a_{s+j+1}$ for $j \geq 1$;

E) let $s = qk + r$, $0 \leq r < k$; then

$$a_j = a_{s+j+1} = a_{qk+r+j+1} = a_{r+j+1}$$

so the minimum period $k$ of the scf expansion of $\sqrt{D}$ is $\leq r+1$; but $r+1 \leq k$ so $r+1 = k$ and $r = k-1$; thus $s = qk + k - 1 = (q+1)k - 1 \equiv -1 \pmod{k}$;

c) this follows immediately from (a) and (b-E);

d) $\sqrt{D} = \frac{(a_0+\sqrt{D})p_{tk-1} + p_{tk-2}}{(a_0+\sqrt{D})q_{tk-1} + q_{tk-2}}$ and multiplication by the denominator on the right and equating rational and irrational parts yields the expressions for $p_{tk-1}$ and $Dq_{tk-1}$;

now multiplying the 1st by $p_{tk-1}$, the 2nd by $q_{tk-1}$ and subtracting yields the result;

e) this is immediate from (d);

f-A) $\sqrt{22} = [4, \overline{1, 2, 4, 2, 1, 8}]$

| | | | | | | | | |
|---|---|---|---|---|---|---|---|---|
| | | 4 | 1 | 2 | 4 | 2 | 1 | 8 |
| 0 | 1 | 4 | 5 | 14 | 61 | 136 | 197 | |
| 1 | 0 | 1 | 1 | 3 | 13 | 29 | 42 | |

$197^2 - 22 \cdot 42^2 = 1$;

B) $\sqrt{13} = [3, \overline{1, 1, 1, 1, 6}]$

| | | | | | | | | | | | |
|---|---|---|---|---|---|---|---|---|---|---|---|
| | | 3 | 1 | 1 | 1 | 1 | 6 | 1 | 1 | 1 | 1 |
| 0 | 1 | 3 | 4 | 7 | 11 | 18 | 119 | 137 | 256 | 393 | 649 |
| 1 | 0 | 1 | 1 | 2 | 3 | 5 | 33 | 38 | 71 | 109 | 180 |

$649^2 - 13 \cdot 180^2 = 1$;

C) $\sqrt{33} = [5, \overline{1, 2, 1, 10}]$

| | | | | | |
|---|---|---|---|---|---|
| | | 5 | 1 | 2 | 1 | 10 |
| 0 | 1 | 5 | 6 | 17 | 23 | |
| 1 | 0 | 1 | 1 | 3 | 4 | |

$23^2 - 33 \cdot 4^2 = 1$;

g) see, for example, the number theory book by Niven and Zuckerman [1966].

18. i) $b = q_n = E(a_1, \cdots, a_n) \geq E_n$ ;

ii) by induction

$\tau < E_2, \quad \tau^2 = 1 + \tau < E_1 + E_2 = E_3$ ;

$\tau^n = \tau^{n-2} + \tau^{n-1} < E_{n-1} + E_n = E_{n+1}$ for $n \geq 2$ ;

iii) $\left(\frac{1+\sqrt{5}}{2}\right)^n < E_n \leq b$ so the result follows by taking logarithms ;

iv) $5 \log \frac{1+\sqrt{5}}{2} > 5(.20898) > 1$ and $\log b < t$ ; thus $n < \frac{\log b}{\log \frac{1+\sqrt{5}}{2}} < 5t$ ;

v) immediate from the above when one observes that $n$ is just the number of divisions required ;

vi) see the references following I #6 .

19. *i*) The square of the distance between the centers of $C(\frac{a}{b})$ and $C(\frac{c}{d})$ is :

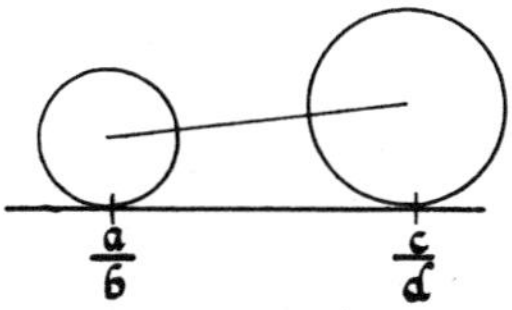


$$\left(\frac{c}{d}-\frac{a}{b}\right)^2+\left(\frac{1}{2d^2}-\frac{1}{2b^2}\right)^2=\frac{1}{4b^4}+\frac{2(bc-ad)^2-1}{2b^2d^2}+\frac{1}{4d^4}$$

$$\geq \frac{1}{4b^4}+\frac{1}{2b^2d^2}+\frac{1}{4d^4}=\left(\frac{1}{2b^2}+\frac{1}{2d^2}\right)^2;$$

thus the circles are disjoint unless $bc-ad=\pm 1$; precisely when $bc-ad=\pm 1$ the circles are tangent and, by #9 (*iii*-*d*) the fractions $\frac{a}{b}, \frac{c}{d}$ are neighboring in $\Phi_m$, where $m = \max\{b, d\}$ ;

*ii*) the point of tangency has :

$$x\text{-coordinate} = \frac{a}{b}+\frac{\frac{1}{2b^2}}{\frac{1}{2b^2}+\frac{1}{2d^2}}\left(\frac{c}{d}-\frac{a}{b}\right)=\frac{ab+cd}{b^2+d^2};$$

$$y\text{-coordinate} = \frac{1}{2b^2}+\frac{\frac{1}{2b^2}}{\frac{1}{2b^2}+\frac{1}{2d^2}}\left(\frac{1}{2d^2}-\frac{1}{2b^2}\right)=\frac{1}{b^2+d^2};$$

*iii*) this is a consequence of each of #10 (*i*) and #14 ;

iv-a) all points of tangency are between $\frac{a}{b}$ and $\frac{c}{d}$ and, consequently, so is the entire curvilinear triangle; thus $\alpha$ must lie between $\frac{a}{b}$ and $\frac{c}{d}$;

b) by (i) and #9 (vi-b);

c) each of $b$ and $d$ is $< b+d = f$;

d) $\frac{a'}{b'} = 1 - \frac{c}{d}$, $\frac{e'}{f'} = 1 - \frac{e}{f}$, $\frac{c'}{d'} = 1 - \frac{a}{b}$, and $b' = d$, $f' = f$, $d' = b$;

e) by proof of (d).

20. i)

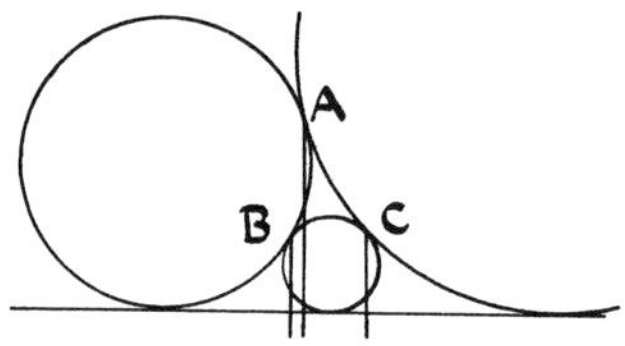


$B \leq A < C$

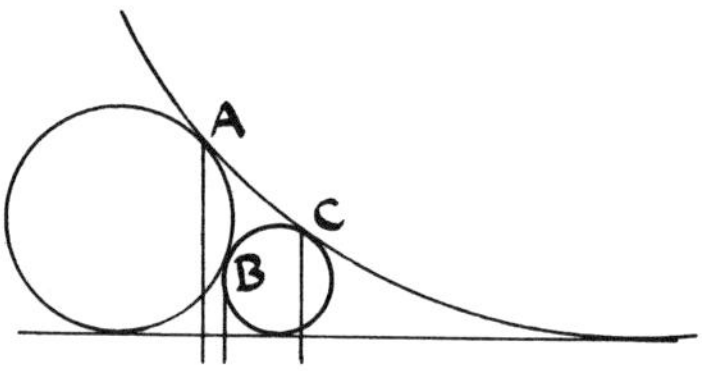


$A \leq B < C$

ii) A and C are points on $C(\frac{c}{d})$ and since $f > b > d$, the point C is farther to the right than is A;

iii-a, b) $B - A = \frac{ab+ef}{b^2+f^2} - \frac{ab+cd}{b^2+d^2}$

$= \frac{b^2 - d^2 - bd}{(b^2+(b+d)^2)(b^2+d^2)} = \frac{t^2 - t - 1}{d^2(t^2+(t+1)^2)(t^2+1)}$,

where $t = \frac{b}{d}$; now $x^2 - x - 1$ has zeros at $\frac{1 \pm \sqrt{5}}{2}$ and $t^2 - t - 1$ is negative precisely between these zeros; thus if $\frac{b}{d} < \frac{1+\sqrt{5}}{2}$ then $B - A < 0$ and if $\frac{b}{d} > \frac{1+\sqrt{5}}{2}$ then $B - A > 0$;

iv-a) $\left|\alpha - \frac{c}{d}\right| < \left|A - \frac{c}{d}\right| = \left|\frac{ab+cd}{b^2+d^2} - \frac{c}{d}\right|$

$= \frac{b}{d}\left|\frac{ad-bc}{b^2+d^2}\right| = \frac{1}{d^2} \cdot \frac{b/d}{(b/d)^2+1} < \frac{1}{\sqrt{5}\,d^2}$

since (putting $t = \frac{b}{d}$) $\frac{t}{t^2+1} < \frac{1}{\sqrt{5}}$ when $t > \frac{1+\sqrt{5}}{2}$;

b) $\left|\alpha - \frac{e}{f}\right| = \left|B - \frac{e}{f}\right| = \left|\frac{ab+ef}{b^2+f^2} - \frac{e}{f}\right|$

$= \frac{b}{f} \cdot \frac{\frac{1}{d^2}}{\left(\frac{b}{d}\right)^2 + \left(\frac{b+d}{d}\right)^2} = \frac{1}{f^2} \cdot \frac{\frac{b}{d}\left(\frac{b}{d}+1\right)}{\left(\frac{b}{d}\right)^2 + \left(\frac{b}{d}+1\right)^2} < \frac{1}{f^2\sqrt{5}}$

since (putting $t = \frac{b}{d}$) $\frac{t(t+1)}{t^2+(t+1)^2} < \frac{1}{\sqrt{5}}$ when $t < \frac{1+\sqrt{5}}{2}$;

v) immediate from (iv) and #19(iii).

21. i) The function $f(x,y) = y - \alpha x$ is continuous, vanishes only on $\mathcal{L}$ and is negative at $(1,0)$ and positive at $(0,1)$;

ii) $\overline{P_{n-2}P_n} = (q_n - q_{n-2},\ p_n - p_{n-2})$
$= (a_n q_{n-1},\ a_n p_{n-1}) = a_n(q_{n-1},\ p_{n-1}) = a_n \overline{OP_{n-1}}$;

iii) the area of $OP_{n-1}P_n$ is the absolute value of $\frac{1}{2}\begin{vmatrix} q_n & p_n & 1 \\ q_{n-1} & p_{n-1} & 1 \\ 0 & 0 & 1 \end{vmatrix} = \pm\frac{1}{2}$; if $(s,t)$ were a lattice point in or on the triangle other than the vertices then the area of the triangle $OP_{n-1}P_n$ would be $\geq$ the sum of the absolute values of the following quantities $\frac{1}{2}\begin{vmatrix} q_{n-1} & p_{n-1} & 1 \\ s & t & 1 \\ 0 & 0 & 1 \end{vmatrix}$, $\frac{1}{2}\begin{vmatrix} q_n & p_n & 1 \\ s & t & 1 \\ 0 & 0 & 1 \end{vmatrix}$ since $(q_{n-1}, p_{n-1}) = (q_n, p_n) = 1$ neither of these is $0$ and each is numerically $\geq \frac{1}{2}$;

iv) immediate from (iii).

22. i-a) $\dfrac{a_{ns}x^{2s}}{a_{n(s-1)}x^{2s-2}} = \dfrac{x^2}{2(2n+2s-1)} < 1$ for large $n$;

b) $$\text{LHS} = \sum_{s=0}^{\infty}\left\{\frac{(n+s)!}{s!\,(2n+2s)!} - \frac{(4n+2)(n+1+s)!}{s!\,(2n+2+2s)!}\right\}x^{2s}$$

$$= 2\sum_{s=1}^{\infty}\frac{(n+s)!}{(s-1)!\,(2n+2s+1)!}x^{2s} = 2x^2\sum_{s=0}^{\infty}\frac{(n+s+1)!}{s!\,(2n+2s+3)!}x^{2s}$$

$$= 4x^2\sum_{s=0}^{\infty}\frac{(n+2+s)!}{s!\,(2n+2s+4)!}x^{2s} = \text{RHS};$$

c) $$f_0(x) = \sum_{s=0}^{\infty}\frac{x^{2s}}{(2s)!} = \frac{e^x+e^{-x}}{2},$$

$$f_1(x) = \frac{1}{2}\sum_{s=0}^{\infty}\frac{x^{2s}}{(2s+1)!} = \frac{1}{2x}\sum_{s=0}^{\infty}\frac{x^{2s+1}}{(2s+1)!} = \frac{1}{2x}\,\frac{e^x-e^{-x}}{2}$$

and, therefore, $\dfrac{f_0(x)}{f_1(x)} = 2x\,\dfrac{e^{2x}+1}{e^{2x}-1}$;

d) $$\frac{e^{2x}+1}{e^{2x}-1} = \frac{1}{2x}\,\frac{f_0(x)}{f_1(x)} = \frac{1}{2x}\left(2+4x^2\frac{f_2(x)}{f_1(x)}\right)$$

$$= \frac{1}{x} + \cfrac{1}{\cfrac{3}{x} + 2x\cfrac{f_3(x)}{f_2(x)}} = \left[\frac{1}{x}, \frac{3}{x} + 2x\frac{f_3(x)}{f_2(x)}\right]$$

$$= \cdots = \left[\frac{1}{x}, \frac{3}{x}, \cdots, \frac{2n-1}{x} + 2x\frac{f_{n+1}(x)}{f_n(x)}\right]$$

$$= \left[\frac{1}{x}, \frac{3}{x}, \cdots, \frac{2n-1}{x} + \cfrac{2x}{4n+2+4x^2\cfrac{f_{n+2}(x)}{f_{n+1}(x)}}\right]$$

$$= \left[\frac{1}{x}, \frac{3}{x}, \cdots, \frac{2n-1}{x}, \frac{2n+1}{x} + 2x\frac{f_{n+2}(x)}{f_{n+1}(x)}\right]$$

$$\to \left[\frac{1}{x}, \frac{3}{x}, \cdots, \frac{2n+1}{x}, \cdots\right],$$

which converges by Seidel's theorem #7(iii);

ii-a) $p_{3n+1} = p_{3n} + p_{3n-1} = 2p_{3n-1} + p_{3n-2}$

$= (4n+1)p_{3n-2} + 2p_{3n-3} = (4n+2)p_{3n-2} + 2p_{3n-3} - p_{3n-2}$

$= (4n+2)p_{3n-2} + p_{3n-3} + p_{3n-4} + p_{3n-5} - p_{3n-2}$

$= (4n+2)p_{3n-2} + p_{3n-5}$

and the same equations are true with all p's replaced by q's ;

b) $\frac{e+1}{e-1} = [2, 6, 10, 14, 18, \cdots]$ and the first few convergents for $\frac{P_n}{Q_n}$ are $\frac{2}{1}, \frac{13}{6}, \frac{132}{61}$ ; for $n = 0, 1, 2$ the quantity $\frac{\frac{1}{2}(p_{3n+1} + q_{3n+1})}{\frac{1}{2}(p_{3n+1} - q_{3n+1})}$ is $\frac{\frac{1}{2}\cdot 4}{\frac{1}{2}\cdot 2}, \frac{\frac{1}{2}\cdot 26}{\frac{1}{2}\cdot 12}, \frac{\frac{1}{2}\cdot 264}{\frac{1}{2}\cdot 122}$ ; thus the result is true for $n = 0, 1, 2$ ; now assuming the result true up to and including $n-1$ we have

$$\frac{p_{3n+1} + q_{3n+1}}{p_{3n+1} - q_{3n+1}} = \frac{(4n+2)(p_{3n-2} + q_{3n-2}) + (p_{3n-5} + q_{3n-5})}{(4n+2)(p_{3n-2} - q_{3n-2}) + (p_{3n-5} - q_{3n-5})}$$

$$= \frac{(4n+2)P_{n-1} + P_{n-2}}{(4n+2)Q_{n-1} + Q_{n-2}} = \frac{P_n}{Q_n} ;$$

we now use the fact that

$$\left(\tfrac{1}{2}(p_{3n+1} + q_{3n+1}), \tfrac{1}{2}(p_{3n+1} - q_{3n+1})\right) = 1$$

for all $n$ ;

c) put $t_n = \frac{p_{3n+1}}{q_{3n+1}}$, so $t_n \to \alpha$ and $\frac{t_n+1}{t_n-1} - \frac{e+1}{e-1} = \frac{2(e-t_n)}{(t_n-1)(e-1)} \to 0$; since $\frac{t_n+1}{t_n-1} = \frac{1+1/t_n}{1-1/t_n} \not\to 1$ we see $\{t_n\}$ is bounded; hence from $\frac{2(e-t_n)}{(t_n-1)(e-1)} \to 0$ we conclude that $t_n \to e$, and uniqueness of limits guarantees

$$\alpha = e\ ;$$

iii-a) put $x = \frac{\sqrt{2}}{2}$ in (i-d);

b) $\sqrt{2}\,[\sqrt{2}, 3\sqrt{2}, 5\sqrt{2}, 7\sqrt{2}, \cdots]$

$$= \sqrt{2}\left(\sqrt{2} + \cfrac{1}{3\sqrt{2} + \cfrac{1}{5\sqrt{2} + \cfrac{1}{7\sqrt{2} \cdots}}}\right)$$

$$= 2 + \cfrac{1}{3 + \cfrac{1}{10 + \cfrac{\sqrt{2}}{7\sqrt{2} + \cfrac{1}{9\sqrt{2} + \cdots}}}}$$

$$= 2 + \cfrac{1}{3 + \cfrac{1}{10 + \cfrac{1}{7 + \cfrac{1}{18 + \cdots}}}}$$

$$= [2, 3, 10, 7, 18, \cdots],$$

and this is non-periodic since the partial quotients are unbounded;

c) if $e^{\sqrt{2}}$ were rational then $\sqrt{2}\,\frac{e^{\sqrt{2}}+1}{e^{\sqrt{2}}-1}$ would be a quadratic irrational and would have a periodic scf expansion in violation of (6).

23. I-i) True for $n=0$; suppose true for $n-1$; then:

$$\begin{pmatrix} a_0 & 1 \\ 1 & 0 \end{pmatrix} \cdots \begin{pmatrix} a_n & 1 \\ 1 & 0 \end{pmatrix} = \begin{pmatrix} p_{n-1} & p_{n-2} \\ q_{n-1} & q_{n-2} \end{pmatrix} \begin{pmatrix} a_n & 1 \\ 1 & 0 \end{pmatrix} =$$

$$\begin{pmatrix} a_n p_{n-1} + p_{n-2} & p_{n-1} \\ a_n q_{n-1} + q_{n-2} & q_{n-1} \end{pmatrix} = \begin{pmatrix} p_n & p_{n-1} \\ q_n & q_{n-1} \end{pmatrix};$$

ii) $\frac{p_{n-1}}{q_{n-1}} \to \alpha$, $\frac{p_n}{q_n} \to \alpha$;

iii) $A_1 \cdots A_n = \begin{pmatrix} k_1 & k_2 \\ k_3 & k_4 \end{pmatrix}$ and $\frac{k_1}{k_3} \to \alpha$;

now $A\,A_1 \cdots A_n = \begin{pmatrix} a & b \\ c & d \end{pmatrix} \begin{pmatrix} k_1 & k_2 \\ k_3 & k_4 \end{pmatrix} =$

$$\begin{pmatrix} ak_1 + bk_3 & ak_2 + bk_4 \\ ck_1 + dk_3 & ck_2 + dk_4 \end{pmatrix} \quad \text{and}$$

$$\frac{ak_1 + bk_3}{ck_1 + dk_3} = \frac{a\frac{k_1}{k_3} + b}{c\frac{k_1}{k_3} + d} \to \frac{a\alpha + b}{c\alpha + d};$$

iv) if $A_1 \cdots A_n = \begin{pmatrix} e & f \\ g & h \end{pmatrix}$ then

$$(k_1 A_1) \cdots (k_n A_n) = \begin{pmatrix} k_1 \cdots k_n e & k_1 \cdots k_n f \\ k_1 \cdots k_n g & k_1 \cdots k_n h \end{pmatrix}$$

and $\dfrac{k_1 \cdots k_n e}{k_1 \cdots k_n g} = \dfrac{e}{g}$ ;

v) subsequences of convergent sequences converge and to the same limit ;

vi) $|K_1(P_n) - K_2(P_n)| = \left|\dfrac{p_n}{q_n} - \dfrac{r_n}{s_n}\right| = \dfrac{|p_n s_n - q_n r_n|}{|q_n s_n|}$

$$= \frac{|\det P_n|}{|q_n s_n|} = \frac{\prod_{r=1}^{n} |\det A_r|}{|q_n s_n|} ;$$

vii-a) from (vi), $\dfrac{1}{|q_n s_n|} = |K_1(P_n) - K_2(P_n)| \to 0$;

b) if either a or c is zero then $K_1(P_n B) = K_1(P_n) \to \alpha$ ; if neither a nor c is 0 then $K_1(P_n B) = \dfrac{a p_n + c r_n}{a q_n + c s_n}$ is a Farey mediant of $\dfrac{p_n}{q_n}$, $\dfrac{r_n}{s_n}$ and, therefore, lies between them; since these two fractions tend to $\alpha$ so also does $K_1(P_n B)$ ;

c) same argument as in (b) ;

d) immediate from (b) and (c) ;

viii) since $BC_1C_2 \cdots C_n = A_1 \cdots A_n B$, this follows immediately from (vii);

ix) verify by direct multiplication of the matrices involved;

x) when $d > 1$ we may use the 1st part of (ix) with $x$ chosen so that $0 < c - xd < d$ (since $ad - bc = \pm 1$ it is clear that $d$ does not divide $c$); the matrix $\begin{pmatrix} b & a - xb \\ d & c - xd \end{pmatrix}$ has determinant $\pm 1$ so we can again apply the same part if $c - xd > 1$; if $c - xd = 1$ we may use the 2nd part of (ix) getting the product of 2 matrices when $a - bc = 1$ and the product of 3 matrices when $bc - a = 1$.

II i) $A_1 = \begin{pmatrix} 1+x & 1 \\ 1 & 1-x \end{pmatrix}$ so

$h_1(x) = 1 = g_1(-x)$, $k_1(x) = 1 - x = 1 + (-x) = f_1(-x)$;

suppose true up to $n$, then

$$\prod_{m=1}^{n+1} A_m = \begin{pmatrix} f_n(x) & g_n(x) \\ g_n(-x) & f_n(-x) \end{pmatrix} \begin{pmatrix} 2n+1+x & 2n+1 \\ 2n+1 & 2n+1-x \end{pmatrix}$$

so

$$f_{n+1}(x) = (2n+1+x) f_n(x) + (2n+1) g_n(x)$$

$$g_{n+1}(x) = (2n+1) f_n(x) + (2n+1-x) g_n(x)$$

$$h_{n+1}(x) = (2n+1+x) g_n(-x) + (2n+1) f_n(-x)$$

$$k_{n+1}(x) = (2n+1) g_n(-x) + (2n+1-x) f_n(-x);$$

clearly $f_{n+1}(-x) = k_{n+1}(x)$ and $g_{n+1}(-x) = h_{n+1}(x)$;

ii) a) $f_{n+1}(x) = (2n+1)(f_n(x) + g_n(x)) + x f_n(x)$

$$= \sum_{k=0}^{n} \frac{(2n-k)!}{(n-k)!\,k!}(2n+1)x^k + \sum_{k=1}^{n+1} \frac{n(2n-k)!}{(n-k+1)!\,(k-1)!} x^k$$

$$= \frac{(2n+1)!}{n!} + \sum_{k=1}^{n} \frac{(2n-k)!}{(n-k)!\,(k-1)!} \left\{ \frac{(n+1)(2n-k+1)}{k(n-k+1)} \right\} x^k + x^{n+1}$$

$$= \sum_{k=0}^{n+1} \frac{(n+1)(2n-k+1)!}{(n-k+1)!\,k!} x^k \quad ;$$

b) $g_{n+1}(x) = (2n+1)(f_n(x) + g_n(x)) - x g_n(x)$

$$= \sum_{k=0}^{n} \frac{(2n-k)!}{(n-k)!\,k!}(2n+1)x^k - \sum_{k=1}^{n+1} \frac{(n-k+1)(2n-k)!}{(n-k+1)!\,(k-1)!} x^k$$

$$= \frac{(2n+1)!}{n!} + \sum_{k=1}^{n} \frac{(2n-k)!}{(n-k)!\,(k-1)!} \left\{ \frac{2n+1}{k} - \frac{n-k+1}{n-k+1} \right\} x^k$$

$$= \sum_{k=0}^{n+1} \frac{(n+1-k)(2n-k+1)!}{(n+1-k)!\,k!} x^k \quad ;$$

iii-a) $$\frac{f_n(x)}{n(n+1)\cdots(2n-1)}$$

$$= \sum_{k=0}^{n} \frac{n(2n-k-1)!\,x^k}{n(n+1)\cdots(2n-1)(n-k)!\,k!}$$ and, for $k>2$,

$$\frac{n(2n-k-1)!}{n(n+1)\cdots(2n-1)(n-k)!\,k!}$$

$$= \frac{1}{k!}\prod_{i=0}^{k-1}\frac{n-i}{2n-i-1} < \frac{1}{k!}\cdot\frac{1}{2^{k-1}} \quad ;$$ since

$$\sum_{k=0}^{\infty}\frac{1}{k!\,2^{k-1}} < \infty, \quad \lim \frac{f_n(x)}{n(n+1)\cdots(2n-1)}$$

$$= \sum_{k=0}^{\infty}\left\{\lim \frac{n(2n-k-1)!}{n(n+1)\cdots(2n-1)(n-k)!\,k!}\right\} x^k$$

$$= \sum_{k=0}^{\infty}\frac{1}{k!}\left(\frac{x}{2}\right)^k = e^{x/2} \quad ;$$

b) same argument as in (a) ;

iv) $$\frac{f_n(x)}{h_n(x)} = \frac{f_n(x)}{g_n(-x)} = \frac{\dfrac{f_n(x)}{n(n+1)\cdots(2n-1)}}{\dfrac{g_n(-x)}{n(n+1)\cdots(2n-1)}}$$

$$\longrightarrow \frac{e^{x/2}}{e^{-x/2}} = e^x,$$

and $\dfrac{g_n(x)}{k_n(x)} = \dfrac{g_n(x)}{f_n(-x)} \longrightarrow e^x$ as above ;

thus $K(A_1 A_2 \cdots)$ exists and is $e^x$ ;

v-a) multiplication on the right side yields the left side ;

b) $[1, k-1, 1, 1, 3k-1, 1, 1, 5k-1, 1, \cdots]$

$$= K\left\{\begin{pmatrix}1&1\\1&0\end{pmatrix}\begin{pmatrix}k-1&1\\1&0\end{pmatrix}\begin{pmatrix}1&1\\1&0\end{pmatrix}\begin{pmatrix}1&1\\1&0\end{pmatrix}\begin{pmatrix}3k-1&1\\1&0\end{pmatrix}\begin{pmatrix}1&1\\1&0\end{pmatrix}\begin{pmatrix}1&1\\1&0\end{pmatrix}\begin{pmatrix}5k-1&1\\1&0\end{pmatrix}\cdots\right\}$$

$$= K\left\{\left(\begin{pmatrix}1&1\\1&0\end{pmatrix}\begin{pmatrix}k-1&1\\1&0\end{pmatrix}\begin{pmatrix}1&1\\1&0\end{pmatrix}\right)\left(\begin{pmatrix}1&1\\1&0\end{pmatrix}\begin{pmatrix}3k-1&1\\1&0\end{pmatrix}\begin{pmatrix}1&1\\1&0\end{pmatrix}\right)\left(\begin{pmatrix}1&1\\1&0\end{pmatrix}\begin{pmatrix}5k-1&1\\1&0\end{pmatrix}\begin{pmatrix}1&1\\1&0\end{pmatrix}\right)\cdots\right\}$$

$$= K\left\{\begin{pmatrix}k+1&k\\k&k-1\end{pmatrix}\begin{pmatrix}3k+1&3k\\3k&3k-1\end{pmatrix}\begin{pmatrix}5k+1&5k\\5k&5k-1\end{pmatrix}\cdots\right\}$$

$$= K\left\{\left(k\begin{pmatrix}1+\frac{1}{k}&1\\1&1-\frac{1}{k}\end{pmatrix}\right)\left(k\begin{pmatrix}3+\frac{1}{k}&3\\3&3-\frac{1}{k}\end{pmatrix}\right)\left(k\begin{pmatrix}5+\frac{1}{k}&5\\5&5-\frac{1}{k}\end{pmatrix}\right)\cdots\right\}$$

$$= K\left\{\begin{pmatrix}1+\frac{1}{k}&1\\1&1-\frac{1}{k}\end{pmatrix}\begin{pmatrix}3+\frac{1}{k}&3\\3&3-\frac{1}{k}\end{pmatrix}\begin{pmatrix}5+\frac{1}{k}&5\\5&5-\frac{1}{k}\end{pmatrix}\cdots\right\} = e^{1/k};$$

we have used I (ii), I (v), II (v-a), I (iv), II (iv) in order ;

c) put $k = 1$ in (b) to obtain

$$[1, 0, 1, 1, 2, 1, 1, 4, 1, 1, 6, 1, \cdots]$$

$$= 1 + \cfrac{1}{0 + \cfrac{1}{[1, 1, 2, 1, 1, 4, 1, 1, 6, 1, \cdots]}}$$

$$= 1 + [1, 1, 2, 1, 1, 4, 1, 1, 6, 1, \cdots]$$

$$= [2, 1, 2, 1, 1, 4, 1, 1, 6, 1, \cdots] .$$

24. I i) $c_1 = \frac{b_1}{a_1} = \frac{a_1 \cdot 0 - b_1(-1)}{a_1 \cdot 1 - b_1 \cdot 0} = \frac{p_1}{q_1}$;

$$c_2 = \frac{b_1}{a_1 - \frac{b_2}{a_2}} = \frac{b_1 a_2}{a_1 a_2 - b_2} = \frac{a_2 b_1 - b_2 \cdot 0}{a_2 a_1 - b_2 \cdot 1} = \frac{p_2}{q_2};$$

suppose true for $n-1$; then

$$c_n = \frac{b_1}{a_1 -} \cdots - \frac{b_n}{a_n} = \frac{b_1}{a_1 -} \cdots \frac{b_{n-1}}{a_{n-1} - \frac{b_n}{a_n}}$$

$$= \frac{(a_{n-1} - \frac{b_n}{a_n}) p_{n-2} - b_{n-1} p_{n-3}}{(a_{n-1} - \frac{b_n}{a_n}) q_{n-2} - b_{n-1} q_{n-3}}$$

$$= \frac{a_n(a_{n-1} p_{n-2} - b_{n-1} p_{n-3}) - b_n p_{n-2}}{a_n(a_{n-1} q_{n-2} - b_{n-1} q_{n-3}) - b_n q_{n-2}}$$

$$= \frac{a_n p_{n-1} - b_n p_{n-2}}{a_n q_{n-1} - b_n q_{n-2}} = \frac{p_n}{q_n};$$

ii) $p_n - p_{n-1} = (a_n - 1) p_{n-1} - b_n p_{n-2}$

$\geq b_n (p_{n-1} - p_{n-2}) \geq \cdots \geq b_0 \cdots b_n$;

consequently $p_n \geq p_{n-1} + b_0 \cdots b_n$

$\geq \cdots \geq b_0 + b_0 b_1 + \cdots + b_0 \cdots b_n$;

clearly if $a_n - 1 = b_n$ for all $n$ strict equality holds everywhere; if $a_j > b_j + 1$ then, for $n \geq j$, $p_j > b_0 + \cdots + b_0 \cdots b_n$; changing all $p_j$ to $q_j$ yields the same result for $q_n$;

iii) $p_0 q_{-1} - p_{-1} q_0 = b_0$, $p_1 q_0 - p_0 q_1 = b_0 b_1$

$$p_n q_{n-1} - p_{n-1} q_n =$$

$$(a_n q_{n-1} - b_n p_{n-2}) q_{n-1} - p_{n-1}(a_n q_{n-1} - b_n q_{n-2})$$

$$= (p_{n-1} q_{n-2} - p_{n-2} q_{n-1}) b_n = b_0 \cdots b_n \,;$$

dividing by $q_{n-1} q_n$ yields the last equality ;

iv) $q_0 - p_0 = 1 = q_{-1} - p_{-1}$,

$$q_1 - p_1 = a_1 - b_1 \geq 1 = q_0 - p_0 \,;$$

assuming true for $n-1$, $n-2$ we have

$$q_n - p_n = a_n(q_{n-1} - p_{n-1}) - b_n(q_{n-2} - p_{n-2})$$

$$= q_{n-1} - p_{n-1} + (a_n - 1)(q_{n-1} - p_{n-1}) - b_n(q_{n-2} - p_{n-2})$$

$$\geq q_{n-1} - p_{n-1} + b_n\{(q_{n-1} - p_{n-1}) - (q_{n-2} - p_{n-2})\}$$

$$\geq q_{n-1} - p_{n-1} \geq 1 \,;$$

if $a_j - 1 = b_j$ for $j \leq n$ we have

$$q_n - p_n = q_{n-1} - p_{n-1} = \cdots = q_0 - p_0 = 1 \,;$$

v) from (iv), after dividing by $q_n$, we see $1 - \frac{p_n}{q_n} \geq \frac{1}{q_n}$ so $\frac{p_n}{q_n} \leq 1 - \frac{1}{q_n} < 1$; from (iii) the sequence $\left\{\frac{p_n}{q_n}\right\}$ is monotone increasing ;

therefore since this sequence is monotone increasing and bounded above by 1 it must converge to a limit $\le 1$;

if $a_j > b_j + 1$ and $n \ge j$ we have

$$\frac{p_n}{q_n} = \frac{b_1}{q_1 q_0} + \cdots + \frac{b_1 \cdots b_n}{q_n q_{n-1}}$$

$$\le \frac{q_1 - q_0}{q_1 q_0} + \cdots + \frac{q_n - q_{n-1}}{q_n q_{n-1}} - \left\{ \frac{q_j - q_{j-1}}{q_j q_{j-1}} - \frac{b_1 \cdots b_j}{q_j q_{j-1}} \right\}$$

$$= \left(\frac{1}{q_0} - \frac{1}{q_1}\right) + \cdots + \left(\frac{1}{q_{n-1}} - \frac{1}{q_n}\right) - \left\{ \frac{q_j - q_{j-1}}{q_j q_{j-1}} - \frac{b_1 \cdots b_j}{q_j q_{j-1}} \right\}$$

$$= \frac{1}{q_0} - \frac{1}{q_n} - \left\{ \frac{q_j - q_{j-1}}{q_j q_{j-1}} - \frac{b_1 \cdots b_j}{q_j q_{j-1}} \right\} ;$$

now, since $\frac{q_j - q_{j-1}}{q_j q_{j-1}} - \frac{b_1 \cdots b_j}{q_j q_{j-1}} > 0$, we have

$$\lim \frac{p_n}{q_n} \le \frac{1}{q_0} - \left\{ \frac{q_j - q_{j-1}}{q_j q_{j-1}} - \frac{b_1 \cdots b_j}{q_j q_{j-1}} \right\} < 1 ;$$

vi) when $a_n = b_n + 1$ for all $n$,

$$\frac{p_n}{q_n} = 1 - \frac{1}{q_n} = 1 - \frac{1}{b_0 + \cdots + b_0 \cdots b_n}$$

and the results follow from this equality.

II *i*) Since $\frac{b_n}{a_n}$ is positive and the convergents to $\alpha_n$ are monotone increasing (see I (*iii*)), we know $0 < \alpha_n$;

since for each $n$ there is a $j > n$ such that $a_j > b_j + 1$, we know by I(v) that $\alpha_n < 1$;

ii) for $m > n$,

$$\alpha_1 = \lim_m \left\{ \frac{b_1}{a_1 -} \cdots \frac{b_m}{a_m} \right\} = \lim_m \left\{ \frac{b_1}{a_1 -} \cdots \frac{b_{n-1}}{a_{n-1} - c_{nm}} \right\}$$
$$= \frac{b_1}{a_1 -} \cdots \frac{b_{n-1}}{a_{n-1} - \lim_m c_{nm}} = \frac{b_1}{a_1 -} \cdots \frac{b_{n-1}}{a_{n-1} - \alpha_n},$$

where $c_{nm} = \frac{b_n}{a_n -} \cdots \frac{b_m}{a_m}$; thus rationality of any $\alpha_j$ implies that of $\alpha_1$ and hence of every $\alpha_j$;

iii) $\alpha_j = \dfrac{b_j}{a_j - \alpha_{j+1}}$ so $\alpha_{j+1} = \dfrac{r a_j - s b_j}{r}$;
by (i), $0 < \alpha_{j+1} < 1$ so $0 < r a_j - s b_j < r$;

iv) otherwise, by (iii), we could construct an infinite strictly monotone decreasing sequence of positive integers;

v) this follows from (iv) if $a_n \geq b_n + 1$ from $n = 1$ on; otherwise, suppose this is true only for $n \geq m$; then $\alpha_m$ is irrational, by (v), hence $\alpha = \frac{b_1}{a_1 -} \cdots \frac{b_{m-1}}{a_{m-1} - \alpha_m}$ $= \frac{(a_{m-1} - \alpha_m)p_{m-2} - b_{m-1}p_{m-3}}{(a_{m-1} - \alpha_m)q_{m-2} - b_{m-1}q_{m-3}}$, from which we find $(\alpha_m - a_{m-1})(p_{m-2} - \alpha q_{m-2}) = -b_{m-1}(p_{m-3} - \alpha q_{m-3})$; the irrationality of $\alpha_m - a_{m-1}$ then says, if $\alpha$ is rational, $p_{m-2} - \alpha q_{m-2} = p_{m-3} - \alpha q_{m-3} = 0$ which implies the false equality

$$\alpha = \frac{p_{m-2}}{q_{m-2}} = \frac{p_{m-3}}{q_{m-3}}.$$

25. *i*) True for $n = 2$; assume true for $n$, then

$$\sum_{k=1}^{n+1} \frac{1}{c_k} = \sum_{k=1}^{n-1} \frac{1}{c_k} + \frac{1}{c_n} + \frac{1}{c_{n+1}} = \sum_{k=1}^{n-1} \frac{1}{c_k} + \cfrac{1}{\cfrac{1}{\cfrac{1}{c_n} + \cfrac{1}{c_{n+1}}}}$$

$$= \frac{1}{c_1 -} \frac{c_1^2}{c_1 + c_2 -} \cdots \cfrac{c_{n-1}^2}{c_{n-1} + \cfrac{1}{\cfrac{1}{c_n} + \cfrac{1}{c_{n+1}}}}$$

$$= \frac{1}{c_1 -} \frac{c_1^2}{c_1 + c_2 -} \cdots \cfrac{c_{n-1}^2}{c_{n-1} + c_n - \left\{c_n - \cfrac{1}{\cfrac{1}{c_n} + \cfrac{1}{c_{n+1}}}\right\}}$$

$$= \frac{1}{c_1 -} \frac{c_1^2}{c_1 + c_2 -} \cdots \cfrac{c_{n-1}^2}{c_{n-1} + c_n - \cfrac{c_n^2}{c_n + c_{n+1}}};$$

ii) as in (i) after noting that

$$\frac{c_{n-2}(c_n+c_{n+1})}{c_{n-1}+(c_n+c_{n+1})} = \frac{c_{n-2}\,c_n}{c_{n-1}+c_n-\dfrac{c_{n-1}\,c_{n+1}}{c_n+c_{n+1}}}$$

iii) clear;

iv) using (i),

$$e = 2! - \cfrac{2!^2}{2!-3! - \cfrac{3!^2}{-3!+4! - \cfrac{4!^2}{4!-5! - \cfrac{5!^2}{-5!+6!-\cdots}}}}$$

$$= 2 + \cfrac{2}{2+\cfrac{3}{3+\cfrac{4}{4+\cfrac{5}{5+\cfrac{6}{6+\cdots}}}}}$$

$$= 2 + \cfrac{1}{1+\cfrac{3}{2\cdot 3+\cfrac{2\cdot 4}{4+\cfrac{5}{5+\cfrac{6}{6+\cdots}}}}}$$

$$= 2 + \cfrac{1}{1+\cfrac{1}{2+\cfrac{2\cdot 4}{3\cdot 4+\cfrac{3\cdot 5}{5+\cfrac{6}{6+\cdots}}}}} = 2 + \cfrac{1}{1+\cfrac{1}{2+\cfrac{2}{3+\cfrac{3}{4+\cdots}}}}$$

v) using (ii)

$$\frac{\pi}{4} = 1 - \frac{1}{3} + \frac{1}{5} - \frac{1}{7} + \cdots$$

$$= \cfrac{1}{1 - \cfrac{-\frac{1}{3}}{1 - \frac{1}{3} - \cfrac{\frac{1}{5}}{-\frac{1}{3} + \frac{1}{5} - \cfrac{\frac{1}{3\cdot 7}}{\frac{1}{5} - \frac{1}{7} - \cdots}}}}$$

$$= \cfrac{1}{1 + \cfrac{1}{2 + \cfrac{3^2}{2 + \cfrac{5^2}{2 + \cfrac{7^2}{2 + \cdots}}}}}$$

26. i) For each fixed $x$

$$\frac{x^{2n}}{2^{2n} n!\, m(m+1)\cdots(m+n-1)} \cdot \frac{2^{2n-2}(n-1)!\, m(m+1)\cdots(m+n-2)}{x^{2n-2}}$$

$$= \frac{x^2}{2^2 n\,(m+n-1)} < 1$$

for $n$ sufficiently large ;

ii) the given assertion is equivalent to

$$f_{m+1}(x) - \frac{x^2}{2^2 m(m+1)} f_{m+2}(x) = f_m(x) ;$$

the left side equals

$$1+\sum_{k=1}^{\infty}\frac{(-1)^k x^{2k}}{2^{2k}k!(m+1)\cdots(m+k)}-\frac{x^2}{2^2 m(m+1)}-\frac{x^2}{2^2 m(m+1)}\sum_{k=1}^{\infty}\frac{(-1)^k x^{2k}}{2^{2k}k!(m+2)\cdots(m+k+1)}$$

$$= 1 + \quad '' \quad - \quad '' \quad -\sum_{k=1}^{\infty}\frac{(-1)^k x^{2(k+1)}}{2^{2(k+1)}k!\,m(m+1)\cdots(m+k+1)}$$

$$= 1 + \quad '' \quad - \quad '' \quad +\sum_{k=2}^{\infty}\frac{(-1)^k x^{2k}}{2^{2k}(k-1)!\,m(m+1)\cdots(m+k)}$$

$$= 1+\sum_{k=1}^{\infty}\frac{(-1)^k x^{2k}}{2^{2k}(k-1)!(m+1)\cdots(m+k)}\left(\frac{1}{k}+\frac{1}{m}\right)$$

$$= 1+\sum_{k=1}^{\infty}\frac{(-1)^k x^{2k}}{2^{2k}k!\,m(m+1)\cdots(m+k-1)} = f_m(x) \, ;$$

the $2^{nd}$ equality follows immediately from the $1^{st}$ ;

iii) $$f_{1/2}(x) = 1+\sum_{k=1}^{\infty}\frac{(-1)^k x^{2k}}{2^{2k}\,k!\,\frac{1}{2}\cdot\frac{3}{2}\cdots\frac{2k-1}{2}}$$

$$= 1+\sum_{k=1}^{\infty}\frac{(-1)^k x^{2k}}{(2k)!} = \sum_{k=0}^{\infty}\frac{(-1)^k x^{2k}}{(2k)!} = \cos x \, ;$$

$$f_{3/2}(x) = 1+\sum_{k=1}^{\infty}\frac{(-1)^k x^{2k}}{2^{2k}\,k!\,\frac{3}{2}\cdots\frac{2k+1}{2}}$$

$$= 1+\sum_{k=1}^{\infty}\frac{(-1)^k x^{2k}}{(2k+1)!} = \sum_{k=0}^{\infty}\frac{(-1)^k x^{2k}}{(2k+1)!} = \frac{\sin x}{x} \, ;$$

thus $$\frac{f_{3/2}(x)}{f_{1/2}(x)} = \frac{\tan x}{x} \, ;$$

iv) in (iii) put $x = \frac{\pi}{4}$ to obtain (successively)

$$\frac{4}{\pi} = \frac{1}{1-}\,\frac{(\pi/4)^2}{3-}\,\frac{(\pi/4)^2}{5-}\,\frac{(\pi/4)^2}{7-}\cdots$$

$$1 = \frac{\pi/4}{1-}\,\frac{(\pi/4)^2}{3-}\,\frac{(\pi/4)^2}{5-}\,\frac{(\pi/4)^2}{7-}\cdots$$

$$= \frac{m}{n-}\,\frac{m^2}{3n-}\,\frac{m^2}{5n-}\,\frac{m^2}{7n-}\cdots\frac{m^2}{(2k-1)n-}\cdots;$$

the right hand side of this last expression satisfies, for $k$ sufficiently large, the condition $(2k-1)n \geq m^2 + 1$ and hence by #24 II (v) is irrational.

27. I. The number of integral polynomials of degree $n$ is countable as are the zeros of such polynomials; hence for each $n$ there are at most countably many algebraic numbers of degree $n$; since this means the set of all algebraic numbers is countable and since the set of real numbers is uncountable there exist transcendental numbers;

II i) one merely divides $f(x)$ by $x-\alpha$ to obtain $g(x)$, which, being of degree $n-1$, does not have $\alpha$ as a zero;

ii) since $g$ is continuous and $g(\alpha) \neq 0$ this is immediate;

iii) choose integers $a, b$ ($b > 0$) such that $\alpha - \delta \le \frac{a}{b} \le \alpha + \delta$; then if $M$ is the maximum absolute value of $g$ on $[\alpha-\delta, \alpha+\delta]$ we have

$$\left|\alpha - \frac{a}{b}\right| = \left|\frac{f(\frac{a}{b})}{g(\frac{a}{b})}\right| \ge \frac{|f(\frac{a}{b})|}{M} \ge \frac{|f(\frac{a}{b})|b^n}{Mb^n};$$

now $f(\frac{a}{b})b^n$ is an integer and is not zero since if so $\alpha$ would not be algebraic of degree $n$; hence $\left|\alpha - \frac{a}{b}\right| \ge \frac{1}{Mb^n}$;

iv) choose $c = \min\left\{\frac{1}{2}\delta, \frac{1}{M+1}\right\}$; then for $\frac{a}{b}$ outside $[\alpha-\delta, \alpha+\delta]$,

$$\left|\alpha - \frac{a}{b}\right| > \delta > c \ge \frac{c}{b^n}$$

and for $\frac{a}{b}$ inside $[\alpha-\delta, \alpha+\delta]$,

$$\left|\alpha - \frac{a}{b}\right| \geq \frac{1}{Mb^n} > \frac{1}{(M+1)b^n} \geq \frac{c}{b^n}.$$

III i) According to II (iv) if $\alpha$ were rational there would be a positive constant $c$ such that $\left|\alpha - \frac{a}{b}\right| \geq \frac{c}{b^n}$ for all integers $a, b$ $(b > 0)$; but with $a = 10^{2^n}\sum_{m=0}^{n} 10^{-2^m}$, $b = 10^{2^n}$ we have

$$\left|\alpha - \frac{a}{b}\right| = \sum_{m=n+1}^{\infty} 10^{-2^m} < 10^{-2^{n+1}+1} = \frac{10^{-2^n+1}}{b}$$

and this, for $n$ sufficiently large, will not be $\geq \frac{c}{b^n}$ for any positive $c$;

an alternative proof is to observe that the decimal expansion of $\alpha$ has infinitely many non-zero digits as well as arbitrarily long blocks of consecutive 0's, hence can not be periodic;

ii) for each $n$, by II (iv) if $\alpha$ is Liouville then $\alpha$ is not algebraic of degree $n$;

iii) let $n$ be a positive integer and $c$ be a positive constant; choose $k > n$ such that $M \cdot 10^{-k!+1} < c$ and put $a = 10^{k!} \sum_{m=0}^{k} a_m 10^{-m!}$, $b = 10^{k!}$; then

$$\left|\alpha - \frac{a}{b}\right| = \left|\sum_{m=k+1}^{\infty} a_m 10^{-m!}\right| \leq M \cdot 10^{-(k+1)!+1} = \frac{M \cdot 10^{-k!+1}}{(10^{k!})^k}$$
$$< \frac{c}{b^k} < \frac{c}{b^n} \;;$$

thus by (ii) $\alpha$ is transcendental.

IV i) Let $n$ and $c$ be given and choose $k > n$ so that $\frac{1}{q_k} < c$; then with $a = p_k$, $b = q_k$ we have

$$\left|\alpha - \frac{a}{b}\right| = \left|\alpha - \frac{p_k}{q_k}\right| < \frac{1}{q_k q_{k+1}} = \frac{1}{q_k(a_{k+1}q_k + q_{k+1})}$$
$$< \frac{1}{a_{k+1} q_k^2} < \frac{1}{q_k^{k+1}} < \frac{c}{q_k^k} = \frac{c}{b^k} < \frac{c}{b^n}$$

and the result follows from III (ii);

ii) take $a_k = 2^{k!}$, giving

$$[1, 2, 2^2, 2^6, 2^{24}, 2^{120}, \ldots];$$

iii) we need to show $a_{k+1} > (2^k a_1 \cdots a_k)^{k-1}$ implies $a_{k+1} > q_k^{k-1}$ (for $k \geq 1$) and this, in turn, would follow from $2^k a_1 \cdots a_k > q_k$ (for $k \geq 1$); for $k=1$, $2a_1 > q_1 = a_1$; if true for $k$ then $2^{k+1} a_1 \cdots a_{k+1} > 2a_{k+1} q_k > a_{k+1} q_k + q_{k-1} = q_{k+1}$.

28. I i) The probability that an arbitrary $x$ in $[0,1]$ is irrational is 1 and that probability is also the indicated sum since every $x$ has exactly 1 integral value for $a_n(x)$; similarly the set of irrational $x$ with given $a_1, \cdots, a_{n-1}$ is the same as the set of irrational $x$ with given $a_1, \cdots, a_{n-1}$ and $a_n(x)$ a positive integer;

ii) the set of $x$ with $a_1(x) = k$ is just the set of $x$ satisfying $\frac{1}{k+1} < x < \frac{1}{k}$ and the probability of $x$ being in here is just $\frac{1}{k} - \frac{1}{k+1} = \frac{1}{k(k+1)}$;

iii) the set of $x$ with $a_1(x) = n$ and $a_2(x) = k$ is just the set of $x$ satisfying

$$\frac{1}{n+\frac{1}{k}} < x = \frac{1}{n+\frac{1}{k+x_2}} < \frac{1}{n+\frac{1}{k+1}} \quad ;$$

thus the probability of $a_2(x) = k$ is just

$$\sum_{n=1}^{\infty}\left(\frac{1}{n+\frac{1}{k+1}} - \frac{1}{n+\frac{1}{k}}\right) = P_{1k}\sum_{n=1}^{\infty}\frac{1}{(n+\frac{1}{k})(n+\frac{1}{k+1})}$$

$$= P_{1k}\left\{\frac{\pi^2}{6} + \sum_{n=1}^{\infty}\left(\frac{1}{(n+\frac{1}{k})(n+\frac{1}{k+1})} - \frac{1}{n^2}\right)\right\}$$

$$= \frac{\pi^2}{6k(k+1)}\left(1 - \frac{6}{\pi^2}\sum_{n=1}^{\infty}\frac{(\frac{1}{k}+\frac{1}{k+1})n + \frac{1}{k(k+1)}}{n^2(n+\frac{1}{k})(n+\frac{1}{k+1})}\right)$$

$$= \frac{\pi^2}{6k(k+1)}(1-\epsilon_k),$$

where $\epsilon_k = \frac{6}{\pi^2}\sum_{n=1}^{\infty}\frac{(\frac{1}{k}+\frac{1}{k+1})n + \frac{1}{k(k+1)}}{n^2(n+\frac{1}{k})(n+\frac{1}{k+1})}$

$$< \frac{6}{k\pi^2}\sum_{n=1}^{\infty}\frac{2}{n^3} \to 0 \text{ as } k \to \infty \quad ;$$

iv) for $k \geq 2$,

$$P_{2k} = P_{1k}\sum_{n=1}^{\infty}\frac{1}{(n+\frac{1}{k})(n+\frac{1}{k+1})} > P_{1k}\sum_{n=1}^{\infty}\frac{1}{(n+\frac{1}{2})(n+\frac{1}{3})}$$

$$> P_{1k}\sum_{n=1}^{\infty}\frac{1}{n(n+1)} = P_{1k} \quad ;$$

$$P_{21} = P_{11}\sum_{n=1}^{\infty}\frac{1}{(n+1)(n+\frac{1}{2})} < P_{11}\sum_{n=1}^{\infty}\frac{1}{n(n+1)} = P_{11}\,.$$

II i) By #3 (xiii-a) the relevant x lie between $[0, a_1, \cdots, a_{n-1}, k]$ and $[0, a_1, \cdots, a_{n-1}, k+1]$; thus

$$p(a_1, \cdots, a_{n-1}, k) = \left| \frac{kp_{n-1}+p_{n-2}}{kq_{n-1}+q_{n-2}} - \frac{(k+1)p_{n-1}+p_{n-2}}{(k+1)q_{n-1}+q_{n-2}} \right|$$

$$= \frac{|p_{n-2}q_{n-1} - p_{n-1}q_{n-2}|}{(kq_{n-1}+q_{n-2})((k+1)q_{n-1}+q_{n-2})}$$

$$= 1 \Big/ q_{n-1}^2 \left(k + \frac{q_{n-2}}{q_{n-1}}\right)\left(k+1+\frac{q_{n-2}}{q_{n-1}}\right);$$

ii) the middle (equality) follows directly from (i); the inequalities follow from the fact that $\frac{k+x}{k+2+x}$ is strictly increasing on $[0,1]$;

iii) the middle expression is $\frac{(1+x)(2+x)}{(k+x)(k+1+x)}$, with $x = \frac{q_{n-2}}{q_{n-1}}$, and is strictly increasing on $[0,1]$;

iv) summing (iii) over all positive k yields the result when one takes account of I(i);

v) multiply the equality in (iii) term by term by the (reciprocal) inequality in (iv), i.e. by $\frac{1}{3} < \frac{P(a_1,\cdots,a_{n-1},1)}{P(a_1,\cdots,a_{n-1})} < \frac{1}{2}$ and one obtains the desired inequality when one observes that

$$P(a_1,\cdots,a_{n-1},k) = P(a_1,\cdots,a_{n-1})\, P_{nk}\ .$$

III i) $$\sum_{k=1}^{M} P_{1k} = \sum_{k=1}^{M} \frac{1}{k(k+1)} = \sum_{k=1}^{M} \left(\frac{1}{k} - \frac{1}{k+1}\right)$$
$$= 1 - \frac{1}{M+1} = \frac{M}{M+1}\ ;$$

ii) $$\sum_{k=1}^{M} P(a_1,\cdots,a_{n-1},k) =$$
$$P(a_1,\cdots,a_{n-1}) - \sum_{k=M+1}^{\infty} P(a_1,\cdots,a_{n-1},k)$$
$$= P(a_1,\cdots,a_{n-1}) - P(a_1,\cdots,a_{n-1})\,\frac{2}{3}\sum_{k=M+1}^{\infty}\frac{1}{k(k+1)}$$
$$= P(a_1,\cdots,a_{n-1})\left(1 - \frac{2}{3(M+1)}\right)\ ;$$

iii) using (ii) iteratively we have

$$\sum_{\substack{1\le a_j\le M\\ 1\le j\le n}} P(a_1,\cdots,a_n) = \sum_{a_1=1}^{M}\cdots\sum_{a_{n-1}=1}^{M}\sum_{k=1}^{M} P(a_1,\cdots,a_{n-1},k)$$
$$< \alpha \sum_{a_1=1}^{M}\cdots\sum_{a_{n-1}=1}^{M} P(a_1,\cdots,a_{n-1})$$
$$< \alpha^2 \sum_{a_1=1}^{M}\cdots\sum_{a_{n-2}=1}^{M} P(a_1,\cdots,a_{n-2})$$
$$< \cdots < \alpha^{n-1}\sum_{a_1=1}^{M} P(a_1) = \alpha^{n-1}\,\frac{M}{M+1}\ ;$$

iv) by (iii), for each M the probability that the partial quotients do not exceed M is 0; thus the probability that they are bounded is also 0.

IV i) Sum II (v) for $k \geq \varphi(t)$ to obtain $\frac{2}{3A} < \sum_{k \geq \varphi(t)} \frac{p(a_1, \cdots, a_{t-1}, k)}{p(a_1, \cdots, a_{t-1})} < \frac{3}{A+1}$, where A is an integer satisfying $A-1 < \varphi(t) \leq A$; now $\frac{2}{3(\varphi(t)+1)} < \frac{2}{3A}$ and $\frac{3}{A+1} \leq \frac{3}{\varphi(t)+1}$ and the result follows;

ii) this follows immediately from (i) & I(i);

iii) the result in (ii) may be written

$$p(a_1, \cdots, a_{t-1})\left(1 - \frac{3}{\varphi(t)+1}\right) < \sum_{1 \leq a_t < \varphi(t)} p(a_1, \cdots, a_{t-1}, a_t)$$

$$< p(a_1, \cdots, a_{t-1})\left(1 - \frac{2}{3(\varphi(t)+1)}\right), \quad t > N;$$

iteration of this result leads successively to

$$p(a_1,\cdots,a_{t-2})\left(1-\frac{3}{\varphi(t-1)+1}\right)\left(1-\frac{3}{\varphi(t)+1}\right)$$

$$<\sum_{1\le a_{t-1}<\varphi(t-1)}\ \sum_{1\le a_t<\varphi(t)} p(a_1,\cdots,a_t)$$

$$<p(a_1,\cdots,a_{t-2})\left(1-\frac{2}{3(\varphi(t-1)+1)}\right)\left(1-\frac{2}{3(\varphi(t)+1)}\right),$$

$$\cdots$$

$$p(a_1,\cdots,a_N)\prod_{j=N+1}^{t}\left(1-\frac{3}{\varphi(j)+1}\right)$$

$$<\sum_{1\le a_{N+1}<\varphi(N+1)}\cdots\sum_{1\le a_t<\varphi(t)} p(a_1,\cdots,a_t)$$

$$<p(a_1,\cdots,a_N)\prod_{j=N+1}^{t}\left(1-\frac{2}{3(\varphi(j)+1)}\right);$$

*iv*) the series $\sum\frac{1}{\varphi(n)}$, $\sum\frac{1}{\varphi(n)+1}$ converge or diverge together; also the series $\sum\frac{1}{\varphi(n)+1}$ and the product $\prod\left(1-\frac{2}{3(\varphi(n)+1)}\right)$ converge or diverge together; thus when $\sum\frac{1}{\varphi(n)}$ diverges so also does $\sum\frac{2}{3(\varphi(n)+1)}$ and, since the terms are all positive and $<1$, this implies $\prod\left(1-\frac{2}{3(\varphi(n)+1)}\right)$ diverges to 0; thus by the left inequality in (*iii*) the probability that a random $x$ with first $N+1$ partial quotients fixed satisfies $a_n(x) < \varphi(n)$ for $n > N$ is 0; since this is true for all choices of $a_1, \cdots, a_N$, and there are only countably many such choices the result follows; when $\Sigma \frac{1}{\varphi(n)}$ converges so also do $\Sigma \frac{1}{3(\varphi(n)+1)}$ and $\Pi\left(1 - \frac{1}{3(\varphi(n)+1)}\right)$; further, given $\epsilon > 0$ there is an $N$ such that for $t > N$ the inequality

$$1 - \epsilon < \prod_{j=N}^{t}\left(1 - \frac{3}{\varphi(j)+1}\right) < 1$$

holds (otherwise the product would diverge to 0); but then, by the inequality in (*iii*),

$$(1-\epsilon)p(a_1, \cdots, a_N) < \Sigma p(a_1, \cdots, a_t) < p(a_1, \cdots, a_N)$$

and, by summing

$$1 - \epsilon < \sum_{a_1} \cdots \sum_{a_N} \sum_{1 \le a_{N+1} < \varphi(N+1)} \cdots \sum_{1 \le a_t < \varphi(t)} p(a_1, \cdots, a_t) < 1$$

and the result follows by observing that this is true for each $\epsilon > 0$.

1. i) For $1 \le s < t \le p_j$ we see that all prime factors of $N_t - N_s = (t-s)p_1 \cdots p_{j-1}$ are smaller than $p_j$; consequently each of $p_j, \cdots, p_n$ divides at most one of $N_s$ and $N_t$;

ii) since $p_{n-1} > 2$ we may clearly take $j = n-1$;

iii) each of the $n-i+1$ primes $p_i, \cdots, p_n$ divides at most one of the $p_i$ numbers $tp_1 \cdots p_{i-1} - 1$, $1 \le t \le p_i$; thus, since $p_i > n-i+1$, there must be one of these numbers divisible by none of the primes $p_i, \cdots, p_n$; since, also, none of the primes $p_1, \cdots, p_{i-1}$ divides any of the numbers the conclusion follows;

iv) if $i \le 4$ then $7 \ge p_i > n-i+1 \ge n-3$ so $n < 10$; hence $i > 4$ by our hypothesis; for

$i = 5$, $p_{i-1} - 2 = 7 - 2 = 5 = i$ so $p_{i-1} - 2 \geq i$; if this last inequality is true for $i = j$ then

$$p_j - 2 \geq p_{j-1} \geq j + 2 > j + 1$$

so it is also true for $i = j + 1$; consequently $p_{i-1} - 2 \geq i$ for all $i \geq 5$; now, using the minimal property of $i$,

$$i \leq p_{i-1} - 2 \leq n - (i - 1) + 1 - 2 < n - i + 1$$

so the number of factors in $p_1 \cdots p_i$ is smaller than the number of factors in $p_{i+1} \cdots p_n$; the desired inequality follows from the fact that $p_1 < p_{i+1}$, $p_2 < p_{i+2}$, $\cdots$, $p_i < p_{2i}$;

v) by (iii), $p_{n+1} < p_1 \cdots p_i$ so, using (iv), $p_{n+1}^2 < p_1 \cdots p_n$.

2. i) Since $j \leq k$ we clearly have $p_j^2 \leq p_k^2 \leq n$ and, since $p_j$ does not divide $n$ we must have $p_j^2 < n$, $(p_j^2, n) = 1$;

ii) by (i), no $p_j$ with $j \leq k$ can fail to divide $n$; consequently $p_1 \cdots p_k | n$ and, therefore, $p_1 \cdots p_k \leq n$;

iii) if there is no composite integer $< n$ and prime to $n$ then, by (ii), $p_1 \cdots p_k \le n$ ; since $n \ge 49$ and $p_k^2 \le n < p_{k+1}^2$ we see that $k \ge 4$ ; therefore, by #1(v), $p_{k+1}^2 < p_1 \cdots p_k \le n < p_{k+1}^2$, which is a contradiction;

iv) by (iii) such an integer must be $< 49$ ; direct checking of the integers from 30 to 48 shows 30 to be the largest integer with the stated property.

3. By #2 (iv), if $n > 30$ there are integers $a, b$ satisfying

$$1 < a \le b , \quad ab < n , \ (ab, n) = 1 \ ;$$

now $a < \sqrt{n}$ and $a \nmid n$ ; this shows no such integer is larger than 30 ; direct checking of the integers from 24 to 30 shows 24 to be the largest integer with the stated property.

4. i) The canonical factorization of $n$ has no primes other than $p_1, \cdots, p_j$ ; hence $n = p_1^{\alpha_1} \cdots p_j^{\alpha_j}$ where the $\alpha_j \geq 0$ ; since each positive integer is of the form $2t + \epsilon$, where $\epsilon = 0$ or $1$, the conclusion follows;

ii) in (i) the number of possible $m$ is $\leq \sqrt{n}$ and the number of possible $p_1^{\epsilon_1} \cdots p_j^{\epsilon_j}$ is $2^j$ ; hence $N_j(n) \leq \sqrt{n}\, 2^j$ ;

iii) put $j = \pi(n)$ in (ii) to obtain

$$n = N_{\pi(n)}(n) \leq \sqrt{n}\, 2^{\pi(n)} \; ;$$

now take natural logs of both sides after dividing by $\sqrt{n}$ ; since $\ln n / 2 \ln 2 \to \infty$ as $n \to \infty$ the number of primes is infinite;

iv) in (iii) replace each $n$ by $p_n$ to obtain

$$n = \pi(p_n) \geq \frac{\ln p_n}{2 \ln 2} \; ;$$

thus, $p_n \leq 4^n$; $p_n \neq 4^n$ so the conclusion follows;

v) if the series converged there would be a $j$ such that $\sum_{n>j} \frac{1}{p_n} < \frac{1}{2}$; then

$$2^j\sqrt{x} \geq N_j(x) \geq x - \sum_{n>j}\left[\frac{x}{p_n}\right] \geq x - \sum_{n>j}\frac{x}{p_n} > \frac{x}{2}$$

and, therefore,

$$\sqrt{x} < 2^{j+1} \text{ for all } x \quad ;$$

this is clearly false so the given series converges.

5. If $k$ is composite and $3 \leq k \leq [x]$ then $\sin\frac{k\pi}{j}$ is $0$ for some $j$, $2 \leq j \leq k-1$, so the $k^{th}$ term in the right hand sum is $0$; if $k$ is prime in the same range then $1-(\sin\frac{k\pi}{j})^2$ is always $<1$ so its $m^{th}$ power $\to 0$ as $m \to \infty$ and the $k^{th}$ term in the right hand sum is $1$; the summand $1$ counts the prime $2$.

6. For fixed $n$ if $s$ is larger than the largest exponent in the prime factorization of $n$ and if $m \geq \vartheta(n)$ then $\frac{(j!)^s\pi}{n}$ is an integer when $j \geq m$ and is not an integer for $0 \leq j < \vartheta(n)$; thus

for s and m so chosen,

$$\lim_{k\to\infty}\sum_{j=0}^{m}\left(1-\left(\cos\frac{(j!)^s\pi}{n}\right)^{2k}\right)=\lim_{k\to\infty}\sum_{j=0}^{\vartheta(n)-1}\left(1-\left(\cos\frac{(j!)^s\pi}{n}\right)^{2k}\right)=\vartheta(n);$$

therefore, the triple limit as stated is just $\vartheta(n)$.

7. i) This follows, using #4 (iv), by comparison with $\sum_{m=k+1}^{\infty}\left(\frac{4}{10}\right)^m$, which is a convergent geometric series;

ii) put $A_t = 10^{\frac{t(t-1)}{2}}\sum_{m=1}^{t-1} p_m 10^{-\frac{m(m+1)}{2}}$,

$$B_t = 10^{\frac{t(t-1)}{2}}\sum_{m=t}^{\infty} p_m 10^{-\frac{m(m+1)}{2}}$$

so that $10^{\frac{t(t-1)}{2}}\beta_0 = A_t + B_t$; clearly $A_t$ is an integer and $0 < B_t = \sum_{j=0}^{\infty} p_{t+j}\, 10^{-(1+j)(t+j/2)}$

$< 4^t \sum_{j=0}^{\infty} 4^j 10^{-t-j} = \left(\frac{4}{10}\right)^t \sum_{j=0}^{\infty}\left(\frac{4}{10}\right)^j = \frac{2}{3}\left(\frac{4}{10}\right)^{t-1} < 1$

for $t \geq 1$; thus $[10^{\frac{n(n+1)}{2}}\beta_0] - 10^n[10^{\frac{n(n-1)}{2}}\beta_0] =$

$A_{n+1} - 10^n A_n = 10^{\frac{n(n+1)}{2}}\sum_{m=1}^{n} p_m 10^{-\frac{m(m+1)}{2}} - 10^{n+\frac{n(n-1)}{2}}\sum_{m=1}^{n-1} p_m 10^{-\frac{m(m+1)}{2}}$

$= p_n$.

8. i, ii) The proof is virtually the same as that given for # 7 .

9. i) Put $f(n) = 10^{\sum_{j=1}^{n}([\log a_j]+2)}$ (base 10 logarithm); then (a) is clearly true; for (b) note that, when $\nu > n$,

$$a_\nu \frac{f(n)}{f(\nu)} = a_\nu 10^{-\sum_{j=n+1}^{\nu}([\log a_j]+2)} \le 10^{-\sum_{j=n+1}^{\nu-1}([\log a_j]+2)} \cdot \frac{a_\nu}{10^{\log a_\nu + 1}}$$

$$= 10^{-\sum_{j=n+1}^{\nu-1}([\log a_j]+2)-1} < 10^{-2(\nu-n)+1}$$

and, therefore,

$$\sum_{\nu=n+1}^{\infty} \frac{a_\nu f(n)}{f(\nu)} < \sum_{\nu=n+1}^{\infty} 10^{-2(\nu-n)+1} < 1 ;$$

ii) $$[f(t)\alpha] = \sum_{\nu=1}^{t} \frac{a_\nu f(t)}{f(\nu)} + \left[\sum_{\nu=t+1}^{\infty} \frac{a_\nu f(t)}{f(\nu)}\right]$$

$$= \sum_{\nu=1}^{t} \frac{a_\nu f(t)}{f(\nu)} \quad \text{so}$$

$$[f(n)\alpha] - \frac{f(n)}{f(n-1)}[f(n-1)\alpha] = \sum_{\nu=1}^{n} \frac{a_\nu f(n)}{f(\nu)} - \frac{f(n)}{f(n-1)} \sum_{\nu=1}^{n-1} \frac{a_\nu f(n-1)}{f(\nu)}$$

$$= a_n ;$$

iii) for #7 put $a_\nu = p_\nu$, $f(n) = 10^{\frac{n(n+1)}{2}}$ ;
for #8 put $a_\nu = p_\nu$, $f(n) = 10^{2^n}$ .

10. i) This follows immediately from

$$\binom{2n}{n} = \left\{\frac{2n(2n-2)\cdots 2}{n!}\right\}\left\{\frac{(2n-1)(2n-3)\cdots 1}{n!}\right\}$$

$$= 2^n\left(2-\tfrac{1}{n}\right)\left(2-\tfrac{1}{n-1}\right)\cdots\left(2-\tfrac{1}{1}\right) \;;$$

ii) by IV #24, the highest power of $p$ in $\binom{2n}{n}$ is

$$\sum_{j=1}^{t_p}\left\{\left[\frac{2n}{p^j}\right]-2\left[\frac{n}{p^j}\right]\right\},$$

which, by IV #13, is $\le t_p$ ; this shows $\binom{2n}{n}$ divides $\prod_{p<2n} p^{t_p}$ ; the other division is clear since no prime $p$, $n<p\le 2n$, is canceled from the numerator of $\binom{2n}{n} = \frac{(2n)!}{n!\,n!}$ when one reduces this expression to an integer ;

iii-a, b) immediate from (i) and (ii) ;

iv) $\sum_{p\le x} \ln p \ge \sum_{\sqrt{x}<p\le x} \ln p \ge \frac{1}{2}(\pi(x)-\sqrt{x})\ln x$ and the rest is immediate ;

v) from (iii-a) we see $2^n < (2n)^{\pi(2n)}$ and, taking logs, this yields $\pi(2n) > \frac{n\ln 2}{\ln 2n}$ ; now let $x \geq 2$ and suppose $2n \leq x < 2n+2$; then

$$\pi(x) \geq \pi(2n) > \frac{n \ln 2}{\ln 2n} > \frac{\frac{x-2}{2} \ln 2}{\ln x} = \left(\frac{1}{2} - \frac{1}{x}\right) \ln 2 \frac{x}{\ln x}$$

$$\geq \frac{\ln 2}{6} \cdot \frac{x}{\ln x} \text{ for } x \geq 3 \;;$$

for $2 \leq x \leq 3$, $\pi(x) \geq 1 > \frac{\ln 2}{3} \cdot \frac{2}{\ln 2} \geq \frac{\ln 2}{3} \cdot \frac{x}{\ln x}$;

hence one may take $A = \frac{\ln 2}{6}$;

vi) from (iii-b), $\sum_{n<p\leq 2n} \ln p < 2n \ln 2$ so $\sum_{p \leq 2n} \ln p < 2n \ln 2 + \sum_{p \leq n} \ln p$; putting $n = 2^{k-1}$ and repeatedly using this last inequality we obtain $\sum_{p \leq 2^k} \ln p < 2^k \ln 2 + \sum_{p \leq 2^{k-1}} \ln p < \cdots <$

$$2^k \ln 2 + 2^{k-1} \ln 2 + \cdots + 2 \ln 2 < 2^{k+1} \;;$$

vii) let $2^{k-1} \leq x < 2^k$; then

$$\sum_{p \leq x} \ln p \leq \sum_{p \leq 2^k} \ln p < 2^{k+1} \leq 4x \;;$$

viii) from (iv) and (vii),

$$\pi(x) \leq \frac{2}{\ln x} \sum_{p \leq x} \ln p + \sqrt{x} \leq 2A \frac{x}{\ln x} + \sqrt{x} < (2A+1) \frac{x}{\ln x},$$

for x sufficiently large; but for bounded x there is clearly a constant B for which

$\pi(x) < B\frac{x}{\ln x}$ since $\frac{x}{\ln x}$ for $x \geq 2$ is bounded away from 0; the conclusion follows;

ix) immediate from (v) and (viii).

11. i) Taking logs in the Chebyshev inequality we have

$\ln A + \ln x - \ln\ln x < \ln \pi(x) < \ln B + \ln x - \ln\ln x$

and dividing by $\ln x$ yields

$$1 + \frac{\ln A - \ln\ln x}{\ln x} < \frac{\ln \pi(x)}{\ln x} < 1 + \frac{\ln B - \ln\ln x}{\ln x} \quad ;$$

the conclusion follows from the fact that the left and right quotients tend to 0 as $x \to \infty$;

ii) multiply the Chebyshev inequality term by term with the inequality in (i);

iii) put $x = p_n$ in (ii);

iv) from (iii) it is immediate that

$$\frac{n \ln n}{B(1+\epsilon)} < p_n < \frac{n \ln n}{A(1-\epsilon)};$$

v) comparison with the series $\sum_{n=2}^{\infty} \frac{1}{(n \ln n)^{\alpha}}$ yields the result.

12. i) Since $\binom{2n-1}{n} = \frac{(2n-1)(2n-2)\cdots(2n-(n-1))}{n!}$ is an integer and no prime between $n$ and $2n$ gets canceled in the division the left inequality is clear; now

$$(1+1)^{2n-1} = \sum_{j=0}^{2n-1} \binom{2n-1}{j} > \binom{2n-1}{n-1} + \binom{2n-1}{n} = 2\binom{2n-1}{n}$$

so $\binom{2n-1}{n} < \frac{1}{2} \cdot 2^{2n-1} = (2^2)^{n-1} = 4^{n-1}$;

ii) it suffices to prove the statement for integral $x$; for $x = 2$, $x = 3$ it is clearly correct; supposing its truth up to and including $n-1$ we have, for $n$ even, $\prod_{p \le n} p = \prod_{p \le n-1} p < 4^{n-1} < 4^n$ and, for $n$ odd (say $n = 2k+1$),

$$\prod_{p \le n} p = \left(\prod_{p < k+1} p\right) P_{k+1} < 4^{k+1} \cdot 4^k = 4^n;$$

iii) $\binom{2n}{n} = \frac{2n(2n-1)\cdots(n+1)}{n!}$ and, since $p \le n < 2p \le 2n < 3p$, we see that $p$, but not $p^2$, divides the numerator and $p$ divides the denominator, thus $p$ does not divide the quotient;

iv-a) we prove the left inequality by induction, it being clearly true for $n=2$; assume true for $n$; then

$$\binom{2n+2}{n+1} = \frac{(2n+2)(2n+1)}{(n+1)^2}\binom{2n}{n} > 2\cdot\frac{2n+1}{n+1}\cdot\frac{4^n}{2\sqrt{n}} = \frac{4^{n+1}}{2\sqrt{n+1}}\cdot\frac{2n+1}{2\sqrt{n^2+n}} > \frac{4^{n+1}}{2\sqrt{n+1}};$$

b) for the right inequality note that

$$\binom{2n}{n} \le \Big(\prod_{p\le\sqrt{2n}}(2n)\Big)\Big(\prod_{p\le\frac{2n}{3}} p\Big)\mathbb{P}_n < (2n)^{\frac{1}{2}\sqrt{2n}}\,4^{\frac{2n}{3}}\,\mathbb{P}_n;$$

v) by (iv), $\mathbb{P}_n > 1$ when $4^{\frac{n}{3}} > 2\sqrt{n}\,(2n)^{\frac{1}{2}\sqrt{2n}}$; raising both sides to the $6^{th}$ power yields the desired result;

vi) using the binomial theorem we note

$$2n = (\sqrt[6]{2n})^6 < (1+[\sqrt[6]{2n}])^6 \le ((1+1)^{\sqrt[6]{2n}})^6 = 2^{6\sqrt[6]{2n}};$$

hence, for $n > 500$,

$$8\cdot(2n)^{3(\sqrt{2n}+1)} < 2^{3+18\sqrt[6]{2n}+18(2n)^{2/3}} < 4^{10(2n)^{2/3}} \le 4^{2n};$$

vii) for $n > 500$ this was proved in (vi) ; now each odd prime in the list 2, 3, 5, 7, 13, 23, 43, 83, 163, 317, 631 is less than twice the preceding prime in the list ; if $n \le 500$ there is a pair of consecutive terms $p, q$ in this list for which $p \le n < q$ ; since $q < 2p \le 2n$ this means $q$ is a prime strictly between $n$ and $2n$ ;

viii) immediate from (vii).

13. i) $(2n)^x < P_n = \prod_{n<p<2n} p < (2n)^{\pi(2n)-\pi(n)}$ and the desired conclusion follows ;

ii) the equality follows by taking logs on both sides of the expression for $(2n)^x$ ; the parenthetical expression is $> 1$ for the values of $n$ specified ;

iii) immediate from (i) and (ii) ;

iv) $$(\pi(2n)-\pi(n))\ln n < \sum_{n<p<2n} \ln p = \ln P_n$$
$$< (n-1)\ln 4 < \frac{7n}{5} \quad \text{for } n \geq 2,$$
where at the 2nd inequality we used # 12 (i) ;

v) direct computation yields

$$\frac{n}{3\ln 2n} < \left\{\begin{matrix} 2 \\ 3 \\ 9 \\ 25 \\ 50 \\ 100 \end{matrix}\right. \leq \pi(2n)-\pi(n) < \frac{7n}{5\ln n} \text{ for } n \text{ in } \left\{\begin{matrix} [6,8] \\ [9,39] \\ [36,150] \\ [135,500] \\ [321,1000] \\ [720,2500] \end{matrix}\right. ;$$

vi) for $n \geq 90$, $\pi(2n)-\pi(n) > \frac{n}{3\ln 2n} > 2$ ; for $6 \leq n < 90$ one sees the truth of the assertion by checking the data given in the proof of (v) ;

vii) by (vi), $\pi(2p_n)-\pi(p_n) \geq 2$ for $p_n \geq 6$ ; since $p_6 = 13 < 2\cdot 7 = 2p_4$ and $p_7 = 17 < 2\cdot 11 = 2p_5$ the result is correct for $n \geq 4$ ;

viii) by (vi), for $n \geq 6$ there are at least 2 primes between $n$ and $2n$; at most one of these may be $\geq 2n-2$; direct checking proves the result for $n=4$ and $n=5$;

ix) this follows from (v) and the fact that $\frac{n}{3\ln 2n}$ increases without bound as $x$ increases.

14. i) This is true for $n = 1, 2, 3, 4, 5$ as can be seen by direct checking; suppose true for some $k$, $k \geq 6$, then we have, using #13 (v),

$$\pi(2^{k+1}) < \pi(2^k) + \frac{7 \cdot 2^k}{5k \ln 2} < \frac{2^{k+1}}{k \ln 2} + \frac{7 \cdot 2^k}{5k \ln 2}$$

$$= \frac{2^{k+2}}{(k+1)\ln 2} \cdot \frac{17(k+1)}{20k} < \frac{2^{k+2}}{(k+1)\ln 2}$$

where only at the last inequality do we need $k \geq 6$;

ii) let $2^{k-1} \leq n < 2^k$, so that

$$\pi(n) \leq \pi(2^k) < \frac{2^{k+1}}{k \ln 2} < \frac{4 \cdot 2^{k-1}}{\ln 2^k} < 4 \frac{n}{\ln n};$$

iii) using #13(v), $\pi(2n) > \pi(n) + \frac{n}{3\ln 2n} > 1 + \frac{n}{3\ln 2n}$;
hence $\pi(n) \geq \pi(2[\frac{n}{2}]) \geq 1 + \frac{[n/2]}{3\ln 2[n/2]} \geq 1 + \frac{n/4}{3\ln n}$

$$= 1 + \frac{n}{12\ln n};$$

iv) this follows from (ii) and (iii) when $x$ is an integer $\geq 2$; since $\frac{1}{12}\left(\frac{x}{\ln x} - \frac{[x]}{\ln[x]}\right) < 1$ and $\frac{x}{\ln x}$ is increasing the result follows from (ii) and (iii) (with some special attention paid to $2 < x < 3$).

15. i) $\pi(mn) - \pi(m) > \frac{mn}{12\ln mn} - \frac{4m}{\ln m}$

$$> \frac{192m}{12\ln m^2} - \frac{4m}{\ln m} = \frac{4m}{\ln m} \geq \frac{4n}{\ln n} > \pi(n);$$

ii) $\pi(mn) - \pi(m) \geq \pi(2m) - \pi(m)$

$$> \frac{m}{3\ln 2m} \geq \frac{4000}{3\ln 8000} > 148 > \frac{4n}{\ln n}.$$

16. i) $\pi(p_m p_n) > \pi(p_m) + \pi(p_n)$

$$= m + n = \pi(p_{m+n});$$

ii) this is true by (i) for $2 \leq n \leq m$, $4 \leq m$; thus we need only check the cases where $n = 1, 2, 3$ and $n \leq m$; but, by #13 (vii),

$$p_1 p_m = 2p_m > p_{m+2},$$

$$p_2 p_m = p_m + 2p_m > p_m + p_{m+1} > p_{m+2} \text{ and}$$

$$p_3 p_m > p_{m-1} + 4p_m > p_{m+1} + 3p_m > p_{m+2} + 2p_m > p_{m+2} + p_{m+1} > p_{m+3};$$

iii) immediate from (ii);

iv) multiply the inequalities in (iii).

17. i) $\prod_{d|n} \prod_{\substack{k=1 \\ (k,d)=1}}^{d} \left(x - e^{\frac{2\pi i k}{d}}\right) = \prod_{d|n} \prod_{\substack{k=1 \\ (k,n/d)=1}}^{n/d} \left(x - e^{\frac{2\pi i k d}{n}}\right)$

$= \prod_{d|n} \prod_{\substack{k=1 \\ (kd,n)=d}}^{n/d} \left(x - e^{\frac{2\pi i k d}{n}}\right) = \prod_{t=0}^{n-1} \left(x - e^{\frac{2\pi i t}{n}}\right) = x^n - 1$;

ii) $F_1(x) = x - 1$ and the proposition is true; assume the proposition is true for all positive integers $< n$; then $\prod_{\substack{d|n \\ d<n}} F_d(x)$ is monic

and integral and, therefore, so also is $F_n(x)$ since $x^n - 1 = F_n(x) \prod_{\substack{d|n \\ d<n}} F_d(x)$; noting that there are exactly $\varphi(n)$ values of $k$ for which $(k, n) = 1$ we see the degree of $F_n(x)$ is $\varphi(n)$;

iii) true for $n = 2$ and the identity

$$x^n - 1 = F_n(x) \prod_{\substack{d|n \\ d<n}} F_d(x)$$

shows the proposition carries over to $n$ from integers $< n$ (note that $F_1(0) = -1$ and $x^n - 1$ is also $-1$ at $x = 0$);

iv) if $p \mid F_n(a)$ then $p \mid a^n - 1$, which implies $(a, p) = 1$;

v-a) suppose $n = qt + r$, $0 \leq r < t$; then

$$1 \equiv a^n = (a^t)^q a^r \equiv a^r \pmod{p};$$

thus $r = 0$ (by definition of $t$) and $t \mid n$;

b) let $c$ be either $p$ or $a+p$; then $p$ divides each of $c^t-1$ and $F_n(c)$; but since $c^t-1=\prod_{d|t} F_d(c)$ and $t|n$, $t<n$, we know $(c^t-1)F_n(c)$, and therefore also $p^2$, divides $c^n-1=\prod_{d|n} F_d(c)$;

$$\text{hence } c^t-1\equiv 0 \pmod{p^2};$$

c) from (b),

$$(a+p)^n-1 = a^n-1+\sum_{j=0}^{n-1}\binom{n}{j}a^j p^{n-j} \equiv na^{n-1}p \pmod{p^2};$$

$$\text{but } (a,p)=1 \text{ so } p|n;$$

d) if $p\nmid n$ then, by (a) and (c), $t=n$; but $a^{p-1}\equiv 1 \pmod p$ and, since $t$ is the smallest number with $a^t\equiv 1 \pmod p$ we must have $t|p-1$;

$$\text{since } t=n,\ n|p-1;$$

vi) that $F_n(nyp_1\cdots p_k)>1$ for $y$ sufficiently large follows from the fact that the leading coefficient of $F_n(x)$ is 1; since

$$nyp_1\cdots p_k\equiv 0 \pmod{np_1\cdots p_k}$$

it is clear that $F_n(nyp_1\cdots p_k)\equiv F_n(0) \pmod{np_1\cdots p_k}$;

the last congruence follows from (iii); finally, let $p$ be any prime divisor of $F_n(nyp_1 \cdots p_k)$; then $p \neq p_j$ for $1 \le j \le k$ and also $p \nmid n$; therefore, by (v-d), $p \equiv 1 \pmod{n}$;

vii) by (vi) no finite collection $p_1, \cdots, p_k$ can exhaust all primes $p$ with $p \equiv 1 \pmod{n}$; the conclusion follows.

18. i) $(\epsilon_m)^{\frac{n}{\alpha}} = e^{\frac{2\pi i m}{n} \cdot \frac{n}{\alpha}} = e^{2\pi i \frac{m}{\alpha}} = 1$ precisely when $\alpha | m$; since the derivative of $x^{\frac{n}{\alpha}} - 1$ is $\frac{n}{\alpha} x^{\frac{n}{\alpha} - 1}$ we see no zero of $x^{\frac{n}{\alpha}} - 1$ is of order $> 1$;

ii) the number of elements in $A_j$ which divide $d$ is just $\binom{s}{j}$; consequently the power of $x - \epsilon_m$ in $q(x)$ is just $\binom{s}{1} + \binom{s}{3} + \binom{s}{5} + \cdots$; note that $x - \epsilon_m$ is not a factor of $F_n(x)$ since $d > 1$; similar reasoning applies to $f(x)$;

iii) when $d = 1$, $x - \epsilon_m$ appears exactly once in each of $F_n(x)$, $x^n - 1$ and in no other factors of $g$ and $f$; thus, since $g$ and $f$ are monic and have the same zeros to the same orders, and since $\binom{s}{1} + \binom{s}{3} + \binom{s}{5} + \cdots = 1 + \binom{s}{2} + \binom{s}{4} + \cdots$ we conclude $f$ and $g$ are identical; the result follows;

iv) the $A_j$ corresponding to $np$ are the same as for $n$; thus, since $x^{\frac{np}{\alpha}} = (x^p)^{n/\alpha}$, the conclusion follows;

v) let $A_j$, $1 \le j \le r+1$, be the sets corresponding to the $A_j$ for $n$; then $F_{np}(x) F_n(x) =$

$$\frac{(x^{np}-1) \prod_{\substack{1 \le j \le r+1 \\ j \text{ odd}}} \prod_{\alpha \in A_j} (x^{\frac{np}{\alpha}} - 1)}{\prod_{\substack{1 \le j \le r+1 \\ j \text{ even}}} \prod_{\alpha \in A_j} (x^{\frac{np}{\alpha}} - 1)} \cdot \frac{(x^n - 1) \prod_{\substack{1 \le j \le r \\ j \text{ odd}}} \prod_{\alpha \in A_j} (x^{\frac{n}{\alpha}} - 1)}{\prod_{\substack{1 \le j \le r \\ j \text{ even}}} \prod_{\alpha \in A_j} (x^{\frac{n}{\alpha}} - 1)};$$

for each $\alpha$ containing the factor $p$ the factor $x^{\frac{np}{\alpha}} - 1$ in the expression for $F_{np}(x)$ is canceled by the factor $x^{\frac{n}{\alpha/p}} - 1$ in the expression for $F_n(x)$; consequently the right side is just $F_n(x^p)$;

vi) from (v), $F_p(x) = \frac{F_1(x^p)}{F_1(x)} = \frac{x^p - 1}{x-1} = x^{p-1} + \cdots + 1$ ;

vii) $F_1(1) = x - 1$ so $F_1(1) = 0$ ; by (iv), $F_{np}(1) = F_n(1)$ ; thus $F_{p^k}(1) = F_{p^{k-1}}(1) = \cdots = F_p(1)$, and this last quantity is p, by (vi); if n has 2 distinct prime factors, say $n = mq^\alpha$ where $m > 1$, $(m, q) = 1$, q prime then, using (iv) and (v),
$F_n(1) = F_{mq^{\alpha-1}}(1) = F_{mq^{\alpha-2}}(1) = \cdots = F_{mq}(1) = \frac{F_m(1)}{F_m(1)} = 1$ ;

viii) this follows immediately from (iii); when d contains a square we do not find $x^{\frac{n}{d}} - 1$ in either numerator or denominator; when $d = 1$ we get the factor $x^n - 1$ and when d is the product of j distinct primes then $x^{\frac{n}{d}} - 1$ is in the numerator when j is even and in the denominator when j is odd ;

ix) using (viii), we see that when $d = \frac{n}{p_i}$, $\mu(d) = 1$ so $x^{p_i} - 1$ is a factor of $F_n(x)$ ;

all other factors than those of this form are congruent to $-1 \pmod{x^{p_t+1}}$ since all exponents of $x$ in these factors exceed $p_t+1$ as a consequence of $p_1+p_2>p_t$; the 2nd congruence follows for similar reasons;

x) for each $i$, $1 \le i \le t-1$, there is a $j$, $0 \le j \le p_t - 1$ for which $x^{p_i} x^j = x^{p_t}$; since there are $t-1$ such $i$ the result follows.

19. i) This is by direct examination;

ii) for $n=7$, each of $12, 13, \cdots, 29$ are in $S_6$ (and in $S_7$) so each of $12+p_7, 13+p_7, \cdots, 29+p_7$ are in $S_7$; if $12, 13, \cdots, 29+p_7+\cdots+p_{n-1}$ are in $S_{n-1}$ (and $S_n$) then $12+p_n, 13+p_n, \cdots, 29+p_7+\cdots+p_n$ are in $S_n$; these blocks overlap since (using #12 viii) $12+p_k<29+p_7+\cdots+p_{k-1}$ for $k \ge 8$;

iii) immediate from (ii);

iv) direct inspection and (iii);

v) direct inspection and (iv).

20.i) $\varphi$, $S$, and unions of open sets are clearly open; if $\mathcal{O}_1, \cdots, \mathcal{O}_t$ are open and $x \in \mathcal{O}_1 \cap \cdots \cap \mathcal{O}_t$ then there are integers $s_1, \cdots, s_t$ such that $x + ns_i \in \mathcal{O}_i$ for all integers $n$; thus $x + ns_1 \cdots s_t \in \mathcal{O}_i$ for all $i$, $1 \leq i \leq t$, and all integers $n$; i.e. $x$ is in an arithmetic progression contained in $\mathcal{O}_1 \cap \cdots \cap \mathcal{O}_t$ and, therefore, this intersection is open;

ii) the complement of an arithmetic progression with difference $d$ is a union of $d-1$ arithmetic progressions with difference $d$; since both sets are open and they mutually exhaust $S$ they must both be closed;

iii) immediate from (ii) ;

iv) this is true since each integer other than $\pm 1$ has a prime factor ;

v) if there were only finitely many primes the set A of (iv) would be closed and consequently would have an open complement ; since the complement $\{-1, 1\}$ is not open we conclude there are infinitely many primes .

21. When $n$ is an odd prime Wilson's theorem tells us $(n-2)! \equiv -(n-1)! \equiv 1 \pmod{n}$ so $\left[\frac{(n-2)!}{n}\right] = \frac{(n-2)!-1}{n}$ is an odd integer ; when $n$ is composite IX#11 tells us $\left[\frac{(n-2)!}{n}\right] = \frac{(n-2)!}{n}$ is an even integer ; therefore when either $n$ or $n+2$ is composite the term of the summand corresponding to $n$ will be 0 since the sine of an even

multiple of $\frac{\pi}{2}$ is 0, while if $n$ and $n+2$ are each prime then the term is the product of the sine of an odd multiple of $\frac{\pi}{2}$ with another sine of an odd multiple of $\frac{\pi}{2}$, hence is $(-1)(-1)=1$; the 2 accounts for 3,5 and 5,7.

22. i) Immediate;

ii) $\sum_{k=1}^{m} F(k) = 1 + \pi(m)$ by (i); now if $m < p_n$ then $1 + \pi(m) \leq n$, while if $m \geq p_n$ then $1 + \pi(m) > n$; since $p_n < 2^{2n}$, by #4 (iv), the sum $\sum_{m=1}^{2^{2n}} \left[\frac{n}{1+\pi(m)}\right]$ counts 1 for each $m < p_n$; thus this sum plus 1 is exactly $p_n$.

## XV Quaternions, Complex Numbers, & Sums of 4 and 2 Squares ~ Solutions

1. i) The determinant of $\begin{pmatrix} a & b \\ -\bar{b} & \bar{a} \end{pmatrix}$ is $|a|^2+|b|^2$ and this is 0 if and only if $a = b = 0$; the upper left elements in the two products of $\begin{pmatrix} 1 & 1+i \\ -1+i & 1 \end{pmatrix}$ and $\begin{pmatrix} i & 1 \\ -1 & -i \end{pmatrix}$ are $-1$ and $2i-1$ so $\mathbb{C}'$ is non-commutative; for $a, b$ real $\begin{pmatrix} a & b \\ -\bar{b} & \bar{a} \end{pmatrix} = \begin{pmatrix} a & b \\ -b & a \end{pmatrix}$; all else is routine verification;

ii) the correspondence is $\begin{pmatrix} a & b \\ -b & a \end{pmatrix} \leftrightarrow a+ib$;

iii) if $a' = a+ib$, $b' = c+id$, where $a, b, c, d$ are in $\mathbb{R}$, then $\begin{pmatrix} a' & b' \\ -\bar{b}' & \bar{a}' \end{pmatrix} = a\begin{pmatrix} 1 & 0 \\ 0 & 1 \end{pmatrix} + b\begin{pmatrix} i & 0 \\ 0 & -i \end{pmatrix} + c\begin{pmatrix} 0 & 1 \\ -1 & 0 \end{pmatrix} + d\begin{pmatrix} 0 & i \\ i & 0 \end{pmatrix}$

and $\begin{pmatrix} a & -b & -c & -d \\ b & a & -d & c \\ c & d & a & -b \\ d & -c & b & a \end{pmatrix} =$

$$a\begin{pmatrix} 1&0&0&0 \\ 0&1&0&0 \\ 0&0&1&0 \\ 0&0&0&1 \end{pmatrix} + b\begin{pmatrix} 0&-1&0&0 \\ 1&0&0&0 \\ 0&0&0&-1 \\ 0&0&1&0 \end{pmatrix} + c\begin{pmatrix} 0&0&-1&0 \\ 0&0&0&1 \\ 1&0&0&0 \\ 0&-1&0&0 \end{pmatrix} + d\begin{pmatrix} 0&0&0&-1 \\ 0&0&-1&0 \\ 0&1&0&0 \\ 1&0&0&0 \end{pmatrix};$$

make the correspondence

$$\begin{pmatrix} 1 & 0 \\ 0 & 1 \end{pmatrix} \leftrightarrow \begin{pmatrix} 1 & 0 & 0 & 0 \\ 0 & 1 & 0 & 0 \\ 0 & 0 & 1 & 0 \\ 0 & 0 & 0 & 1 \end{pmatrix}, \quad \begin{pmatrix} i & 0 \\ 0 & -i \end{pmatrix} \leftrightarrow \begin{pmatrix} 0 & -1 & 0 & 0 \\ 1 & 0 & 0 & 0 \\ 0 & 0 & 0 & -1 \\ 0 & 0 & 1 & 0 \end{pmatrix},$$

$$\begin{pmatrix} 0 & 1 \\ -1 & 0 \end{pmatrix} \leftrightarrow \begin{pmatrix} 0 & 0 & -1 & 0 \\ 0 & 0 & 0 & 1 \\ 1 & 0 & 0 & 0 \\ 0 & -1 & 0 & 0 \end{pmatrix}, \quad \begin{pmatrix} 0 & i \\ i & 0 \end{pmatrix} \leftrightarrow \begin{pmatrix} 0 & 0 & 0 & -1 \\ 0 & 0 & -1 & 0 \\ 0 & 1 & 0 & 0 \\ 1 & 0 & 0 & 0 \end{pmatrix};$$

the assertion is now a matter of routine verification;

iv) one needs only show the 4 $4 \times 4$ matrices of $R''$ which are displayed in the proof of (iii) are independent and this is clear;

v) that this set is a subfield is clear and the non-commutativity is proved by the example given in the proof of (i);

vi) this follows from associativity and distributivity properties of addition and multiplication.

2\. i) $\alpha - \bar{\alpha} = 2(ib + jc + kd) = 0$ if and only if $b = c = d = 0$ since $i, j, k$ are independent;

ii) $N\alpha = a^2 + b^2 + c^2 + d^2$

$$= a^2 + (-b)^2 + (-c)^2 + (-d)^2 = N\bar{\alpha}\ ;$$

$$T\alpha = 2a = (a + ib + jc + kd)$$
$$+ (a - ib - jc - kd) = \alpha + \bar{\alpha}\ ;$$

iii) $N\alpha = a^2 + b^2 + c^2 + d^2 = 0$ if and only if $a = b = c = d = 0$ ; i.e. if and only if $\alpha = 0$ ;

iv) direct verification ;

v) $\overline{\alpha \cdot 1} = \bar{\alpha} = \bar{1} \cdot \bar{\alpha}$ ,

$\overline{\alpha i} = \overline{ia - b - kc + jd} = -b - ia - jd + kc =$ $(-i)(a - ib - jc - kd) = \bar{i}\,\bar{\alpha}$ ; similarly $\overline{\alpha j} = \bar{j}\,\bar{\alpha}$, $\overline{\alpha k} = \bar{k}\,\bar{\alpha}$ ; now use (iv) and distributivity to obtain $\overline{\alpha\beta} = \bar{\beta}\,\bar{\alpha}$ ;

vi) clear ;

vii) $N(\alpha\beta) = (\alpha\beta)(\overline{\alpha\beta}) = (\alpha\beta)(\bar{\beta}\,\bar{\alpha})$
$= \alpha(N\beta)\bar{\alpha} = (\alpha\bar{\alpha})(N\beta) = (N\alpha)(N\beta)$ ;

viii) substitution yields the result immediately.

3. i) Let $f(\alpha)=0$, where $f$ is a monic integral polynomial; since $\alpha$ also satisfies the principal equation there is a monic integral irreducible polynomial $g$, of degree 1 or 2, satisfying $g(\alpha)=0$; if $\alpha$ is not rational then $g(x)=0$ is the principal equation, while if $\alpha$ is rational $g^2(x)=0$ is the principal equation; in either case $T\alpha$ and $N\alpha$ are integers; the reverse direction is clear by # 2 (viii) ;

ii) clearly $\mathcal{L} \subset \mathcal{H}$ and $\rho \in \mathcal{H} \setminus \mathcal{L}$; if $\beta = \rho + \alpha$, $\alpha = a + ib + jc + kd \in \mathcal{L}$ then
$T\beta = \rho + \alpha + \bar{\rho} + \bar{\alpha} = 1 + \alpha + \bar{\alpha}$
$= 1 + T\alpha = 1 + 2a \in \mathbb{Z}$ ,

$$N\beta = (\rho+\alpha)(\bar\rho+\bar\alpha) = \rho\bar\rho + \alpha\bar\rho + \bar\alpha\bar{\bar\rho} + \alpha\bar\alpha$$
$$= \left(\tfrac{1}{4}+\tfrac{1}{4}+\tfrac{1}{4}+\tfrac{1}{4}\right) + (a-b-c-d) + (a^2+b^2+c^2+d^2) \in \mathbb{Z}$$

so, by (i), $\mathcal{H} \subset I$ ; now $\frac{3}{5}i + \frac{4}{5}j$ is not in $\mathcal{H}$ but $T(\frac{3}{5}i + \frac{4}{5}j) = 0 \in \mathbb{Z}$, $N(\frac{3}{5}i + \frac{4}{5}j) = \frac{9}{25} + \frac{16}{25} = 1 \in \mathbb{Z}$ so $\frac{3}{5}i + \frac{4}{5}j \in I$ ;

iii) when $A = \mathcal{L}$ this is obvious ; when $A = \mathcal{H}$ note first that $\bar\rho = \rho + (-i-j-k) \in \mathcal{H}$ so if $\alpha = \rho + \beta \in \mathcal{H}$, $\beta \in \mathcal{L}$, then $\bar\alpha = \bar\rho + \bar\beta \in \mathcal{H}$; further, $i(\rho+\beta) = \rho + (-1-j+i\beta)$ so when $\rho+\beta \in \mathcal{H}$, $\beta \in \mathcal{L}$, so also is $i(\rho+\beta)$ in $\mathcal{H}$ ; similar arguments work for $j\alpha$ and $k\alpha$ ;

iv) put $\alpha = \frac{3}{5}i + \frac{4}{5}j$ ; by the proof of (ii) we know $\alpha \in I$ ; now $i\alpha = -\frac{3}{5} + \frac{4}{5}k$ and $T(i\alpha) = -\frac{6}{5} \notin \mathbb{Z}$ ; thus by (i) $i\alpha \notin I$ ;

v) as we have seen in (iv) $I$ is not closed under multiplication and is, therefore, not an integral domain ; using (iii) it is routine verification to prove $\mathcal{L}$ and $\mathcal{H}$ are integral domains ;

vi) if $\alpha = a + ib + jc + kd$ is in an integral domain of I which contains $\mathcal{H}$ then since $\alpha$, $-i\alpha, -j\alpha, -k\alpha$ are all in I, we know $T\alpha = 2a \in \mathbb{Z}$, $T(-i\alpha) = 2b \in \mathbb{Z}$, $T(-j\alpha) = 2c \in \mathbb{Z}$, $T(-k\alpha) = 2d \in \mathbb{Z}$ ; further $N\alpha = a^2 + b^2 + c^2 + d^2 \in \mathbb{Z}$ ; if any one of $a, b, c, d$ is half an odd integer this last fact implies they are all halves of odd integers ; since $a, b, c, d$ are all integers or all halves of odd integers we see $\alpha \in \mathcal{H}$ ; this shows $\mathcal{H}$ is maximal in I ;

vii) $a\rho + ib + jc + kd$ is in $\mathcal{L}$ and hence in $\mathcal{H}$ when $a$ is even ; when $a$ is odd this equals $\rho + \beta$, where $\beta = (a-1)\rho + ib + jc + kd \in \mathcal{L}$ ; on the other hand $\rho + \alpha = (1 + 2a)\rho + i(b-a) + j(c-a) + k(d-a)$.

4. i) Each element of $\mathcal{L}$ is its own left associate; if $\alpha \in \mathcal{H} \setminus \mathcal{L}$ we may write

$$\alpha = 2(a + ib + jc + kd) + \alpha_1 \text{, where}$$

$\alpha_1 = \frac{1}{2}(e + if + jg + kh)$; $e, f, g, h$ are all $\pm 1$, $a, b, c, d$ are in $\mathbb{Z}$; since $N\alpha_1 = 1$, $\alpha\overline{\alpha}_1$ is a left associate of $\alpha$; finally

$$\alpha\overline{\alpha}_1 = (a + ib + jc + kd)(e - if - jg - kh) + 1 \in \mathcal{L};$$

ii) let $\alpha = a\rho + ib + jc + kd$ and try for $\beta = e\rho + if + jg + kh$; then $\alpha - m\beta =$

$$\frac{a-me}{2} + i\frac{a-me+2b-2mf}{2} + j\frac{a-me+2c-2mg}{2} + k\frac{a-me+2d-2mh}{2};$$

now choose $e, f, g, h$ so that each of the following four quantities falls between $-\frac{m}{2}$ and $\frac{m}{2}$ inclusive: $a - me$, $\frac{a-me+2b}{2} - mf$, $\frac{a-me+2c}{2} - mg$, $\frac{a-me+2d}{2} - mh$; with these choices, $N(\alpha - m\beta) \le \frac{m^2}{16} + \frac{3m^2}{4} < m^2$;

iii) let $m = \beta\overline{\beta}$ and choose $\gamma_1$ so that $N(\overline{\beta}\alpha - m\gamma_1) < m^2$; put $\delta_1 = \alpha - \beta\gamma_1$;

then $\alpha = \beta\gamma_1 + \delta_1$ and $mN\delta_1 = N(\bar{\beta}\delta_1) = N(\bar{\beta}\alpha - m\gamma_1) < m^2$; consequently $N\delta_1 < m = N\beta$; the other part goes the same way;

iv) under the hypothesis each of $\delta, \delta_1$ is a left divisor of the other; thus $\delta = \delta_1\delta'$, $\delta_1 = \delta\delta'' = \delta_1\delta'\delta''$ and $\delta'\delta'' = 1$; this means $\delta$ and $\delta_1$ are left associates; on the other hand if $\delta$ is a left gcd of $\alpha$ and $\beta$ and $\delta_1 = \delta\nu$ then from $\gamma = \delta\gamma_1$ we find $\gamma = \delta_1\nu^{-1}\gamma_1$; thus $\delta_1$ left divides all numbers left divisible by $\delta$, so $\delta_1$ is a left gcd of $\alpha$ and $\beta$;

v) let $\delta$ be an element of minimal positive norm in A; further, let $\delta_1$ be a common left divisor of $\alpha$ and $\beta$; then, by (iii),

$$\alpha = \delta\alpha_1 + \alpha_2 \quad , \quad N\alpha_2 < N\delta ;$$

$$\beta = \delta\beta_1 + \beta_2 \quad , \quad N\beta_2 < N\delta ;$$

$$\delta_1 = \delta\delta' + \delta'' \quad , \quad N\delta'' < N\delta ;$$

since $\alpha - \delta\alpha_1$, $\beta - \delta\beta_1$ are in A we must have $N\alpha_2 = N\beta_2 = 0$ so $\alpha_2 = \beta_2 = 0$ and $\delta$ is a common left divisor of $\alpha$ and $\beta$; also since $\delta_1$ is a left divisor of $\alpha$ we know $\alpha$ is a left divisor of $\delta_1$ so $\delta_1 - \delta\delta'$ is in A which implies $\delta'' = 0$ and, therefore, $\delta$ is a left divisor of $\delta_1$;

vi) if $\alpha = a + ib + jc + kd$ and $N\alpha = 1$ then $a^2 + b^2 + c^2 + d^2 = 1$; since a, b, c, d are either all integers or all halves of odd integers the units are precisely those quaternions corresponding to (a, b, c, d) one of: $(\pm 1, 0, 0, 0)$, $(0, \pm 1, 0, 0)$, $(0, 0, \pm 1, 0)$, $(0, 0, 0, \pm 1)$, $(\pm \frac{1}{2}, \pm \frac{1}{2}, \pm \frac{1}{2}, \pm \frac{1}{2})$, and there are just 24 of these; that every unit is a two-sided divisor of each element of $\mathcal{H}$ follows from (iii) and the fact that no element of $\mathcal{H}$ has positive norm smaller than 1; as for the units in $\mathcal{L}$, this is clear since $a^2+b^2+c^2+d^2=1$ for integers $a, b, c, d$ only if one of these is $\pm 1$ and all the others are $0$ ;

vii) if $(\alpha, n) = \alpha\mu + n\nu$ were a unit then
$$1 = (\alpha\mu + n\nu)(\overline{\alpha\mu + n\nu}) = (\alpha\mu + n\nu)(\bar{\mu}\bar{\alpha} + \bar{\nu}n)$$
$\equiv 0 \pmod{n}$, contrary to fact.

5. i) If $\alpha$ is a prime and $\beta = \mu\alpha\nu$, where $\mu$ and $\nu$ are units then $N\beta = N\alpha > 1$ so $\beta$ is not a unit ; if $\beta = \gamma\delta$, $N\gamma > 1$, $N\delta > 1$ then $\alpha = (\mu^{-1}\gamma)(\delta\nu^{-1})$, $N(\mu^{-1}\gamma) = N\gamma > 1$, $N(\delta\nu^{-1}) = N\delta > 1$ and $\alpha$ would not be prime ; thus $\beta$ is prime ;

ii) if $\alpha$ is not a prime in $\mathcal{H}$, $\alpha = \beta\gamma$, $N\beta > 1$, $N\gamma > 1$ ; but then $N\alpha = N\beta N\gamma$ is not a rational prime ;

iii) let $\pi = (\alpha, p)$, $\alpha = \pi\alpha_1$, $p = \pi p_1$; by #4(vii), $N\pi > 1$ so, from $p^2 = Np = N\pi\, Np_1$, either $N\pi = p$ or $N\pi = p^2$; in the 1st case we are done; in the other case $Np_1 = 1$ and, therefore, $\alpha = \pi\alpha_1 = p\bar{p}_1\alpha_1$ which means $\alpha$ is divisible by $p$; this contradicts the hypothesis that $\alpha$ is primitive;

iv) for $p$ an odd rational prime there are rational integers $m$ and $n$ for which $1 + m^2 + n^2$ is divisible by $p$ (see e.g. XI #14(i)); let $\alpha = 1 + im + jn$; then $(p, \alpha)$ is prime and $N((p, \alpha)) = p$; also $2 = (1 + i)(1 - i)$;

v) in view of (ii) we need only prove one direction; let $\alpha$ be a prime in $\mathcal{H}$ and suppose $p$ is a rational prime dividing $N\alpha$; then if $\delta = (\alpha, p)$ we must have $\alpha = \delta\alpha_1$ and by (iii), $N\delta = p$ ; since $\alpha$ is prime in $\mathcal{H}$ this means $N\alpha_1 = 1$ so $N\alpha = N\delta = p$ ;

vi) since $2 \mid N\alpha$, $\alpha = a + ib + jc + kd \in \mathcal{L}$ and $a + b + c + d \equiv N\alpha \equiv 0 \pmod 2$;
now put $\beta = \frac{1}{2}((a+b) + i(b-a) + j(c+d) + k(d-c))$ and observe that $\alpha = (1+i)\beta$ ;

vii) by (vi), $\alpha = (1+i)^s \gamma$, where $N\gamma$ is odd; $(1+i)^2 = 2i$ so $(1+i)^s = (1+i)^r n\nu$, where $n$ is a rational integer, $\nu$ is a unit, and $r = 0$ or $1$; now $\gamma = t\delta$, where $t \in \mathbb{Z}$ and $\delta$ is primitive; thus $\alpha = (1+i)^r m\nu\delta = (1+i)^r m\beta\nu$, where $\nu\delta = \beta\nu$ and $N\beta = N\delta = N\gamma$ is odd; if $\beta$ is not in $\mathcal{L}$, by #4(i) we can make it be in $\mathcal{L}$ by altering the unit $\nu$ ;

viii) by (iii) if $\pi_1 = (\gamma, p_1)$ then $N\pi_1 = p_1$ ; putting $\gamma = \pi_1\gamma_2$ we see that $p_2$ divides $N\gamma_2$

so, again by (iii), if $\pi_2 = (\gamma_2, p_2)$ then $N\pi_2 = p_2$; repeating yields $\gamma = \pi_1 \cdots \pi_s$; now, by (vii), $\alpha = (1+i)^r \beta \nu$ for some primitive $\beta$ of odd norm in $\mathcal{L}$ and $\nu$ some unit; by the 1st part (with $\gamma = \beta$) we see $\alpha = (1+i)^r \pi_1 \cdots \pi_s$ ($\pi_s$ is the earlier $\pi_s$ multiplied on the right by $\nu$);

ix) let $\beta = \tau_1 \cdots \tau_s$; then $N\beta = p_1 \cdots p_s$ and $\pi_1 = (\beta, p_1)$; since $\tau_1$ divides each of $\beta$ and $p_1$ so $\tau_1$ divides $\pi_1$; since $\tau_1$ and $\pi_1$ are primes they must be associates; repeat with

$$\beta_2 = \tau_2 \cdots \tau_s, N\beta_2 = p_2 \cdots p_s, \text{etc.};$$

x) immediate from (vii)-(ix);

xi) $7 = (1+i+j+2k)(1-i-j-2k)$

$$= (1-i-j+2k)(1+i+j-2k).$$

6. i) All quaternions of the form $(1+i)\nu$, where $\nu$ is a unit have norm 2 and, by #4 (vi), there are 24 of them; on the other hand all quaternions of norm 2 are in $\mathcal{C}$ and $a^2+b^2+c^2+d^2=2$ implies precisely 2 of the $a, b, c, d$ are $\pm 1$; the number of such quaternions in $\mathcal{C}$ is $4\binom{4}{2}=24$;

ii-a) for each quadruple $A, B, C, D$ satisfying (1) there is exactly one quadruple $A_1, B_1, C_1, D_1$ satisfying (2), namely, let $A_1=A$, $D_1=D$ and $B_1, C_1$ be chosen so that $-aA+B_1\equiv B$, $-bA+C_1\equiv C \pmod p$; the reverse direction is clear;

b) we may choose $a, b$ so that $a^2+b^2+1$ is divisible by $p$ (see XI #14 (i)); then using (2) we have $B^2+C^2+D^2\equiv 2A(aB+bC)-(1+a^2+b^2)A^2$

$$\equiv 2A(aB+bC) \pmod p;$$

c) from (3) we see, when $B \equiv C \pmod{p}$, that $D \equiv 0 \pmod{p}$; hence the $p$ solutions are obtained from $A$, since all $p$ possible values of $A$ will be suitable; when $B \not\equiv C \pmod{p}$ and $a, b$ are as in (6) it is clear that $(B, C, p) = (a, b, p) = 1$; thus, without loss of generality, $(B, p) = (b, p) = 1$; then there is a unique $E$ such that $B \equiv bE \pmod{p}$ and, for this $E$, we have $C \equiv -aE \pmod{p}$; hence $0 \equiv B^2 + C^2 + D^2 \equiv (b^2 + a^2)E^2 + D^2 \equiv -E^2 + D^2 \pmod{p}$; therefore $D \equiv \pm E \pmod{p}$; with $A$ arbitrary, $E$ any non-zero value ($B \not\equiv C \pmod{p}$) and $D \equiv \pm E \pmod{p}$, we find altogether $2p(p-1)$ solutions;

d) as in the proof of (c) we note that $(a, b, p) = 1$ so that we may assume without loss of generality that $(b, p) = 1$; thus for each of the $p$ possible values of $B$ there are

$p-1$ values for $C$ such that $aB+bC \not\equiv 0 \pmod p$; for each of these $p^2-p$ choices for $B, C$ and every one of the $p$ possible choices for $D$ there is a unique $A$; consequently there are exactly $p(p^2-p)$ quadruples $A, B, C, D$ satisfying (3) ;

e) by (c) and (d) the number of solutions is $p+2p(p-1)+p(p^2-p)=(p^2-1)(p+1)+1$ .

iii-a) If the assertion were false then $p$ would divide each of $a_0^2+a_1^2, a_0^2+a_2^2, a_0^2+a_3^2$ and then, since it also divides $N\alpha = a_0^2+a_1^2+a_2^2+a_3^2$, it would divide $2a_0^2$; thus $p$ would divide $a_0$ and, thence, also $a_1, a_2, a_3$ so would divide $\alpha$, contrary to supposition ;

b-1) $\beta+\gamma i_{\nu+1} = a_0 + a_\nu i_\nu + a_{\nu+1} i_{\nu+1} + a_{\nu+2} i_\nu i_{\nu+1}$

$$= a_0 + a_\nu i_\nu + a_{\nu+1} i_{\nu+1} + a_{\nu+2} i_{\nu+2} = \alpha ;$$

similarly the expression for $x$ is correct ;

2) this is clear since $i_\nu$ commutes with itself and with scalars ;

3) for $a, b$ scalars $(a+bi_\nu)i_{\nu+1} =$
$ai_{\nu+1}+bi_\nu i_{\nu+1} = ai_{\nu+1} - bi_{\nu+1}i_\nu = i_{\nu+1}(a-bi_\nu)$ ;

c) let $\nu$ be as in (a);

$\alpha x = (\beta\eta - \gamma\bar{\xi}) + (\beta\xi + \gamma\bar{\eta})i_{\nu+1}$ ;

since $\beta\eta - \gamma\bar{\xi}$ is of the form $a+bi_\nu$ and $(\beta\xi+\gamma\bar{\eta})i_{\nu+1}$ is of the form $ci_{\nu+1}+di_{\nu+2}$, with $a, b, c, d$ scalars, we see that when $p$ divides $\alpha x$ it must also divide each of $\beta\eta-\gamma\bar{\xi}$ and $\beta\xi+\gamma\bar{\eta}$ ; on the other hand if $p$ divides $\beta\eta-\gamma\bar{\xi}$ then $\beta\eta \equiv \gamma\bar{\xi} \pmod p$ so, multiplying by $\bar{\beta}\bar{\gamma}$ yields $\beta\bar{\beta}\eta\bar{\gamma} \equiv \gamma\bar{\gamma}\bar{\beta}\bar{\xi} \pmod p$ ; but since $\beta\bar{\beta} \not\equiv 0 \pmod p$ (since $p$ does not divide $a_0^2+a_\nu^2$) and $\beta\bar{\beta} \equiv -\gamma\bar{\gamma} \pmod p$ (since $p$ does divide $N\alpha$) we see $\eta\bar{\gamma} \equiv -\bar{\beta}\bar{\xi} \pmod p$ and, therefore,

$\beta\delta \equiv -\gamma\bar{\eta} \pmod{p}$; hence if $p$ divides $\beta\eta - \gamma\bar{\delta}$ it also divides $\beta\delta + \gamma\bar{\eta}$, and, therefore, divides $\alpha x$;

d) by (c), $\alpha x \equiv 0 \pmod{p}$ is equivalent to $\beta\eta \equiv \gamma\bar{\delta} \pmod{p}$, which in turn, after multiplying by $\bar{\beta}$, is equivalent to $(N\beta)\eta \equiv \bar{\beta}\gamma\bar{\delta} \pmod{p}$; now $\beta$ and $\gamma$ are fixed and, since $p$ does not divide $\beta$, each of the $p^2$ possible $\delta$ values yields a unique $\eta$; thus there are exactly $p^2$ solutions

$$x = \eta + \delta i_{v+1} \text{ of } \alpha x \equiv 0 \pmod{p}.$$

iv) Each prime $\pi$ in $\mathcal{L}$ with $N\pi = p$ gives rise to one of the $(p^2-1)(p+1)$ non-trivial solutions of $N\alpha \equiv 0 \pmod{p}$, $\alpha \not\equiv 0 \pmod{p}$; conversely to each such $\alpha$ there is (see #5 (iii)) a prime $\pi$ in $\mathcal{L}$ with $N\pi = p$; since

$\bar{\pi} x \equiv 0 \pmod p$ has (see (iii-d)) exactly $p^2-1$ non-trivial solutions each $\pi$ must arise from exactly $p^2-1$ different $\alpha$; thus there are $p+1$ $\left(=\frac{(p^2-1)(p+1)}{p^2-1}\right)$ distinct $\pi$.

7. *i*) The presence of such a pair of consecutive factors means that

$$N(\pi_{\nu\eta}\pi_{\nu,\eta+1}) = N(\pi_{\nu\eta}\bar{\pi}_{\nu\eta}) = p_\nu$$

so $p_\nu$ divides $\alpha$ and $\alpha$ is not primitive;

*ii*) the proof is by induction; the proposition is vacuously true for the case when the number of factors is 1; suppose the proposition true for $t-1$ factors and let $\alpha = \pi_1 \cdots \pi_t \equiv 0 \pmod p$ for some prime $p$ in $\mathbb{Z}$; if $\pi_2 \cdots \pi_t$ is not primitive the conclusion is true by (*i*) and the induction hypothesis; otherwise

$$\overline{\pi}_1 \pi_1 \pi_2 \cdots \pi_t = (N\pi_1)\pi_2 \cdots \pi_t \equiv 0 \pmod{p}$$

and, therefore, $N\pi_1 \equiv 0 \pmod{p}$; but then $N\pi_1 = p$; hence

$$p\,\pi_2 \cdots \pi_t = \overline{\pi}_1 \pi_1 \pi_2 \cdots \pi_t = \overline{\pi}_1 p \beta,$$

for suitable $\beta$; but then $\pi_2 \cdots \pi_t = \overline{\pi}_1 \beta$ and, by #5 (x), $\pi_2$ and $\overline{\pi}_1$ are associates;

*iii*) in the product $\pi_{\nu 1} \cdots \pi_{\nu \alpha_\nu}$, there are $p_\nu + 1$ choices for $\pi_{\nu 1}$ by #6 (*iv*), and for each other $\pi_{\nu j}$ there are only $p_\nu$ choices; since, by (*ii*), consecutive factors may not be conjugates; the multiplier 8 comes from the possible 8 factors which are the units in $\mathcal{L}$;

*iv*) by (*ii*) a non-primitive $\alpha$ will have consecutive factors one of which is an associate of the other; each square which is a divisor of $m$ arises from a collection of disjoint such pairs ; thus if $d^2 | m$ then associated with this $d$ there are, by (iii), $8\frac{m}{d^2}\prod_{p|\frac{m}{d^2}}(1+\frac{1}{p})$ primitive elements of $\mathcal{C}$ of norm $\frac{m}{d^2}$, each of which when multiplied by $d^2$ yields an $\alpha$ in $\mathcal{C}$ of norm $m$ ; the total number of such $\alpha$ is just the left hand side of the indicated expression ; now, for each $\delta | m$ if we let $\delta'$ be the largest square factor of $\frac{m}{\delta}$ we find

$$\sigma(m) = \sum_{\delta|m} \delta = \sum_{\delta'|m} \sum_{t|\frac{m}{\delta'}} \frac{m}{\delta' t} = \sum_{d^2|m} \frac{m}{d^2} \sum_{t|\frac{m}{d^2}} \frac{1}{t}$$

$$= \sum_{d^2|m} \frac{m}{d^2} \prod_{p|\frac{m}{d^2}} \left(1+\frac{1}{p}\right),$$

where $t$ is squarefree throughout ;

v) there are 24 elements of $\mathcal{C}$ of norm 2 (see #6(i)) and they are merely permuted by multiplication by the 8 units of $\mathcal{C}$ ; thus, by (iv), we have $24\sigma^{\circ}(n)$ such $\alpha$ ;

vi) since $N\alpha$ is a sum of 4 squares for every $\alpha$ in $\mathcal{L}$ the desired number of solutions may be obtained from (iv) and (v); let $M$ be the number of divisors of $n$ not divisible by 4; if $n$ is odd then $M = \sigma(m)$ by (iv) and the number of $\alpha$ is $8M$; if $n$ is even then $M = \sigma^\circ(n) + 2\sigma^\circ(n) = 3\sigma^\circ(n)$ and by (v) the number of $\alpha$ is $8M$.

8. We sketch the argument in 7 steps.

i) G has exactly 4 units;

ii) #5 (iv) is replaced by: rational primes of the form $4k+3$ are primes in G; no other rational primes are primes in G but each is the norm of a prime in G;

iii) #5(v) is replaced by: $\alpha$ primitive and prime in G implies $N\alpha$ is prime in $\mathbb{Z}$;

iv) #5 (viii) remains the same but it should be noted that if $\alpha$ is primitive and an odd prime divides $N\alpha$ then that prime is of the form $4k+1$; this implies that $N\alpha$, for $\alpha$ in $G$, can never have an odd power of a $4k+3$ prime in its canonical prime factorization;

v) #5 (x) is replaced by: if $\alpha$ in $G$ is such that $N\alpha = 2^r q_1^2 \cdots q_t^2 p_1 \cdots p_s$, where the $q_\nu$ are $4k+3$ primes and the $p_\nu$ are $4k+1$ primes then there exist unique, up to associates, primes $\pi_1, \cdots, \pi_s$ in $G$ such that $\alpha = (1+i)^r q_1 \cdots q_t \pi_1 \cdots \pi_s$, $N\pi_\nu = p_\nu$ for $1 \le \nu \le s$;

vi) the methods of #6 may be used to prove: the number of distinct, up to associates, solutions of $N\alpha \equiv 0 \pmod p$, where $p$ is an odd prime in $Z$, is $0$ or $p$ depending on whether $p$ is a $4k+3$ or a

$4k+1$ prime, while the number of solutions of $N\alpha = p$, again up to associates, is $1$;

vii) from $n = 2^r q_1^2 \cdots q_t^2 p_1 \cdots p_s$, where the $q_\nu$ and $p_\nu$ are as in (v), we may write

$$n = (1+i)^r (1-i)^r \prod q^{2\nu} \prod (a+ib)^\nu (a-ib)^\nu$$
$$= A^2 + B^2 = (A+iB)(A-iB)$$

and so, by unique factorization

$$A+iB = i^\alpha (1+i)^{r_1} (1-i)^{r-r_1} \prod q^{\nu_1} \prod (a+ib)^{\nu_1} (a-ib)^{\nu-\nu_1}$$
$$A-iB = (-i)^\alpha (1+i)^{r-r_1} (1-i)^{r_1} \prod q^{\nu_2} \prod (a+ib)^{\nu-\nu_1} (a-ib)^{\nu_1},$$

where $\nu_1 + \nu_2 = 2\nu$; from this we see $\nu_1 = \nu_2$ and, since $1+i$ and $1-i$ are associates, the number of possible pairs $A, B$ is just the number $\prod(\nu+1)$ of divisors of $\prod p^\nu$;

the sum of all odd divisors of $n$ is

$$\prod (1+q+\cdots+q^{2\nu}) \prod (1+p+\cdots+p^\nu);$$

replacing each $q$ by $-1$ and each $p$ by $1$ yields, on the one hand $\prod(\nu+1)$, and, on the other hand, $d_1(m) - d_3(m)$.

9. i - a, b) These follow from the fact that the product of 2 odd numbers is of the form $4k+3$ if and only if exactly one of them is of the form $4k+3$;

c) let $(a,b)=1$; then, using (a), (b), & #8,

$$f_2(ab) = \tfrac{1}{4} r_2(ab) = d_1(ab) - d_3(ab)$$
$$= d_1(a)d_1(b) + d_3(a)d_3(b) - d_3(a)d_1(b) - d_1(a)d_3(b)$$
$$= (d_1(a) - d_3(a))(d_1(b) - d_3(b)) = \tfrac{1}{4} r_2(a)\, \tfrac{1}{4} r_2(b)$$
$$= f_2(a) f_2(b);$$

ii - a) this is a restatement of Jacobi's theorem;

b) let $(a,b)=1$; then using (a) and the multiplicativity of $\sigma$ we have:

for $ab$ odd,

$$f_4(ab) = \tfrac{1}{8} r_4(ab) = \sigma(ab) = \sigma(a)\sigma(b)$$
$$= \tfrac{1}{8} r_4(a)\, \tfrac{1}{8} r_4(b) = f_4(a) f_4(b)$$

while for $ab$ even (without loss of generality take $a$ even, $b$ odd)

$$f_4(ab) = \frac{1}{8} r_4(ab) = 3\sigma^\circ(ab) = 3\sigma^\circ(a)\sigma^\circ(b)$$
$$= 3 \cdot \frac{1}{24} r_4(a) \frac{1}{8} r_4(b) = f_4(a) f_4(b) ;$$

iii-a) if $2 = a_1^2 + \cdots + a_s^2$ then $s \geq 2$ and all $a_j$ except for 2 must be 0; the 2 which are not 0 are each either $\pm 1$; thus the total number of solutions is 4 times the number of pairs of the $a_j$ which may be taken to be non-zero;

b) similar to (a);

c) if $6 = a_1^2 + \cdots + a_s^2$ then $s \geq 3$ and either 6 of the $a_j$ are $\pm 1$ or 2 of the $a_j$ are $\pm 1$ and 1 other is $\pm 2$; the total number of ways of doing this is

$$2^6 \binom{s}{6} + 3 \cdot 2^3 \binom{s}{3} = 64 \binom{s}{6} + 24 \binom{s}{3} ;$$

d) $f_s(6) - f_s(2)f_s(3) =$

$$\tfrac{1}{2s}\left(64\binom{s}{6} + 24\binom{s}{3} - 32\binom{s}{2}\binom{s}{3}\right)$$

$$= \tfrac{2}{45}s(s-1)(s-2)(s-4)(s-8) \;;$$

e) if $f_s$ were multiplicative then $f_s(6) - f_s(2)f_s(3)$ would have to be 0; but, by (d), this can happen only for $s = 1, 2, 4, 8$.

## XVI Brun's Theorem ~ Solutions

1. Let $n, n+2$ be primes, $1 < n < \sqrt{x}$; the number of such pairs with $n \le z$ is less than or equal to $z$ and each pair with $n > z$ has $(a_n, R) = 1$ so contributes 1 to $S$. Hence $\pi_2(x) \le z + S$.

2. Suppose $a_n = dd'$, where $d \mid R$ and $(d', R) = 1$; if $d = 1$ then $(a_n, R) = 1$ and $n$ contributes 1 to $S_1$ and 0 to $S_\delta$, $\delta \ne 1$, so $n$ contributes 1 to the right hand side of the given expression; on the other hand if $d = p_1 \cdots p_j$, $1 \le j \le 2k$, then $n$ contributes 1 to each of the $\binom{j}{i}$ terms $S_\delta$, where $\delta \mid d$, $\nu(\delta) = i$; thus $n$ contributes

$$1 - \binom{j}{1} + \binom{j}{2} - \binom{j}{3} + \cdots + (-1)^j \binom{j}{j} = (1-1)^j = 0$$

to the right hand side.

3. When $p$ is odd exactly two, namely the last and third last, of the numbers

$$a_1 = 1\cdot 3, \cdots, a_p = p(p+2)$$

are divisible by $p$, while when $p=2$ only the last is divisible by $p$; thus when $\nu(\delta)=1$ the expressions for $\rho(\delta)$ are correct; assume them to be correct for $\nu(\delta)=n$ and examine $\nu(\delta p)$, where $\delta p \mid R, p\nmid\delta$, and, as we may assume without loss of generality, $p$ is odd; now $pi+j$, $1\le i\le\delta$, $1\le j\le p$ is a complete system of residues modulo $\delta p$ so we wish to find the number of solutions of

$$(pi+j)(pi+j+2) \equiv 0 \pmod{\delta p};$$

the number of solutions of this congruence is equal to the number of solutions of the system

$$(pi+j)(pi+j+2) \equiv 0 \pmod{p},$$
$$(pi+j)(pi+j+2) \equiv 0 \pmod{\delta};$$

from the 1st congruence we see that

$$j(j+2) \equiv 0 \pmod{p}$$

and hence there are exactly 2 values of $j$ possible; for each of these 2 values of $j$ the numbers $pi+j$, $1\le i\le\delta$, constitute a complete system of residues modulo $\delta$ so the number of solutions of the 2nd congruence is $\rho(\delta)$; therefore the number of solutions of the system is $2\rho(\delta)$; this gives $\rho(\delta p) = 2^{\nu(\delta p)-\epsilon}$, where $\epsilon$ is 0 or 1 depending on whether $\delta p$ is odd or even; this completes the induction.

4. Let $x = q\delta + r$, $0 \leq r < \delta$; then the numbers $n$, $1 \leq n \leq [x]$ fall either into one of $q$ complete systems of residues modulo $\delta$ or into a remaining partial system of residues modulo $\delta$; the total number of $n$ with $\delta \mid a_n$ is $S_\delta$ while the number in each complete system of residues is $\rho(\delta)$; thus there is a $\vartheta_1$, $0 \leq \vartheta_1 < 1$, such that

$$S_\delta = q\,\rho(\delta) + \vartheta_1\,\rho(\delta)$$
$$= \left(\frac{x}{\delta} + \vartheta_1 - \frac{r}{\delta}\right)\rho(\delta);$$

putting $\vartheta = \vartheta_1 - \frac{r}{\delta}$ yields the desired result.

5. Using #2 and #4 we find

$$S = \sum_{\delta | R,\, \nu(\delta) \le 2k} (-1)^{\nu(\delta)} \left( \frac{x}{\delta} + \vartheta \right) \rho(\delta)$$

$$= x \sum_{\delta | R,\, \nu(\delta) \le 2k} \frac{(-1)^{\nu(\delta)} \rho(\delta)}{\delta} + \sum_{\delta | R,\, \nu(\delta) \le 2k} \vartheta (-1)^{\nu(\delta)} \rho(\delta)$$

$$= x \sum_{\delta | R} \frac{(-1)^{\nu(\delta)} \rho(\delta)}{\delta} - x \sum_{\delta | R,\, \nu(\delta) > 2k} \frac{(-1)^{\nu(\delta)} \rho(\delta)}{\delta} + \sum_{\delta | R,\, \nu(\delta) \le 2k} \vartheta (-1)^{\nu(\delta)} \rho(\delta)$$

$$\le x \left( \sum_{\delta | R} \frac{(-1)^{\nu(\delta)} \rho(\delta)}{\delta} + \sum_{\delta | R,\, \nu(\delta) > 2k} \frac{\rho(\delta)}{\delta} \right) + \sum_{\delta | R,\, \nu(\delta) \le 2k} \rho(\delta)$$

$$= x (T_1 + T_2) + T_3 \;;$$

the alternate expression for $T_1$ is a direct consequence of #3 as is the inequality in the expression for $T_2$; finally, since the number of $\delta$ such that $\nu(\delta) = j$ and $\delta | R$ is just $\binom{\nu(R)}{j}$, where $\nu(R) = \pi(z)$, we have

$$T_3 = \sum_{j=0}^{2k} \sum_{\delta | R,\, \nu(\delta) = j} \rho(\delta) \le \sum_{j=0}^{2k} \binom{\pi(z)}{j} 2^j .$$

6. By XIV#11(iv), there is a constant $c$ such that $c\, n \ln n < p_n$ for all $n$; thus

$$\sum_{p \le t} \frac{1}{p} \le \sum_{n=2}^{\pi(t)} \frac{1}{c\, n \ln n} \le \frac{1}{c} \left( 1 + \int_2^t \frac{1}{x \ln x} dx \right) < A \ln \ln t$$

for suitable positive $A$ with $e A \ln 2 > 1$.

7. The last 2 inequalities are clearly true for $x$ sufficiently large; the 1st follows from the fact that as $x \to \infty$ so also does $z$ plus the fact that the left side is of the order of $\ln\ln z$ while the right side (see XIV #10 (ix)) is of the order of $\frac{z}{\ln z}$.

8. Using #7,

$$4 < \ln z = \frac{\ln x}{6eA\ln\ln x} < \frac{\ln x}{18} < \ln\sqrt{x}$$

so $2 < z < \sqrt{x}$; further, $2k < \pi(z) = \nu(R)$.

9. In (a) the 1st inequality follows from #5(a) and $1-\frac{2}{p} < (1-\frac{1}{p})^2$, the 2nd inequality from $\prod_{p|R}(1-\frac{1}{p})^{-1} = \prod_{p|R}(1+\frac{1}{p}+\frac{1}{p^2}+\cdots) \geq \sum_{m=1}^{z}\frac{1}{m}$, and the 3rd inequality from the fact $\sum_{m=1}^{z}\frac{1}{m} > \int_1^z \frac{1}{z}\,dz$;

in (b) the 1st inequality follows from #5(b) and $\sum_{\substack{\mathfrak{J} \\ \nu(\mathfrak{J})=j}} \frac{1}{\mathfrak{J}} \leq (\sum_{p\leq z}\frac{1}{p})^j \frac{1}{j!}$, the 2nd inequality from #6 and the fact that $\frac{j^j}{j!} < 1+j+\frac{j^2}{2!}+\cdots+\frac{j^j}{j!}+\cdots = e^j$, the $3^{rd}$ inequality from $2eA\ln\ln z < k < \frac{j}{2}$ and $\sum_{j=2k+1}^{\infty}\left(\frac{1}{2}\right)^j = 2^{-2k}$, the $4^{th}$ inequality from #6 and

$$k\ln 2 > 2eA\ln 2\ln\ln z > \ln\ln z\ ;$$

in (c) the $1^{st}$ inequality follows from #5(c) and $\binom{\pi(z)}{j} \le \frac{(\pi(z))^j}{j!} \le \frac{(\pi(z))^{2k}}{j!}$, the $2^{nd}$ and $3^{rd}$ inequalities from

$$\sum_{j=0}^{2k}\frac{2^j}{j!} < \sum_{j=0}^{\infty}\frac{2^j}{j!} = e^2 < 9 \text{ and } \pi(z) < z\ .$$

10. From the definitions of $z$ and $k$ given in #7 and #8 we see

$\ln z = \frac{\ln x}{6eA\ln\ln x}$, $\ln\ln z = \ln\ln x - \ln\ln\ln x - \ln 6eA$

and, therefore,

$$\frac{1}{(\ln z)^2} = (6eA)^2\left(\frac{\ln\ln x}{\ln x}\right)^2,$$

$$z^{2k} = x^{2k\alpha} < x^{\frac{2eA\ln\ln z}{3eA\ln\ln x}} = x^{\frac{2}{3} - \frac{2\ln\ln\ln x}{3\ln\ln x} - \frac{2\ln 6eA}{3\ln\ln x}} = x^{\beta}\ ;$$

thus, using #1, 5, 8, 9 we find

$$\pi_2(x) \le z + S \le z + x(T_1 + T_2) + T_3 \le \sqrt{x} + x\left(\frac{B+1}{(\ln z)^2}\right) + 9z^{2k}$$

$$\le \sqrt{x} + x(6eA)^2(B+1)\left(\frac{\ln\ln x}{\ln x}\right)^2 + x^{\beta} < Cx\left(\frac{\ln\ln x}{\ln x}\right)^2$$

for suitable positive $C$ and $x$ sufficiently large.

11. For $x$ sufficiently large

$$\pi_2(x) < Cx\left(\frac{\ln\ln x}{\ln x}\right)^2 < \frac{Cx}{(\ln x)^{3/2}}\ ;$$

thus if $p_n$ is the $n^{th}$ prime for which $p_n+2$ is prime then

$$n = \pi_2(p_n) < \frac{C\,p_n}{(\ln p_n)^{3/2}} < \frac{C\,p_n}{(\ln n)^{3/2}}$$

so $\frac{1}{p_n} < \frac{C}{n(\ln n)^{3/2}}$ and the result in (a) follows immediately by the comparison

$$\Sigma\frac{1}{p_n} < \Sigma\frac{C}{n(\ln n)^{3/2}}$$

and the convergence of the right hand series;

(b) follows from (a) by splitting $\Sigma\frac{1}{p}$ into the sum in (a) and a sum dominated by the sum in (a).

## XVII Quadratic Residues ~ Solutions

1. i) Multiplying the 1st congruence by $4a$ leads to the equivalent congruence

$$4af(x) \equiv 0 \pmod{4am};$$

since $4af(x) = (2ax+b)^2 - D$ the result follows;

ii) suppose (a) is true and $x_0^2 \equiv a \pmod{m}$, where $a = a'\delta$, $m = m'\delta$. Then, since $\delta^2 \mid x_0^2$, $x_0 = \delta y_0$ for suitable $y_0$. Thus $y_0^2 \equiv a't \pmod{m't}$ and, therefore, $t$ divides $y_0^2$. But $t$ is squarefree so $t$ divides $y_0$ and $y_0 = ts$ for suitable $s$. This means $ts^2 \equiv a' \pmod{m'}$. Since any common prime factor of $t$ and $m'$ would divide $a'$ and $(a', m') = 1$, we must have $(t, m') = 1$. Hence $ts^2 \equiv a' \pmod{m'}$ and $(ts)^2 \equiv ta' \pmod{m'}$ are equivalent. This proves (a) implies (b).

Suppose now (b) is true and $x_0^2 \equiv ta' \pmod{m'}$. Since $(t,m')=1$ there is an $s$ such that $x_0 \equiv ts \pmod{m'}$. Thus $t^2s^2 \equiv ta' \pmod{m'}$ and, since $(t,m')=1$ this last congruence implies (indeed, is equivalent to) the congruence $ts^2 \equiv a' \pmod{m'}$. Multiplying by $t\delta^2 (=d)$ we obtain $(t\delta s)^2 \equiv a \pmod{m}$. Thus (b) implies (a).

2. i) If $x_0^2 \equiv 12 \pmod{45}$ then $3 \mid x_0$; but then since $3^2 \mid 45$, we would have to have $3^2 \mid 12$, contrary to fact;

ii) taking $a = 252 = 2^2 \cdot 3^2 \cdot 7$, $m = 3^2 \cdot 5^2 \cdot 7$ in #1 (ii) we find $t = 7$, $d = 3^2 \cdot 7$ and $(t, \frac{m}{d}) = (7,25) = 1$; thus $x^2 \equiv 252 \pmod{1575}$ is equivalent to

$$x^2 \equiv 28 \equiv 3 \pmod{25}.$$

3. i-a) In this case the congruence becomes $x^2 \equiv 1 \pmod{2}$; thus, modulo 2, there is exactly 1 solution;

b) in this case $x^2 \equiv a \pmod 4$ ; since $(a,4)=1$, modulo 4, $x$ must be either 1 or 3 ; in either case $x^2 \equiv 1 \pmod 4$ ; thus $a \equiv 1 \pmod 4$ ; on the other hand if $a \equiv 1 \pmod 4$ then $x=1, x=3$ are both solutions ;

c) see the proof of (b) ;

d) since $a$ must be odd so also must $x$ be odd ; but the square of every odd number is congruent to 1 (mod 8) (see IX # 8(i)); on the other hand if $a \equiv 1 \pmod 8$ then all possible odd $x$ satisfy $x^2 \equiv a \pmod 8$ since each of 1, 3, 5, 7 satisfies the congruence ;

e) it is clear that solvability of $(*_{\alpha+1})$ implies solvability of $(*_\alpha)$ ; thus, suppose $(*_\alpha)$ is solvable and that $x_0$ is a solution ; then $x_0$ is odd and $x_0^2 = a + 2^\alpha \cdot t$ for suitable $t$ ; choosing s so that

$t + x_0 s \equiv 0 \pmod 2$ we see that

$$(x_0 + s \cdot 2^{\alpha-1})^2 = x_0^2 + x_0 s\, 2^{\alpha} + s^2 \cdot 2^{2\alpha-2}$$

$$= a + (t + x_0 s) 2^{\alpha} + s^2\, 2^{2\alpha-2} \equiv a \pmod{2^{\alpha+1}},$$

where the last congruence is true since

$$2\alpha - 2 = \alpha + (\alpha - 2) \geq \alpha + 1 \text{ when } \alpha \geq 3;$$

thus solvability of $(*_\alpha)$ implies solvability of $(*_{\alpha+1})$;

f) if $x_0^2 \equiv a \equiv y_0^2 \pmod{2^\alpha}$ then

$$(x_0 + y_0)(x_0 - y_0) \equiv 0 \pmod{2^\alpha};$$

since $x_0, y_0$ are odd exactly one of the even integers $x_0 + y_0$, $x_0 - y_0$ is divisible by 4; hence $x_0 \equiv \pm y_0 \pmod{2^{\alpha-1}}$; thus there are at most the 4 solutions $x_0, -x_0, x_0 + 2^{\alpha-1}, -x_0 + 2^{\alpha-1}$; since, when $x_0$ is a solution, all of these are solutions and since they are pairwise incongruent modulo $2^\alpha$ this proves there are exactly 4 solutions;

g) this follows from (a)-(f).

ii-a) from $x^2 \equiv a \equiv y^2 \pmod p$ we conclude $x \equiv \pm y \pmod p$; thus there are exactly the 2 incongruent solutions $x, -x$ when $x$ is a solution;

b) let $x_0^2 = a + kp^\alpha$ and note that $(2x_0, p) = 1$; thus there is exactly one, modulo $p$, value of $t$ such that $k + 2x_0 t \equiv 0 \pmod p$ and for this $t$ we have $(x_0 + tp^\alpha)^2 = x_0^2 + 2x_0 t p^\alpha + t^2 p^{2\alpha} =$

$$a + (k + 2x_0 t)p^\alpha + t^2 p^{2\alpha} \equiv a \pmod{p^{\alpha+1}};$$

c) this follows immediately from (b);

d) if $x^2 \equiv a \equiv y^2 \pmod{p^\alpha}$ then, since $p$ divides neither $x$ nor $y$, the quantities $x + y$, $x - y$ are not both divisible by $p$; hence $x \equiv \pm y \pmod{p^\alpha}$; this shows there are at most 2 solutions; since $-x$ also is a solution when $x$ is a solution and since $x \not\equiv -x \pmod p$ this completes the proof;

iii) since every solution of the system is a solution of each individual congruence, $N>0$ implies $N_j>0$ for all $j$, $1\le j\le r$; different, modulo $m_1\cdots m_r$, solutions of the system lead to different $r$-tuples of solutions to the individual congruences since if $x\equiv x'\pmod{m_j}$ for all $j$, $1\le j\le r$, then $x\equiv x'\pmod{m_1\cdots m_m}$; hence $N\le N_1\cdots N_r$; on the other hand if $x_1,\cdots,x_r$ is an $r$-tuple solution for the $r$ congruences (i.e. $x_j$ is a solution to the $j^{\text{th}}$ congruence) then if $t_j\frac{m_1\cdots m_r}{m_j}\equiv 1\pmod{m_j}$, $1\le j\le r$, we have $x=x_1t_1\frac{m_1\cdots m_r}{m_1}+\cdots+x_rt_r\frac{m_1\cdots m_r}{m_r}$ a solution of the system; finally if $x_1',\cdots,x_r'$ is a different $r$-tuple solution for the $r$ congruences, $x'=x_1't_1\frac{m_1\cdots m_r}{m_1}+\cdots+x_r't_r\frac{m_1\cdots m_r}{m_r}$ is not congruent, modulo $m_1\cdots m_r$, to $x$ since otherwise for all $j$, $1\le j\le r$, $x_j\equiv x_j'\pmod{m_j}$; hence $N_1\cdots N_r\le N$; the conclusion follows;

iv-a) this follows from (iii), (i-b), & (i-d);
b) this follows from (iii), (i-g), (ii-d) & (a).

4. i) By Fermat's theorem $(a^{\frac{p-1}{2}}+1)(a^{\frac{p-1}{2}}-1) \equiv 0 \pmod p$; consequently $a^{\frac{p-1}{2}} \equiv \pm 1 \pmod p$; since $p$ is odd not both are possible since that would imply $2 \equiv 0 \pmod p$;

ii) from $x_0^2 \equiv a \pmod p$ we find
$$a^{\frac{p-1}{2}} \equiv x_0^{p-1} \equiv 1 \pmod p;$$

iii-a) by the remainder theorem in algebra we know $z^{\frac{p-1}{2}}-1 = (z-a)q(z) + a^{\frac{p-1}{2}}-1$, where $q$ is a polynomial of degree $\frac{p-1}{2}-1 = \frac{p-3}{2}$; replacing $z$ by $x^2$ yields the result;

b) immediate from (a) and Fermat's theorem;

iv) if $a$ is a qr of $p$ then, by (ii),

$$a^{\frac{p-1}{2}} \equiv 1 \pmod{p};$$

on the other hand if $a^{\frac{p-1}{2}} \equiv 1 \pmod{p}$ then, by (iii-b), since $g(x^2)$ is of degree $p-3$ in $x$ and hence not always congruent to 0 modulo $p$, there is an $x$ for which $x^2 - a \equiv 0 \pmod{p}$; i.e. $a$ is a qr of $p$; the rest follows from (i).

5. i) Since $x = a$ satisfies $x^2 \equiv a^2 \pmod{p}$ it is clear that $\left(\frac{a^2}{p}\right) = 1$;

ii) immediate from #4 (iv) and the definition of $\left(\frac{a}{p}\right)$;

iii) $\left(\frac{a}{p}\right) \equiv a^{\frac{p-1}{2}} \equiv b^{\frac{p-1}{2}} \equiv \left(\frac{b}{p}\right) \pmod{p}$; since $\left(\frac{a}{p}\right)$ and $\left(\frac{b}{p}\right)$ are $\pm 1$ and since $1 \not\equiv -1 \pmod{p}$ this means $\left(\frac{a}{p}\right) = \left(\frac{b}{p}\right)$;

iv) $\left(\frac{ab}{p}\right) \equiv (ab)^{\frac{p-1}{2}} \equiv a^{\frac{p-1}{2}} \cdot b^{\frac{p-1}{2}} \equiv \left(\frac{a}{p}\right)\left(\frac{b}{p}\right) \pmod{p}$;

as in (iii), $\left(\frac{ab}{p}\right) = \left(\frac{a}{p}\right)\left(\frac{b}{p}\right)$ ;

v) a) if $a^2 \equiv b^2 \pmod{p}$ then $a \equiv \pm b \pmod{p}$, which cannot happen because of the conditions on a and b ;

b) for $0 < a \leq p-1$ there is a non-zero number c, $-\frac{p-1}{2} \leq c \leq \frac{p-1}{2}$ with $a \equiv c \pmod{p}$; putting $b = |c|$ yields the desired result ;

c) immediate from (b) and the meaning of $\left(\frac{a}{p}\right) = 1$ ;

vi) $\left(\frac{-1}{p}\right) \equiv (-1)^{\frac{p-1}{2}} \pmod{p}$, by (ii); since both sides are $\pm 1$ and p is odd this implies the equality.

6. i) From #5 (i) ;

ii) from #5 (iii) ;

iii) from #5(iv);

iv) by the proof of IX #20 (i-d) we know a polynomial of degree $n$ may have no more than $n$ zeros modulo $p$; thus $x^{\frac{p-1}{2}} \equiv 1$ and $x^{\frac{p-1}{2}} \equiv -1$ (mod $p$) each has at most $\frac{p-1}{2}$ solutions; since all the integers $1, 2, \cdots, p-1$ satisfy exactly one of these congruences each must have exactly $\frac{p-1}{2}$ solutions; the conclusion now follows from #4(iv).

7. i) Let $n = p_1^{\alpha_1} \cdots p_k^{\alpha_k}$; since $\left(\frac{n}{p}\right) = -1$ we know $p \nmid n$ so, by #5(iv), $\left(\frac{n}{p}\right) = \left(\frac{p_1}{p}\right)^{\alpha_1} \cdots \left(\frac{p_k}{p}\right)^{\alpha_k} = -1$; thus for some $j$, $\alpha_j$ is odd and $\left(\frac{p_j}{p}\right) = -1$; this implies, for such a $j$, $\sum_{d \mid p_j^{\alpha_j}} \left(\frac{d}{p}\right) = 0$; but then, see VIII #27,

$$\sum_{d \mid n} d^{\frac{p-1}{2}} = \prod_{i=1}^{k} \sum_{d \mid p_i^{\alpha_i}} d^{\frac{p-1}{2}} \equiv \prod_{i=1}^{k} \sum_{d \mid p_i^{\alpha_i}} \left(\frac{d}{p}\right) = 0 \pmod{p},$$

where we have also used #5(ii);

ii) this is merely a rephrasing of #6(iv).

iii-a) since, by #5(iv), $\left(\frac{ab}{p}\right) = \left(\frac{a}{p}\right)\left(\frac{b}{p}\right)$ not all three of these Legendre symbols may be $-1$ ;

b) since $x^6 - 11x^4 + 36x^2 - 36 = (x^2-2)(x^2-3)(x^2-6)$ and since, by (a), one at least of 2, 3, 6 is a qr modulo any prime other than 2 and 3 the conclusion follows as soon as we observe that 2 and 3 divide the value of this polynomial when $x = 0$ ;

iv) from $ax_0^2 \equiv -by_0^2 \pmod p$, using #5(i), #5(iii), and #5(iv) we see that

$$\left(\frac{a}{p}\right) = \left(\frac{ax_0^2}{p}\right) = \left(\frac{-by_0^2}{p}\right) = \left(\frac{-b}{p}\right) .$$

8. i) By #5(v) every qr of p is congruent to exactly one of $1^2, 2^2, \cdots, \left(\frac{p-1}{2}\right)^2$ ; since, by #6(iv), there are exactly $\frac{p-1}{2}$ qr of p we see, using Wilson's theorem,

$$\prod_{\substack{a \\ (\frac{a}{p})=1}} a = 1^2 \cdot 2^2 \cdots \left(\tfrac{p-1}{2}\right)^2 \equiv (-1)^{\frac{p-1}{2}} 1 \cdot 2 \cdots \tfrac{p-1}{2} \cdot \left(-\tfrac{p-1}{2}\right) \cdots (-1)$$

$$\equiv (-1)^{\frac{p-1}{2}} (p-1)! \equiv \left(\tfrac{-1}{p}\right)(p-1)! \equiv -\left(\tfrac{-1}{p}\right) \pmod p;$$

the other conclusion follows from

$$-\left(\tfrac{-1}{p}\right) \prod_{\substack{a \\ (\frac{a}{p})=-1}} a \equiv \Big(\prod_{\substack{a \\ (\frac{a}{p})=1}} a\Big)\Big(\prod_{\substack{a \\ (\frac{a}{p})=-1}} a\Big) \equiv (p-1)! \equiv -1 \pmod p;$$

ii-a) when $p \equiv 1 \pmod 4$, $\frac{p-1}{2}$ is even; consequently, when $a^{\frac{p-1}{2}} \equiv 1 \pmod p$ so also is $(-a)^{\frac{p-1}{2}} \equiv 1 \pmod p$; this implies the desired result;

b) when $p \equiv 3 \pmod 4$, $\frac{p-1}{2}$ is odd; consequently when $a^{\frac{p-1}{2}} \equiv 1 \pmod p$ then $(-a)^{\frac{p-1}{2}} \equiv -1 \pmod p$; this implies the desired result;

c) immediate from (a);

d) the 1st part of (i) combined with (a) and (b) yield these results.

9. i) $|a_i| = |a_j|$ implies $ia \equiv a_i = \pm a_j \equiv \pm ja \pmod{p}$ which implies $(i \pm j)a$ is divisible by $p$; but $1 \le i < j \le \frac{p-1}{2}$ precludes $i \pm j$ being divisible by $p$ and we are given $p \nmid a$; the conclusion follows;

ii) the equality follows immediately from (i) and the congruence follows from the congruences

$$a_i \equiv ia \pmod{p}, \quad 1 \le i \le \tfrac{p-1}{2};$$

iii) immediate from (ii).

10. i) Since $\nu$ is the number of numbers among $2, 4, \cdots, 2j, \cdots, p-1$ which have negative least absolute residues the result is clear;

ii) by IV #11 and (i), $\nu = \left[\frac{p}{2}\right] - \left[\frac{p}{4}\right]$;
if $p = 8k + i$ then

$$\left[\tfrac{p}{2}\right] - \left[\tfrac{p}{4}\right] \equiv \left[\tfrac{i}{2}\right] - \left[\tfrac{i}{4}\right] \equiv \begin{cases} 0 \pmod{2} \text{ if } i = 1 \text{ or } 7; \\ 1 \pmod{2} \text{ if } i = 3 \text{ or } 5; \end{cases}$$

iii) this follows from, when $p = 8k + i$,

$$\frac{p^2-1}{8} = \frac{64k^2+16ki+i^2-1}{8} \equiv \frac{i^2-1}{8} \equiv \begin{cases} 0 \text{ if } i = \pm 1 \\ 1 \text{ if } i = \pm 3 \end{cases} \equiv \nu;$$

iv) this follows from (ii) (or (iii)) and the Lemma of Gauss.

11. i) Since $\nu$ is the number of numbers among $3, 6, \cdots, 3j, \cdots, \frac{3}{2}(p-1)$ which have negative least absolute residues and since only those between $\frac{p}{2}$ and $p$ have this property the result is clear;

ii) the proof is similar to the proof of #10(ii);

iii) this follows from (ii) and the Lemma of Gauss.

12. i) The number of $5, 10, \cdots, 5j, \cdots, \frac{5}{2}(p-1)$ with negative least absolute residues modulo $p$ is just the number of $j$ indicated;

ii) this follows from (i) ;

iii) this follows from (ii) and the Lemma of Gauss .

13. i) When $p \equiv \pm q \pmod{4a}$ neither $p$ nor $q$ divides $a$ so $\left(\frac{a}{p}\right) = (-1)^{\nu}$, $\left(\frac{a}{q}\right) = (-1)^{\mathcal{N}}$, where

$$\nu = \sum_{\substack{i=1 \\ i \text{ even}}}^{a} \left\{ \left[ i\frac{p}{2a} \right] - \left[ (i-1)\frac{p}{2a} \right] \right\}, \quad \mathcal{N} = \sum_{\substack{i=1 \\ i \text{ even}}}^{a} \left\{ \left[ i\frac{q}{2a} \right] - \left[ (i-1)\frac{q}{2a} \right] \right\};$$

if $p = \pm q + 4at$ then

$$\nu = \sum_{\substack{i=1 \\ i \text{ even}}}^{a} \left\{ \left[ \frac{\pm iq}{2a} \right] + 2it - \left[ \frac{\pm(i-1)q}{2a} \right] \mp 2(i-1)t \right\} \equiv$$

$$\sum_{\substack{i=1 \\ i \text{ even}}}^{a} \left\{ \left[ \frac{\pm iq}{2a} \right] - \left[ \frac{\pm(i-1)q}{2a} \right] \right\} \pmod 2;$$

since for even $j$ from 1 to $a$ the quantity $\frac{jq}{2a}$ is not an integer (if $\frac{jq}{2a} = s$, since $j \leq a$, $s \leq \frac{q}{2}$, so $jq = 2as$ is impossible as $q$ does not divide any of $2, a, s$) the last sum is $\mathcal{N}$; thus $\nu \equiv \mathcal{N} \pmod 2$ and the result is proved;

ii) if $p \equiv q \pmod 4$ then $p - q = 4a$ for suitable $a$ and $(a, pq) = 1$; if $p \not\equiv q \pmod 4$ then one of p, q is of the form 4k+1 and the other of the form 4k+3 ; thus their sum is of the form 4a with $(a, pq) = 1$ ;

iii) making use of (i), (ii), and the properties in #5 we have : for some a, $p = \pm q + 4a$, and, therefore,

$$\left(\frac{p}{q}\right) = \left(\frac{\pm q + 4a}{q}\right) = \left(\frac{4a}{q}\right) = \left(\frac{a}{q}\right) = \left(\frac{a}{p}\right) = \left(\frac{4a}{p}\right) = \left(\frac{-p+4a}{p}\right) = \left(\frac{\mp q}{p}\right);$$

iv) from (iii),

$$\left(\frac{p}{q}\right)\left(\frac{q}{p}\right) = \begin{cases} \left(\frac{-1}{p}\right) & \text{if } p \equiv q \pmod 4 \\ 1 & \text{otherwise} \end{cases} =$$

$$\begin{cases} (-1)^{\frac{p-1}{2}} & \text{if } p \equiv q \pmod 4 \\ 1 & \text{otherwise} \end{cases} = (-1)^{\frac{p-1}{2} \cdot \frac{q-1}{2}}$$

14. By the reciprocity law $\left(\frac{p}{q}\right)\left(\frac{q}{p}\right)$ is 1, meaning $\left(\frac{p}{q}\right) = \left(\frac{q}{p}\right)$ except when each of p and q is of the form 4k+3 in which case $\left(\frac{p}{q}\right) = -\left(\frac{q}{p}\right)$.

15. i) This is obvious ;

ii) this is true since for each $p_j$ the Legendre symbols $\left(\frac{1}{p_j}\right)$ and $\left(\frac{a^2}{p_j}\right)$ are equal to 1 ;

iii) when a is a qr of m it is a qr of every $p_j$ and hence all $\left(\frac{a}{p_j}\right)=1$ which implies $\left(\frac{a}{m}\right)=1$ ;

iv) $\left(\frac{2}{9}\right)=\left(\frac{2}{3}\right)^2=1$ but $x^2 \equiv 2 \pmod 9$ is not solvable ;

v) this follows from the corresponding property for the Legendre symbol since when $a \equiv b \pmod m$, $a \equiv b \pmod p$ for every prime factor p of m ;

vi) immediate from the definition ;

vii) follows from the corresponding property of the Legendre symbol;

viii) write $m = q_1 \cdots q_t$, where the primes $q_j$ are not necessarily distinct; then, since each $q_j - 1$ is even and the product of 2 or more such factors is a multiple of 4, we see that

$$m - 1 = ((q_1 - 1) + 1) \cdots ((q_t - 1) + 1) - 1$$
$$= 4k + (q_1 - 1) + \cdots + (q_t - 1)$$

for suitable $k$; thus

$$\left(\frac{-1}{m}\right) = \left(\frac{-1}{q_1}\right) \cdots \left(\frac{-1}{q_t}\right) = (-1)^{\frac{(q_1 - 1) + \cdots + (q_t - 1)}{2}} = (-1)^{\frac{m-1}{2}};$$

ix) write $m$ as in the proof of (viii) and note that, since the square of an odd number is congruent to 1 modulo 8,

$$m^2 - 1 = ((q_1^2 - 1) + 1) \cdots ((q_t^2 - 1) + 1) - 1$$
$$= 64k + (q_1^2 - 1) + \cdots + (q_t^2 - 1)$$

for suitable $k$; thus

$$\left(\frac{2}{m}\right) = \left(\frac{2}{q_1}\right) \cdots \left(\frac{2}{q_t}\right) = (-1)^{\frac{(q_1^2 - 1) + \cdots + (q_t^2 - 1)}{8}} = (-1)^{\frac{m^2-1}{8}};$$

x) write $m = q_1 \cdots q_t$, $n = p_1 \cdots p_k$; then

$$\left(\frac{n}{m}\right)\left(\frac{m}{n}\right) = \left(\frac{n}{q_1}\right)\cdots\left(\frac{n}{q_t}\right)\left(\frac{m}{p_1}\right)\cdots\left(\frac{m}{p_k}\right)$$

$$= \prod_{i=1}^{t}\prod_{j=1}^{k}\left(\frac{p_j}{q_i}\right)\left(\frac{q_i}{p_j}\right) = (-1)^{\sum_{i=1}^{t}\sum_{j=1}^{k}\frac{p_j-1}{2}\cdot\frac{q_i-1}{2}}$$

$$= (-1)^{\left(\sum_{i=1}^{t}\frac{q_i-1}{2}\right)\left(\sum_{j=1}^{k}\frac{p_j-1}{2}\right)} = (-1)^{\frac{m-1}{2}\cdot\frac{n-1}{2}}$$

16.i) Making heavy use of #15 we have

a) $\left(\frac{89}{197}\right) = \left(\frac{197}{89}\right) = \left(\frac{19}{89}\right) = \left(\frac{89}{19}\right) = \left(\frac{13}{19}\right) = \left(\frac{19}{13}\right) = \left(\frac{6}{13}\right)$

$= \left(\frac{2}{13}\right)\left(\frac{3}{13}\right) = -\left(\frac{3}{13}\right) = -\left(\frac{13}{3}\right) = -\left(\frac{1}{3}\right) = -1$ ;

b) $\left(\frac{1050}{1573}\right) = \left(\frac{2}{1573}\right)\left(\frac{525}{1573}\right) = -\left(\frac{525}{1573}\right)$

$= -\left(\frac{1573}{525}\right) = -\left(\frac{-2}{525}\right) = -\left(\frac{2}{525}\right) = 1$ ;

c) $(12345, 6789) \neq 1$ so $\left(\frac{12345}{6789}\right)$ is not defined ;

ii-a,b) since $89 = 4\cdot 22 + 1$ and 89 and 197 are prime #14 tells us that (a) and (b) are both solvable or both insolvable ; by (i-a) we see (a) is not solvable so (b) is also not solvable ;

c) this is solvable because 1050 is a qr of 11 and 13 and hence of all prime factors of 1573;

d) this is not solvable because 1573 is a qnr of the prime factor 5 of 1050;

e) using #1 (ii-b) we see $d = 3 = t$ so $\frac{m}{d} = \frac{219}{3} = 73$ and the congruence is solvable if and only if $(3, 73) = 1$ and $x^2 \equiv 111 \equiv 38 \pmod{73}$ is solvable; but $(3, 73)$ is equal to 1, 73 is a prime and $\left(\frac{38}{73}\right) = 1$; thus the congruence in (e) is solvable;

f) again using #1 (ii-b) we see $d = 3 = t$ so $\frac{m}{d} = \frac{111}{3} = 37$ and the congruence is solvable if and only if $(3, 37) = 1$ and $x^2 \equiv 219 \equiv 34 \pmod{37}$ is solvable; again, as in (e) the modulus is a prime and $\left(\frac{34}{37}\right) = 1$; thus the congruence in (f) is solvable.

(This problem shows that all possible situations may occur for pairs of congruences of the type

$$x^2 \equiv a \pmod{b} \quad , \quad x^2 \equiv b \pmod{a} \, ;$$

either both are insolvable, exactly one is solvable, or both are solvable.)

17. i) By #1(i), $f(x) \equiv 0 \pmod{p}$ if and only if $(2ax+b)^2 \equiv D \pmod{4am}$; i.e. if and only if $(2x+1)^2 \equiv -163 \pmod{4p}$; but, making heavy use of the Jacobi symbol and #15, we see $\left(\frac{-163}{p}\right) = -1$ for all primes $< 41$ so the conclusion follows;

ii) this is immediate from (i);

iii) all of the values

$$f(-40), f(-39), \cdots, f(-1), f(0), f(1), \cdots, f(39)$$

are positive and smaller than $41^2$ so all are prime.

18. *i*) If $Z_D a = Z_D b$ then $Da \equiv Db \pmod{p}$ and, therefore, $a \equiv b \pmod{p}$; similarly for $T_D$ ;

*ii*) $T_D(T_D a) = T_D(\widetilde{Da^{-1}}) = T_D(Da^{-1}) = \widetilde{D(Da^{-1})^{-1}}$
$= \widetilde{DD^{-1}a} = \widetilde{a} = a$, for all $a$ in $A$ ;

*iii*) $T_D T_1(a) = T_D(\widetilde{a^{-1}}) = T_D\, a^{-1} = \widetilde{D(a^{-1})^{-1}} = \widetilde{Da}$
$= Z_D a$ ;

*iv*) $T_D x = x$ implies $\widetilde{Dx^{-1}} = x$ which, in turn, implies $x^2 \equiv D \pmod{p}$; by #3 (*ii*-a) the conclusion follows ;

*v*) since $T_D$ is an involution its cyclic representation, as a permutation, consists of $\alpha_D$ cycles of length 1 and $\frac{\varphi(p)-\alpha_D}{2}$ transpositions (cycles of length 2) ; therefore $\operatorname{sgn} T_D = (-1)^{\frac{\varphi(p)-\alpha_D}{2}}$ ; from (*iii*) we find
$\operatorname{sgn} Z_D = (\operatorname{sgn} T_D)(\operatorname{sgn} T_1) = (-1)^{\varphi(p) - \frac{\alpha_D+\alpha_1}{2}} = (-1)^{\frac{\alpha_D+\alpha_1}{2}}$ ;

vi) this follows from (iii) since 1 is a qr of p ;

vii) by (v) and (vi), $\operatorname{sgn} Z_D = (-1)^{\frac{\alpha_D+2}{2}}$ ; when D is a qr of p this is 1 and when D is a qnr of p this is -1, as we see by noting the value of $\alpha_D$ from (iii);

viii) this follows from (vii) when we take

$$A = \{1, 2, \cdots, p-1\} \;;$$

ix) taking $A = \{-\frac{p-1}{2}, \cdots, -1, 1, 2, \cdots, \frac{p-1}{2}\}$ then $Z_{-1}$ gives the permutation

$$\frac{p-1}{2}, \cdots, 1, -1, -2, \cdots, -\frac{p-1}{2}$$

and clearly $\frac{p-1}{2}$ transpositions lead back to the original order of A; hence $(\frac{-1}{p}) = \operatorname{sgn} Z_{-1} = (-1)^{\frac{p-1}{2}}$ ; taking $A = \{1, 2, \cdots, p-1\}$ then $Z_2$ gives the permutation $2, 4, \cdots, p-1, 1, 3, \cdots, p-2$ and the number of transpositions putting it back in the original order is $\frac{p-1}{2} + (\frac{p-1}{2} - 1) + \cdots + 1 = \frac{(\frac{p-1}{2})(\frac{p-1}{2}+1)}{2} = \frac{p^2-1}{8}$ so $(\frac{2}{p}) = \operatorname{sgn} Z_2 = (-1)^{\frac{p^2-1}{8}}$ .

19. i) Since whenever $a \in A$ so also is $-a \in A$, we have from $a' < a''$ and $\widetilde{Da'} > \widetilde{Da''}$ the following: $-a'' < -a'$ and $\widetilde{-Da''} = -\widetilde{Da''} > -\widetilde{Da'} = \widetilde{-Da'}$ ;

ii) $a', a''$ and $-a'', -a'$ are different unless

$$a' = -a'' \;;$$

iii) $\operatorname{sgn} Z_D = (-1)^\alpha$ where $\alpha$ is the number of inversions of the form $-a, a$ ; for exactly such inversions $a \in A^+$ and $\widetilde{Da} = -a \in A^-$ ;

iv) take $D = q$ and $D = p$ respectively in the above and use (iii) as well as #18 (vii) ;

v) if $qx - py = 0$ then $qx = py$ so $p$ would divide $x$, which is not possible since $1 \le x \le \frac{p-1}{2}$ ; thus, since $qx - py \ne 0$ the four inequalities exhaust all possibilities ;

vi) the mapping $x, y \leftrightarrow \frac{p+1}{2} - x, \frac{q+1}{2} - y$ proves the 1st assertion; the 2nd assertion follows from (iv);

vii) by (vi), $\mu + \nu \equiv$ total number of pairs $x, y$ in (v) $\equiv \frac{p-1}{2} \cdot \frac{q-1}{2} \pmod 2$; now use (iv).

## XVIII Exponents, Primitive Roots, & Power Residues ~ Solutions

1. i) Let $s-t = qP(a)+r$, $0 \le r < P(a)$; then $a^s = a^{t+qP(a)+r} = a^t(a^{P(a)})^q a^r \equiv a^t a^r \equiv a^t \pmod m$; since $(a^t, m) = 1$ this implies $a^r \equiv 1 \pmod m$; if $r \neq 0$ then $P(a)$ would not be the exponent of $a$; consequently $r = 0$ and $P(a)$ divides $s-t$;

ii) (a) and (b) follow from (i) by, respectively, putting $t=0$ and $s = \varphi(m)$, $t=0$; in (c) the given quantities are clearly solutions of the stated congruence; if $a^i \equiv a^j \pmod m$ then by (i), $P(a) \mid i-j$; but this may not happen for $i, j$ among $1, 2, \cdots, P(a)$ unless $i = j$;

iii) $(a^k)^t \equiv 1 \pmod m$ if and only if $kt \equiv 0 \pmod{P(a)}$; this last congruence is equivalent to the one obtained by dividing all parts by $(k, P(a))$ and then dividing all parts, except the modulus, by the new coefficient of $t$, which is prime to the modulus and hence results in an equivalent congruence; the result is $t \equiv 0 \pmod{\frac{P(a)}{(k, P(a))}}$; since the least possible positive value for $t$ is clearly the modulus in this last congruence we have the desired conclusion;

iv-a) if $t \nmid \varphi(m)$ then $t$ can be the exponent mod $m$ of no integer since all such exponents must divide $\varphi(m)$, as seen in (ii-b);

b) every number prime to $p$ has some exponent which is a divisor of $\varphi(m)$; since no number has more than one exponent the sum on the left is clearly just the number of integers, mod $m$, prime to $m$, i.e. $\varphi(m)$;

c) let $P(a) = t$; then $a, a^2, \cdots, a^t$ are prime to $m$ and their exponents are given by (iii); the number of $k$ in (iii) for which $P(a^k) = P(a) = t$ is just the number of $k$ for which $(k, P(a)) = 1$, i.e.

$\varphi(P(a)) = \varphi(t)$; thus $\psi(t)$ is at least as large as $\varphi(t)$;

v-a) in this case, when $P_p(a) = t$, then $a, a^2, \cdots, a^t$ are all the solutions of $x^t \equiv 1 \pmod p$; hence by (iii) exactly $\varphi(t)$ of them have exponent $t$;

b) by (iv-b, c) and VIII #19 we have

$$p-1 = \sum_{t \mid p-1} \psi_p(t) \leq \sum_{t \mid p-1} \varphi(t) = p-1 ;$$

thus $\sum_{t \mid p-1} \psi_p(t) = \sum_{t \mid p-1} \varphi(t)$; this with the inequality in (iv-c) yields the desired conclusion;

v) immediate from (v-b) when one takes $t = p-1$.

2. i) $\varphi(p) = p-1$, $p-1 \mid P_{p^\alpha}(q)$, and $P_{p^\alpha}(q) \mid \varphi(p^\alpha) = p^{\alpha-1}(p-1)$; thus $P_{p^\alpha}(q) = p^\beta(p-1)$ for some $\beta$, $0 \leq \beta \leq \alpha - 1$; since $q^{\frac{\varphi(p^\alpha)}{p}} \not\equiv 1 \pmod{p^\alpha}$ this means $\beta = \alpha - 1$ and the conclusion follows;

ii) since $(g+p)^{p-1} - g^{p-1} \equiv -g^{p-2}p \not\equiv 0 \pmod{p^2}$ we see that not both $(g+p)^{p-1}$ and $g^{p-1}$ are congruent to $1 \pmod{p^2}$; the result now follows from (i) since we know $g$ and $g+p$ are primitive roots of $p$ ;

iii) using IV#24 it is easy to see ( $j \geq 2$ ) that the highest power of $p$ in $\binom{p^\beta}{j}$ is $\geq \beta+2-j$; thus, if $g$ is a primitive root of $p^2$ then $g^{p-1} = 1+qp$, where $p \nmid q$; hence $g^{\frac{\varphi(p^\alpha)}{p}} = (g^{p-1})^{p^{\alpha-2}}$
$= (1+qp)^{p^{\alpha-2}} \equiv 1+qp^{\alpha-1} \pmod{p^\alpha}$;
the result now follows from (i);

iv) $a^t \equiv 1 \pmod{2p^\alpha}$ implies $a^t \equiv 1 \pmod{p^\alpha}$ and $a$ is odd; if $a$ is a primitive root of $p^\alpha$ then the largest $t$ could be is $\varphi(p^\alpha) = \varphi(2p^\alpha)$;
this completes the proof;

v) this follows from (ii), (iii), (iv), and #1(vi);

vi) for all other numbers the function $\chi(m)$, introduced in IX#9 is smaller than $\varphi(m)$ and the conclusion follows then from that problem part (i); for powers of 2 larger than 2 the conclusion also follows from IX#8(6).

3. i) If $q$ is a primitive root of $p^{\alpha+1}$ but not of $p^\alpha$ then for some $t$, $0<t<\varphi(p^\alpha)$, $q^t \equiv 1 \pmod{p^\alpha}$; but then $q^t = 1+sp^\alpha$ and, therefore, $q^{tp} = (1+sp^\alpha)^p \equiv 1 \pmod{p^{\alpha+1}}$; since $tp < p\,\varphi(p^\alpha) = p^\alpha(p-1) = \varphi(p^{\alpha+1})$ this contradicts $q$ being a primitive root of $p^{\alpha+1}$;

ii) the numbers in question are all primitive roots of $p$ and, defining $q$ by $g^{p-1} = 1+qp$, we see that $((1+sp)g)^{\frac{\varphi(p^2)}{p}} \equiv (1+qp)(1-sp) \pmod{p^2}$; except when $s \equiv q \pmod p$ this right hand side is not congruent to $1 \pmod{p^2}$; consequently by #2(i) the conclusion follows;

iii) by #1 (vi) there are $\varphi(p-1)$ primitive roots of $p$; by (ii), to each of these there are $p-1$ primitive roots of $p^2$; hence the number of primitive roots of $p^2$ is

$$(p-1)\varphi(p-1) = \varphi(p(p-1)) = \varphi(\varphi(p^2));$$

iv) by (i) and #2 (iii) if $q$ is a primitive root of $p^\alpha$, $\alpha \geq 2$, then $q$ is a primitive root of $p^2$, and, thus, a primitive root of $p^{\alpha+1}$; but then $q, q+p^\alpha, \cdots, q+(p-1)p^\alpha$ are all primitive roots of $p^{\alpha+1}$ and exhaust those congruent to $q$ (again by (i));

v) by (iii) there are $\varphi(\varphi(p^2))$ primitive roots of $p^2$ and by iterated use of (iv) there are then $p^{\alpha-2}\varphi(\varphi(p^2))$ primitive roots of $p^\alpha$; but $\varphi(\varphi(p^\alpha)) = \varphi(p^{\alpha-1}(p-1)) = \varphi(p^{\alpha-1})\varphi(p-1)$

$$= p^{\alpha-2}(p-1)\varphi(p-1) = p^{\alpha-2}\varphi(\varphi(p^2));$$

vi) this follows from (v), #2 (iv), and direct checking for 2 and 4.

4. i) By #1 (iii), $P(a^u) = \frac{P(a)}{(u, P(a))} = \frac{uv}{(u, uv)} = \frac{uv}{u} = v$;

ii) let $P(a) = u$, $P(b) = v$, $P(ab) = w$; then $1 \equiv (ab)^{uv} \equiv (ab)^{wu} \equiv b^{wu} \equiv (ab)^{wv} \equiv a^{wv} \pmod{m}$; hence each of $u$ and $v$ divides $w$ which, in turn, divides $uv$;

iii) with $m = 7$, $a = 3$, $b = 5$ we have $P(3 \cdot 5) = 1$ but $[P(3), P(5)] = [6, 6] = 6$;

iv) let $P(a) = u$, $P(b) = v$ and choose $u_0, v_0$ such that $u_0 | u$, $v_0 | v$, $(u_0, v_0) = 1$, $u_0 v_0 = [u, v]$ (this can be done for example by using the prime factorizations of $u$ and $v$); put $c = a^{\frac{u}{u_0}} \cdot b^{\frac{v}{v_0}}$; then, using #1 (iii),

$$P(a^{\frac{u}{u_0}}) = \frac{P(a)}{(\frac{u}{u_0}, P(a))} = \frac{u}{(\frac{u}{u_0}, u)} = u_0,$$

and similarly, $P(b^{\frac{v}{v_0}}) = v_0$; thus, by (ii),

$$P(a^{\frac{u}{u_0}}, b^{\frac{v}{v_0}}) = u_0 v_0 = [u, v] ;$$

v) let $P$ be the largest exponent and suppose $(a, m) = 1$ ; then, by (iv), there is an integer c such that $P(c) = [P, P(a)]$ ; but this means

$$P \leq P(c) \leq P \text{ so } P = [P, P(a)] \text{ and } P(a) | P ;$$

vi) let $P$ be the largest exponent of the prime $p$ ; then, using (v), $x^P \equiv 1 \pmod p$ has $\varphi(p)$ solutions so $\varphi(p) \leq P$ ; but $P | \varphi(p)$ so $P \leq \varphi(p)$ ; this means $P = \varphi(p)$ and $p$ has a primitive root ;

vii) if $g$ is a primitive root of $p$ then $g^t$ is also for all $t$ satisfying $(t, p-1) = 1$ since, by #1 (iii), $P(g^t) = \frac{P(g)}{(t, P(g))} = \frac{\varphi(p)}{(t, \varphi(p))} = \frac{p-1}{(t, p-1)} = p-1.$

5. i) If $a$ has order $u$ and $b$ has order $v$ choose $u_0, v_0, c$ as in the proof of #4 (iv); then the order of $c$ is $[u, v]$;

ii) the proof is like that of #4 (v);

iii) if the group $G$ is of order $q$ and the maximal element in $A_G$ is $P$ then $P \leq q$ and $x^P = 1$ has $q$ solutions; but in a field the equation $x^P = 1$ has at most $P$ solutions; thus $P = q$ and the desired result follows;

iv) immediate from (iii);

v) a cyclic group of $p-1$ elements has $\varphi(p-1)$ generators; similarly we can deduce #3 (v) from (iv).

6. i) Let $q$ be a primitive root of $p$;
if $x^n \equiv a \pmod{p}$ and $x \equiv q^s \pmod{p}$ then

$$a^{\frac{p-1}{d}} \equiv 1 \pmod{p};$$

on the other hand, if $a^{\frac{p-1}{d}} \equiv 1 \pmod{p}$ and $a \equiv q^t \pmod{p}$ then $q^{t\frac{p-1}{d}} \equiv 1 \pmod{p}$ so $d \mid t$ and $t \equiv ns \pmod{p-1}$ is solvable for $s$ yielding $q^t \equiv q^{ns} \equiv a \pmod{p}$, so $x^n \equiv a \pmod{p}$ has $x = q^s$ as a solution;

ii) this follows from (i) by taking $n = \delta$;

iii) since the number of $\delta^{th}$ power residues modulo $p$ cannot exceed $\frac{p-1}{\delta} < p-1$, see (ii), the conclusion must be true;

iv-a) let $t$ be the exponent of $B_i$ modulo $p$; then, since $B_i^{p_i^{\alpha_i}} = A_i^{p-1} \equiv 1 \pmod{p}$, we know

$t \mid p_i^{\alpha_i}$; but, since $A_i^{\frac{p-1}{p_i}} \not\equiv 1 \pmod{p}$ we know that $t < p_i^{\alpha_i}$ is false; thus $t = p_i^{\alpha_i}$;

6) since for $i \neq j$, $(P(B_i), P(B_j)) = 1$ we may use the finite extension of #4 (ii) to obtain

$$P(B_1 \cdots B_k) = P(B_1) \cdots P(B_k) = p_1^{\alpha_1} \cdots p_k^{\alpha_k}$$
$$= p - 1 .$$

7. i) we first find $A_1, A_2, A_3$ so that

$$A_1^{\frac{42}{2}} \not\equiv 1 , \; A_2^{\frac{42}{3}} \not\equiv 1 , \; A_3^{\frac{42}{7}} \not\equiv 1 \pmod{43} ;$$

where the 2, 3, 7 are the prime factors of 42, $42 = 2 \cdot 3 \cdot 7$; it is easy to see we may take

$$A_1 = 2 , \; A_2 = 7 , \; A_3 = 3$$

and that $2^{21} \equiv -1$, $7^{14} \equiv 6$, $3^6 \equiv -2 \pmod{43}$; since all $\alpha_i = 1$ we find

$B_1 = 2^{21}$, $B_2 = 7^{14}$, $B_3 = 3^6$ so $B_1 B_2 B_3 \equiv 12 \pmod{43}$;

therefore 12 is a primitive root modulo 43;

ii) using 12 we have the table:

| power of 12 → | 1 | 2 | 3 | 4 | 5 | 6 |
|---|---|---|---|---|---|---|
| no. ≡ to (mod 43) → | **12** | 15 | 8 | 10 | **-9** | 21 |
| exponent → | 42 | 21 | 14 | 21 | 42 | 7 |

| 7 | 8 | 9 | 10 | 11 | 12 | 13 | 14 | 15 | 16 | 17 | 18 |
|---|---|---|---|---|---|---|---|---|---|---|---|
| -6 | 14 | -4 | -5 | **-17** | 11 | **3** | -7 | 2 | -19 | **-13** | -27 |
| 6 | 21 | 14 | 21 | 42 | 7 | 42 | 21 | 14 | 21 | 42 | 21 |

| 19 | 20 | 21 | 22 | 23 | 24 | 25 | 26 | 27 | 28 | 29 | 30 |
|---|---|---|---|---|---|---|---|---|---|---|---|
| **20** | -18 | -1 | -12 | **-15** | -8 | **-10** | 9 | -21 | 6 | **-14** | 4 |
| 42 | 21 | 2 | 21 | 42 | 7 | 42 | 21 | 14 | 3 | 42 | 7 |

| 31 | 32 | 33 | 34 | 35 | 36 | 37 | 38 | 39 | 40 | 41 | 42 |
|---|---|---|---|---|---|---|---|---|---|---|---|
| **5** | 17 | -11 | -3 | 7 | -2 | **19** | 13 | -16 | -20 | **18** | 1 |
| 42 | 21 | 14 | 21 | 6 | 7 | 42 | 21 | 14 | 21 | 42 | 1 |

we have circled the primitive roots; the least is 3 and the least absolute is 3; for an easier method to carry out these computations, see ch. 8 of Uspensky & Heaslet, *Elementary Number Theory*.

8. i) When $(3, p-1) = 1$ all integers $a$, prime to $p$, are cubic residues of $p$ (see #6(i)); thus since $(3, 4) = (3, 10) = 1$ all integers are cubic residues modulo 5 and 11;

ii) this follows as in the proof of (i) since $(5, 6) = 1$;

iii) when $n$ is odd, $(n, 4) = (n, 16) = 1$; thus the conclusion follows from #6(i);

iv) if all $a$, $p \nmid a$, are $n^{th}$ power residues modulo $p$ then $(n, p-1) = 1$ since otherwise no such $a$ would have exponent $p-1$ and this contradicts the existence of primitive roots for all odd primes; thus for all odd $n$, $(n, p-1) = 1$; this means $p-1$ must be a power of 2 and, hence, $p$ is a Fermat prime; the other direction is clear since when $n$ is odd and $p$ is a Fermat prime we always have $(n, p-1) = 1$;

v) by XIV # 17 (vii) given $n > 1$ there are infinitely many primes $p$ of the form $nk+1$; for each of these $(n, p-1) = n$ and therefore if all integers were $n^{th}$ power residues then such $p$ would have no primitive roots contrary to fact.

9. If $p-1 \mid n$ then $j^n \equiv 1 \pmod{p}$ so the left side is congruent to $p-1$ and hence to $-1 \bmod p$; otherwise let $q$ be a primitive root modulo $p$ and note that $1, 2, \cdots, p-1$ are congruent, in some order, to $q, q^2, \cdots, q^{p-1}$; hence

$$1^n + 2^n + \cdots + (p-1)^n \equiv q^n + q^{2n} + \cdots + q^{(p-1)n} \pmod{p};$$

now, since $p-1 \nmid n$, $1 - q^n \not\equiv 0 \pmod{p}$ so there is an $x_0$ such that $(1-q^n)x_0 \equiv 1 \pmod{p}$ and $x_0 \not\equiv 0 \pmod{p}$; now

$$1^n + \cdots + (p-1)^n \equiv q^n + \cdots + q^{(p-1)n}$$

$$\equiv x_0(1-q^n)(q^n + \cdots + q^{(p-1)n}) \equiv x_0 q^n (1 - q^{(p-1)n})$$

$$\equiv 0 \pmod{p}.$$

10. If $a^t \equiv 1 \pmod{a^n - 1}$ then $t \geq n$; since $a^n \equiv 1 \pmod{a^n - 1}$ we see that $n$ is the exponent of $a \pmod{a^n - 1}$; the conclusion now follows from #1 (ii-b).

11. Let $g$ be a primitive root modulo $p$; then the product in question is just, using VIII #16,

$$\prod_{\substack{t \\ (t,p-1)=1}} g^t = g^{\frac{p-1}{2}\varphi(p-1)} \equiv 1 \pmod{p},$$

since $\varphi(p-1)$ is even and $g^{p-1} \equiv 1 \pmod{p}$.

12. If $g$ is a primitive root modulo $m$, $m > 2$, then

$$\prod_{\substack{n=1 \\ (n,m)=1}}^{m} n = \prod_{t=1}^{\varphi(m)} g^t \equiv g^{(\varphi(m)+1)\frac{\varphi(m)}{2}} \equiv g^{\frac{\varphi(m)}{2}} \pmod{m};$$

since $(g^{\frac{\varphi(m)}{2}} + 1)(g^{\frac{\varphi(m)}{2}} - 1) \equiv 0 \pmod{m}$ and $g^{\frac{\varphi(m)}{2}} - 1 \not\equiv 0 \pmod{m}$ we have the desired conclusion; on the other hand when $m$ has no primitive root we may follow the lines of the proof of IX #12 except that now it is not true that $x^2 \equiv 1 \pmod{m}$ has only the two solutions $1$ and $m-1$ since the number of solutions is given in XVII #3 (iv) and is

$2^k$ if $4 \nmid m$; $2^{k+1}$ if $4 | m$, $8 \nmid m$; $2^{k+2}$ if $8 | m$,

where $k$ is the number of distinct odd prime factors of $m$; if $n$ has no primitive root then, by #2 (vi), either

$k = 0$, $8 | m$ or $k = 1$, $4 | m$ or $k \geq 2$ ;

in all of these cases the number of solutions is divisible by 4; noting that the solutions of $x^2 \equiv 1 \pmod{m}$ fall into pairs $x, -x$ with a product $-x^2 \equiv -1 \pmod{m}$ we see that the product of all solutions is $(-1)^s$, where $s$ is the number of pairs; but $s$ is even since the total number of solutions of $x^2 \equiv 1 \pmod{m}$ is divisible by 4; the proof is now finished in an identical way to that of IX #12. (It should be noted that the method of the 2nd half of the proof to this problem is also applicable to the 1st half - namely, to the case where m does have a primitive root.)

13. (A) First we show that $Q$ as defined is $< x$;

Case 1. $t$ even; if $2^{\frac{1}{2}t} \equiv -1 \pmod{2x+1}$ then $Q = \frac{1}{2}t - 1 < \frac{1}{2}t \le \frac{1}{2}\varphi(2x+1) \le \frac{1}{2}(2x) = x$; if $2^{\frac{1}{2}t} \not\equiv -1 \pmod{2x+1}$ then $2x+1$ is not a prime, for if it were $(2^{\frac{1}{2}t}+1)(2^{\frac{1}{2}t}-1) = 2^t - 1 \equiv 0 \pmod{2x+1}$ and $2^{\frac{1}{2}t} \equiv 1 \pmod{2x+1}$; thus $t \mid \varphi(2x+1) < 2x$

$$\text{so } Q = t - 1 < x\ ;$$

Case 2. $t$ odd; here $t \le \frac{1}{2}\varphi(2x+1) \le x$ so

$$Q = t - 1 < x\ ;$$

(B) suppose now that $t$ is even and $2^{\frac{1}{2}t} \equiv -1 \pmod{2x+1}$; then $2^{\frac{1}{2}t} = -1 + q(2x+1)$ for some odd $q$, say $q = 2s+1$ (for $x = 1$ the assertion in the problem is obvious); then

$$2^Q = 2^{\frac{1}{2}t-1} = \frac{-1+(2s+1)(2x+1)}{2} = (2s+1)x + s$$

is in the progression; further, if $2^\alpha = (2s+1)x + s$ for some $s$ and if $\alpha$ is as small as possible we have $2^{\alpha+1} \equiv 2x \equiv -1 \pmod{2x+1}$ so $2^{2\alpha+2} \equiv 1 \pmod{2x+1}$; thus $\alpha \le \frac{1}{2}t - 1$ and $t \mid 2\alpha + 2$ so $\alpha = \frac{1}{2}t - 1$ and $2^Q$ is the smallest power of 2 in the progression;

(C) suppose finally that $t$ does not satisfy the conditions of (B); then $2^{a+1} = 2^t = 1 + q(2x+1)$ for some odd $q$, say $q = 2s+1$; thus

$$2^a - 1 = -1 + \frac{1+(2s+1)(2x+1)}{2} = (2s+1)x + s$$

is in the progression; further, if $2^\alpha - 1 = (2s+1)x + s$ and $\alpha$ is as small as possible then

$$2^{\alpha+1} = 2 + 2x \equiv 1 \pmod{2x+1};$$

thus $\alpha \le t-1$ and $t \mid \alpha+1$ so $\alpha = t-1$.

14. If $2^a$ were in the sequence then $a > 3$ since 1, 2, 4, 8 are not in the sequence; if $5 + 7x = 2^a$, $a > 3$, then $x$ is odd, say $x = 2x_1 + 1$, which yields $6 + 7x_1 = 2^{a-1}$ and $x_1$ is even, say $x_1 = 2x_2$, which yields $3 + 7x_2 = 2^{a-2}$ and $x_2$ is odd, say $x_2 = 2x_3 + 1$, which yields $5 + 7x_3 = 2^{a-3}$; thus if $2^a$, $a > 3$, is in the sequence then $2^{a-3}$ is also in the sequence; iteration leads to a contradiction; if $5 + 7x = 2^a - 1$ then $a > 2$, since 0, 1, 3 are not in the sequence; from $5 + 7x = 2^a - 1$ we

see $x$ is even, say $x = 2x_1$, which yields $3 + 7x_1 = 2^{a-1}$ where $x_1$ is odd, say $x_1 = 2x_2 + 1$, which yields $5 + 7x_2 = 2^{a-2}$, which we know is impossible by the first part of the argument.

15. i) The identity is clear as is the congruence modulo $2^{n+2}$; noting that $5^{2^j} + 1 \equiv 2 \pmod 4$ we see that none of the factors $5^{2^j} + 1$ are divisible by 4 and the incongruence assertion follows;

ii) this is an immediate consequence of (i);

iii) clearly $1, 5, 5^2, \cdots, 5^{2^{\alpha-2}-1}$ and $-1, -5, -5^2, \cdots, -5^{2^{\alpha-2}-1}$ are pairwise incongruent modulo $2^\alpha$, among themselves; further, each of the 1st group of numbers is congruent modulo 8 to either 1 or 5, while each of the 2nd group of numbers is congruent modulo 8 to either 3 or 7;

thus no member of the 1st group is congruent to a member of the 2nd group modulo 8, and, therefore, since $\alpha > 2$, modulo $2^\alpha$.

# XIX Special Primes and the Lucas-Lehmer Theorem - Solutions

1\. i) $\left(\frac{3}{F_n}\right) = (-1)^{\frac{3-1}{2}\cdot\frac{F_n-1}{2}}\left(\frac{F_n}{3}\right) = \left(\frac{F_n}{3}\right) = \left(\frac{2}{3}\right) = (-1)^{\frac{3^2-1}{8}} = -1$, where we have used the fact that

$$F_n \equiv (-1)^{2^n} + 1 \equiv 2 \pmod 3 ;$$

ii) by hypothesis $3^{F_n-1} \equiv 1 \pmod{F_n}$ so $P_{F_n}(3) \mid F_n - 1$ and this implies $P_{F_n}(3)$ is a power of 2; but this power cannot be smaller than $F_n - 1$ itself since $3^{\frac{F_n-1}{2}} \not\equiv 1 \pmod{F_n}$; consequently $P_{F_n}(3) = F_n - 1$; finally,

$$F_n - 1 = P_{F_n}(3) \leq \varphi(F_n) \leq F_n - 1$$

so $\varphi(F_n) = F_n - 1$ and $F_n$ must be a prime;

iii) when $F_n$ is prime then, by (i), $\left(\frac{3}{F_n}\right) = -1$ and, by Euler's criterion, see XVII #5 (ii), this implies $3^{\frac{F_n-1}{2}} \equiv -1 \pmod p$; the opposite implication is precisely the statement in (ii).

2. i) $2^{2^n} \equiv -1 \pmod{p}$ so $2^{2^{n+1}} \equiv 1 \pmod{p}$ and, therefore, $P_p(2) \mid 2^{n+1}$; if $P_p(2) = 2^m$, $m < n+1$, then $2^{2^n} \equiv 1 \pmod{p}$ contrary to fact;

ii) since $P_p(2) \mid \varphi(p)$ we know $2^{n+1} \mid p-1$; hence $p^2 - 1 = (p+1)(p-1)$ is divisible by at least the $4^{th}$ power of 2 (recall that $n > 1$ and $p$ is odd); therefore $\left(\frac{2}{p}\right) = (-1)^{\frac{p^2-1}{8}} \equiv 1 \pmod{p}$ and the result follows;

iii) since, by (ii), $\left(\frac{2}{p}\right) = 1$ Euler's criterion tells us $2^{\frac{p-1}{2}} \equiv 1 \pmod{p}$; thus $P_p(2) = 2^{n+1}$ divides $\frac{p-1}{2}$;

iv) from (i)-(iii);

v) since the $F_n$ are pairwise relatively prime, see III #9 (iv-b), prime factors of different $F_n$ are different; now let $p_1, p_2, p_3, \cdots$ be a sequence of prime numbers such that $p_n | F_n$ ; then, by (iv), $p_n$ is in the stated sequence for all values of $n \geq k-2$.

3. i) $F_5 = 4,294,967,297$ with possible prime divisors of the form $1+2^7 \cdot t$, $1 \leq t \leq 2^{25}$; calculation shows $1+2^7 \cdot t$ is not prime for $t=1,3,4$ but is prime for $t=2,5$; for $t=2$ the resulting prime 257 does not divide $F_5$ while for $t=5$ the resulting prime 641 does divide $F_5$ with quotient 6,700,417 ;
(the total calculating time here was $\leq$ 10 minutes;)

ii) ( we follow Sierpinski [1964] )
compute $r_0$ where $r_0 \equiv 3^{2^7} \pmod{F_7}$ ;
now compute $r_1, r_2, \cdots, r_{120}$ successively, where $r_{j-1}^2 \equiv r_j \pmod{F_7}$ ; then, since

$$3^{\frac{F_n-1}{2}} = 3^{2^{127}} \equiv r_{120} \not\equiv -1 \pmod{F_7},$$

we see that $F_7$ is not prime .

4. i) If $n = ab$, $a > 1$, $b > 1$ then $2^a - 1 \mid 2^{ab} - 1$ ;

ii-a) $\left(\frac{2}{q}\right) = (-1)^{\frac{q^2-1}{8}} = (-1)^{\frac{p(p+1)}{2}} = 1$ ;

b) if $q \mid 2^p + 1$ then $2^p = 2^{\frac{q-1}{2}} \equiv -1 \pmod{q}$,

which says $\left(\frac{2}{q}\right) = -1$ in contradiction to (a) ;

c) since $2^{q-1} + 1 = (2^{\frac{q-1}{2}} + 1)(2^{\frac{q-1}{2}} - 1)$

$= (2^p + 1)(2^p - 1) \equiv 0 \pmod{q}$

and since, by (b), $q \nmid 2^p + 1$, $q$ must divide

$M_p = 2^p - 1$ ;

iii) these are all immediate consequences of (ii-c);

iv) each of the examples in (iii) is a counter-example to the converse of (i).

5. i) $U_1 = 1$, $V_1 = 2$ ; suppose now the assertions are correct for $n$ ; then $U_{n+1} =$

$\frac{1}{2\sqrt{3}}\{(1+\sqrt{3})^n(1+\sqrt{3}) - (1-\sqrt{3})^n(1-\sqrt{3})\} =$

$\frac{1}{2\sqrt{3}}\{(1+\sqrt{3})^n - (1-\sqrt{3})^n + \sqrt{3}V_n\} = U_n + \frac{1}{2}V_n$,

$$V_{n+1} = (1+\sqrt{3})^n(1+\sqrt{3}) + (1-\sqrt{3})^n(1-\sqrt{3})$$
$$= (1+\sqrt{3})^n + (1-\sqrt{3})^n + \sqrt{3}\left\{(1+\sqrt{3})^n - (1-\sqrt{3})^n\right\} = V_n + 6U_n;$$

the assertions now follow for $n+1$;

ii-a) multiply out the right side and cancel terms;

b-e) similar to (a);

f) put $m=n$ in (c) and use (e);

iii-a) $U_p = \frac{1}{2\sqrt{3}}\left(\sum_{j=0}^{p}\binom{p}{j}\sqrt{3}^{\,j} - \sum_{j=0}^{p}\binom{p}{j}(-1)^j\sqrt{3}^{\,j}\right)$
$= \sum_{\substack{j=0 \\ j \text{ odd}}}^{p}\binom{p}{j}\sqrt{3}^{\,j-1} \equiv 3^{\frac{p-1}{2}} \equiv \left(\frac{3}{p}\right) \pmod{p}$;

b) similar to (a);

c) put $n=p$ in the expression for $U_{n+1}$ in (i) and put $m=1$, $n=p-1$ into (ii-b) to obtain

$$2U_{p+1} = 2U_p + V_p, \quad 4U_{p-1} = -2U_p + V_p;$$

multiply these equations to obtain

$$8U_{p+1}U_{p-1} = -4U_p^2 + V_p^2 \equiv -4+4 \equiv 0 \pmod{p},$$

where we used (a) and (b) at the 2nd last congruence;

iv-a) by (ii-a) ;

b) by (ii-b) ;

c-1) by (iii-c) ;

2) if $n = q\omega_p + r$, $0 \le r < \omega_p$, then, by (a) and (b), if $r \ne 0$ then $r = n - q\omega_p$ is in $S_p$; but this contradicts the minimality of $\omega_p$; hence

$r = 0$ and, therefore, $\omega_p | n$ ;

v-a) from (ii-c) with $m = 2^p - 1$, $n = 1$ we have, using (iii-a, b),

$2V_{2^p} = 2V_{2^p-1} + 12U_{2^p-1} \equiv 4 + 12\left(\frac{3}{2^p-1}\right) \pmod{2^p - 1}$;

now $2^5 \equiv 8 \pmod{12}$ and, therefore, as we see by induction, all odd integers $s$, larger than 3, satisfy $2^s \equiv 8 \pmod{12}$; hence for such $s$, $2^s - 1 \equiv -5 \pmod{12}$, and, for those which are prime, we have, by XVII #11 (iii), $\left(\frac{3}{2^s-1}\right) = -1$; consequently $\left(\frac{3}{2^p-1}\right) = -1$ and

$2V_{2^p} \equiv -8 \pmod{2^p-1}$ ;

b) this follows by setting $n = 2^{p-1}$ in (ii-e);

c) immediate from (b) and (a);

d) immediate from (c) since

$$2^{\frac{M_p-1}{2}} \equiv \left(\frac{2}{M_p}\right) \equiv 1 \pmod{M_p};$$

vi-a) since $M_q = 2^q - 1 \equiv (-1)^q - 1 \equiv -2 \pmod 3$,

we see $p \neq 3$;

b) from (ii-d) we see $U_{2^q} = U_{2^{q-1}} V_{2^{q-1}}$ so from $p \mid V_{2^{q-1}}$ we conclude $p \mid U_{2^q}$;

c) by (b), $2^q$ is in $S_p$ so, by (iv-c-2), $\omega_p \mid 2^q$;

d) if $\omega_p \mid 2^{q-1}$ then, by (iv-c-2), $2^{q-1}$ is in $S_p$ so $p \mid U_{2^{q-1}}$ and, therefore, by (ii-f) we conclude

$p \mid (-2)^{2^{q-1}} + 2$, contrary to fact;

e) immediate from (c) and (d);

f) from (iv-c-1) we know $\omega_p = 2^q \leq p+1$; thus $M_q = 2^q - 1 \leq p$; but $p \leq M_q$ so $M_q = p$;

vii) $V_2 = 8 = 2^{2^{1-1}} \cdot 4$ so put $s_1 = 4$; suppose $s_1, \cdots, s_k$ have been correctly chosen; then, using (ii-e),

$$V_{2^{k+1}} = V_{2^k}^2 + (-2)^{2^k+1} = 2^{2^k} s_k^2 + (-2)^{2^k+1}$$

$$= 2^{2^k}(s_k^2 - 2) \;;$$

so we put $s_{k+1} = s_k^2 - 2$ ;

viii) if $M_p$ is prime then, by (v-d), $M_p | V_{2^{p-1}}$ and, therefore, by (vii), $M_p | s_{p-1}$ ; on the other hand, any odd prime divisor of $M_p$ must, when $M_p$ divides $V_{2^{p-1}}$, which is true when $M_p | s_{p-1}$, by (vi), equal $M_p$ ; i.e. $M_p$ is itself a prime ;

ix) we may note that all $s_k$ are even so we may suppress a factor of 2 in each term ; this yields

$$t_1 = \frac{s_1}{2} = 2 \;,$$

$$t_{k+1} = \frac{s_{k+1}}{2} = 2\left(\frac{s_k}{2}\right)^2 - 1 = 2t_k^2 - 1 .$$

## XX Pell Equation ~ Solutions

1\. i-a) Since $\alpha y$, for $y$ an integer, is never an integer it must lie between two consecutive integers; let $x$ be the larger of these integers;

b) for each of the $m+1$ values of $y$, $0 \le y \le m$, we select $x$ as in (a); the $m+1$ resulting numbers must contain a pair within $\frac{1}{m}$ of each other and their difference yields the desired result;

c) this follows from (b) since $\frac{1}{m} \le \frac{1}{y}$;

ii-a) $|x+y\sqrt{D}| = |x-y\sqrt{D}+2y\sqrt{D}|$

$$\le |x-y\sqrt{D}| + 2y\sqrt{D} < \tfrac{1}{y} + 2y\sqrt{D};$$

multiplying this inequality by $|x-y\sqrt{D}| < \frac{1}{y}$ yields $|x^2-Dy^2| < \frac{1}{y^2} + 2\sqrt{D} \le 1 + 2\sqrt{D}$;

b) since infinitely many pairs $x, y$ lead to $|x^2-Dy^2| \le 1+2\sqrt{D}$ there must be some infinite number of $x^2-Dy^2$ having the same integral value;

c) modulo $k$ the integers $x, y$ are all in $k^2$ pairs of residue classes; since there are infinitely many such pairs $x, y$ some two are in the same pair of classes;

d-1) $x_3 = x_1 x_2 - D y_1 y_2 \equiv x_1^2 - D y_1^2$
$= k \equiv 0 \pmod{k}$;

$y_3 = x_2 y_1 - x_1 y_2 \equiv x_1 y_1 - x_1 y_1 \equiv 0 \pmod{k}$;

2) if $y_3 = 0$ then $x_2 = \frac{x_1 y_2}{y_1}$ so

$k = x_2^2 - D y_2^2 = \frac{y_2^2}{y_1^2}(x_1^2 - D y_1^2) = \left(\frac{y_2}{y_1}\right)^2 k$

and, therefore, $y_1 = y_2$; but then $x_1 = x_2$ and the two pairs are not distinct;

3) $x_3^2 - D y_3^2 = (x_1^2 - D y_1^2)(x_2^2 - D y_2^2) = k^2$;

e) by (d-1) and (3) we see that $\left(\frac{x_3}{k}\right)^2 - D\left(\frac{y_3}{k}\right)^2 = 1$ is a solution.

2. i) $x_3^2 - D y_3^2 = (x_3 + y_3\sqrt{D})(x_3 - y_3\sqrt{D})$
$= (x_1^2 - D y_1^2)(x_2^2 - D y_2^2) = 1$;

ii) this is clear since the $x$ and $y$ appear in (1) only to the 2nd power and, therefore, either one or both of $x, y$ may be changed to their negative ;

iii) from $1 < x + y\sqrt{D}$ we have
$0 < x - y\sqrt{D} = \frac{1}{x+y\sqrt{D}} < 1$ and $-1 < -x + y\sqrt{D} < 0$ ;
adding each of these new inequalities to the given inequality yields

$$1 < 2x \text{ and } 0 < 2y\sqrt{D}$$

from which our conclusion is immediate ;

iv) let $x + y\sqrt{D}$ be a positive solution; then for suitable $k$ we have

$$(x_0 + y_0\sqrt{D})^k \leq x + y\sqrt{D} < (x_0 + y_0\sqrt{D})^{k+1} ;$$

multiplying by $(x_0 + y_0\sqrt{D})^{-k}$ yields

$$1 \leq x' + y'\sqrt{D} = (x + y\sqrt{D})(x_0 + y_0\sqrt{D})^{-k} < x_0 + y_0\sqrt{D} ;$$

if $x' + y'\sqrt{D}$ is not 1 then, since by (iii), $x' > 0$, $y' > 0$ we would have a positive solution smaller than the smallest positive solution; hence $x'+y'\sqrt{D}=1$ and, therefore, $x+y\sqrt{D}=(x_0+y_0\sqrt{D})^k$;

v) this follows from (iii), (iv), and the fact that $x_0-y_0\sqrt{D}=\frac{1}{x_0+y_0\sqrt{D}}$.

3. i) $x^2-3y^2=-1$ is not solvable because if it were the congruence $x^2\equiv -1 \pmod 3$ would be solvable; but 0, 1, 2 fail to satisfy the congruence;

ii) this parallels the details of #2;

iii) if $\alpha+\beta\sqrt{D}$ is the fundamental solution of (1) then $(\alpha+\beta\sqrt{D})(x'+y'\sqrt{D})$ is a solution of (2) and hence equals $(x'+y'\sqrt{D})^{2k+1}$ for some positive integer $k$; thus $\alpha+\beta\sqrt{D}=(x'+y'\sqrt{D})^{2k}$ and its fundamental character dictates that $k=1$.

4. i) If $x^2 - Dy^2 = 1$ then $(\sigma x)^2 - D(\sigma y)^2 = \sigma^2$;

ii) direct computation;

iii) consider the equation $x^2 - 58y^2 = 9$ with
$$x_1 + y_1\sqrt{D} = x_2 + y_2\sqrt{D} = 61 + 8\sqrt{58};$$

iv) when $\sigma^2 | D$ then $\sigma | x$ as we see from (3); the rest is easily seen by computation;

v-a) this is clear;

b) $4D \equiv \sigma^2 \pmod{4\sigma^2}$ yields $D \equiv \rho^2 \pmod{4\rho^2}$; hence $D = D'\rho^2$ and $D' \equiv 1 \pmod 4$;

c) from $x^2 - Dy^2 = x^2 - D'\rho^2 y^2 = \sigma^2 = 4\rho^2$ we see $\rho | x$; hence $(\frac{x}{\rho})^2 - D'y^2 = 4$; but, by (b), $D'$ is odd so $\frac{x}{\rho}$ and $y$ have the same parity;

d) $$x_3 = \frac{x_1x_2 + Dy_1y_2}{\sigma} = \rho^2\,\frac{x_1'x_2' + D'y_1y_2}{2\rho}$$
$$= \rho\,\frac{x_1'x_2' + D'y_1y_2}{2},$$
$$y_3 = \frac{x_1'y_2 + x_2'y_1}{2}$$ ; now

observe that $x_i'$ and $y_i$ have the same parity for $i = 1, 2$ and the conclusion follows since D is odd ;

vi) if $1 < \frac{x + y\sqrt{D}}{\sigma} \leq \frac{x_0 + y_0\sqrt{D}}{\sigma} = \alpha$ then

$$\frac{1}{\alpha} \leq \frac{x - y\sqrt{D}}{\sigma} < 1 , \quad -1 < \frac{-x + y\sqrt{D}}{\sigma} \leq -\frac{1}{\alpha}$$

so $\frac{2x}{\sigma} > 1$ , $\frac{2y\sqrt{D}}{\sigma} > 0$ .

5. i) By direct computation ;

ii) if $x + y\sqrt{D}$ is a rational solution put

$$r = \begin{cases} \frac{1-x}{y} & \text{if } y \neq 0 \\ 0 & \text{if } y = 0 \end{cases} ;$$

then $\frac{D + r^2}{D - r^2} = x$ and $\frac{-2r}{D - r^2} = y$ ;

iii) $x = 8$, $y = 3$ ;

iv) direct computation .

6. i) The fundamental solutions to these equations are, respectively, $1+\sqrt{2}$ and $(1+\sqrt{2})^2 = 3+2\sqrt{2}$; thus all solutions are given by, respectively, $(1+\sqrt{2})^{2n+1} = (1+\sqrt{2})(3+2\sqrt{2})^n$ and $(3+2\sqrt{2})^n$;

ii) if $s_n = a + (a+1)$ then $a^2 + (a+1)^2 = h_n^2$ so $s_n^2 + 1 = (2a+1)^2 + 1 = 2(a^2 + (a+1)^2) = 2h_n^2$ and, therefore, $s_n + h_n\sqrt{2} = (1+\sqrt{2})(3+2\sqrt{2})^n$, where we have used (i) in the form $s_n^2 - 2h_n^2 = -1$;

iii)

$$((2x+1)+z\sqrt{2})(3+2\sqrt{2})$$
$$=(6x+4z+3)+(4x+3z+2)\sqrt{2}$$

so the "next Pythagorean triple" following $(x, x+1, z)$ is $(3x+2z+1, 3x+2z+2, 4x+3z+2)$;

iv) follows from (i)–(iii);

v) (3, 4, 5), (20, 21, 29), (119, 120, 169), (696, 697, 985).

## XXI Weyl's Theorem on Uniform Distribution ~ Solutions

1 i-a, b) These are clear from the definition of $\chi_{[a,b]}$;

ii) by (a) and (b) if $\frac{1}{n}\sum_{m=1}^{n} f(s_m) \to \int_0^1 f$ then $\lim_{n\to\infty} \frac{n(a,b)}{n} = b-a$ and, therefore, $\{s_n\}$ is uniformly distributed;

iii) immediate from the definitions.

2. Let $\epsilon > 0$ be given and choose $g, h, n$ satisfying the given conditions and $\int g - \epsilon \le \frac{1}{n}\sum_{m=1}^{n} g(s_m)$, $\frac{1}{n}\sum_{m=1}^{n} h(s_m) \le \int h + \epsilon$; then

$$\int f - 2\epsilon < \int g - \epsilon \le \frac{1}{n}\sum_{m=1}^{n} f(s_m) \le \int h + \epsilon < \int f + 2\epsilon,$$

and, therefore,

$$\left|\frac{1}{n}\sum_{m=1}^{n} f(s_m) - \int f\right| < 2\epsilon;$$

this proves $f(s_n) \rightsquigarrow \int f$.

3. Using #1 (ii), (iii) we see that because of the additive and homogeneous properties of the arithmetic mean and of the integral the result is true whenever $f$ is a step function; since for any Riemann integrable $f$ the hypotheses of #2 are realizable with $g$ and $h$ step functions the desired conclusion follows immediately from #2.

4. i) This is clear since when $u$ and $v$ are real,

$$\int (u+iv) = \int u + i \int v \;;$$

ii) by (i), $f(s_n) \rightsquigarrow \int_0^1 e^{2\pi i k x}\, dx = 0$ .

5. i) $P$ implies $e^{2\pi i k s_n} \rightsquigarrow 0$ which, in turn, implies each of

(*) $\cos 2\pi k s_n \rightsquigarrow 0$ and $\sin 2\pi k s_n \rightsquigarrow 0$ ;

but $T(x) = \sum_{k=1}^{q} (a_k \cos 2\pi k x + b_k \sin 2\pi k x)$ and, therefore, $\frac{1}{n}\sum_{m=1}^{n} T(s_m) =$

$$\sum_{k=1}^{q} a_k \left\{\frac{1}{n}\sum_{m=1}^{n} \cos 2\pi k s_m\right\} + \sum_{k=1}^{q} b_k \left\{\frac{1}{n}\sum_{m=1}^{n} \sin 2\pi k s_m\right\} \;;$$

hence, since the expressions in the curly brackets tend to 0, we conclude $T(s_n) \sim 0$;

ii) to go the other way we see that if $T(s_n) \sim 0$ for all trigonometric polynomials with zero constant term then both assertions of (*) are true and, thus, $e^{2\pi i k s_n} \sim 0$;

iii) immediate from (i) and (ii);

iv) if $\tau(x) = a_0 + T(x)$ and $T(s_n) \sim 0$ then

$\frac{1}{n}\sum_{m=1}^{n} \tau(s_m) = a_0 + \frac{1}{n}\sum_{m=1}^{n} T(s_m) \to a_0 + \int T = \int \tau$;

on the other hand, since all trigonometric polynomials certainly include those with zero constant term the opposite direction is obvious;

v) for $f$ continuous and $\epsilon > 0$ there exists a trigonometric polynomial $\tau$ such that $|f - \tau| < \frac{\epsilon}{2}$;

putting $g = \tau - \frac{\epsilon}{2}$, $h = \tau + \frac{\epsilon}{2}$ we see that all the conditions of #2 are satisfied so we may conclude $f(s_n) \rightsquigarrow \int f$;

vi) given a characteristic function $\chi$ of a subinterval of $[0,1]$ we may choose continuous $g$ and $h$ satisfying the conditions of #2 (we are using (v) here) and, therefore, by #2 $\chi(s_n) \rightsquigarrow \int \chi$; but then the conclusion follows by #1 (ii);

vii) this is a restatement of #4 and (vi).

6. i) $\sum_{m=1}^{n} e^{2\pi i k s_m} = \sum_{m=1}^{n} e^{2\pi i k m \alpha} = e^{2\pi i k \alpha}\left(\frac{1-e^{2\pi i k n \alpha}}{1-e^{2\pi i k \alpha}}\right)$;
since the right side is bounded as a function of $n$ we see that on dividing it by $n$ the resulting expression tends to 0 as $n \to \infty$; since this is true for all $k \geq 0$, the sequence $\{s_n\}$ is uniformly distributed, by Weyl's theorem, #5 (vii);

ii) $\frac{1}{n}\sum_{m=1}^{n} e^{2\pi i k s_m} = \frac{1}{n}\sum_{m=1}^{n} e^{2\pi i k(m\alpha+\beta)} = \frac{1}{n} e^{2\pi i k\beta}\left(\sum_{m=1}^{n} e^{2\pi i k m\alpha}\right)$,
and since the sum on the right again tends to 0 as $n \to \infty$ we conclude, as in (i), that $\{s_n\}$ is uniformly distributed.

7. i) $n(a,b) = n(0,b) - n(0,a)$ and, therefore,
$\lim_{n\to\infty} \frac{n(a,b)}{n} = \lim_{n\to\infty} \frac{n(0,b)-n(0,a)}{n} = \lim_{n\to\infty} \frac{n(0,b)}{n} - \lim_{n\to\infty} \frac{n(0,a)}{n} = b - a$;
the opposite direction is clear;

ii) suppose $0 < a < 1$, $0 < \epsilon < \min\{a, 1-a\}$;
then, for $n > N > \frac{1}{\epsilon}$,
$n_\beta(0, a-\epsilon) - N_\beta(0, a-\epsilon) \le n_\alpha(0,a) \le n_\beta(0, a+\epsilon) + N$;
dividing throughout by $n$ and allowing $n$ to increase without bound we find $\frac{n_\alpha(0,a)}{n} - a$ may be made as small as we wish if only we choose n sufficiently large; consequently $\lim_{n\to\infty} \frac{n_\alpha(0,a)}{n}$ exists and equals $a$; the conclusion now follows from (i);

iii) it is obvious that if some enumeration of S is uniformly distributed then S is dense in [0,1]; on the other hand suppose S is dense in [0,1] and $\{\beta_n\}$ is any uniformly distributed sequence; choose a sequence $\{\alpha_n\}$ from S such that $|\alpha_n - \beta_n| < \frac{1}{n}$ and the $\alpha_n$ are distinct; now enumerate $S - \{\alpha_n\}$ to get $\{\gamma_n\}$; finally, place $\gamma_j$ into the $\{\alpha_n\}$ sequence so as to have $\gamma_j$ occupy the $j^2$ th position in the resulting sequence; clearly since $\{\alpha_n\}$ is uniformly distributed, by (ii), so also will the new sequence with the $\gamma$ inserted.

8. i) Study the diagram:

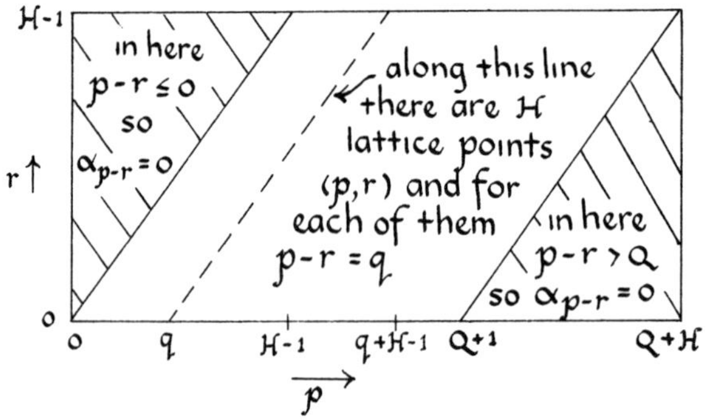

ii) note that if $p-r=p-s=q$ then $r=s$ and one obtains $q$ for the pairs $(p,r)$ equal to: $(q,0),(q+1,1),(q+2,2),\cdots(q+H-1,H-1)$; thus there are exactly $H$ such pairs; if $p-r\neq p-s$ then, say, $p-r=q$, $p-s=q+h$, $h>0$; then $1\le q\le Q-h$ and $p-s=p-r+h$ so $r-s=h$; thus, terms $\alpha_q\bar{\alpha}_{q+h}$ are obtained from the pairs $(r,s)$ equal to: $(h,0),(h+1,1),\cdots,(H-1,H-1-h)$; thus there are exactly $H-h$ such pairs; the same argument works for $\bar{\alpha}_q\alpha_{q+h}$; from these observations the desired conclusion follows.

9. Using #8 and the Schwarz' inequality we have

$$H^2\Big|\sum_{1\le q\le Q}\alpha_q\Big|^2=\Big|H\sum_{1\le q\le Q}\alpha_q\Big|^2=\Big|\sum_{0<p<H+Q}\Big(1\cdot\sum_{0\le r<H}\alpha_{p-r}\Big)\Big|^2$$

$$\le\Big(\sum_{0<p<H+Q}1^2\Big)\Big(\sum_{0<p<H+Q}\Big|\sum_{0\le r<H}\alpha_{p-r}\Big|^2\Big)$$

$$=(H+Q-1)\sum_{\substack{p,r,s\\0<p<H+Q\\0\le r<H,\,0\le s<H}}\alpha_{p-r}\bar{\alpha}_{p-s}$$

$$=(H+Q-1)\Big\{H\sum_{1\le q\le Q}|\alpha_q|^2+\sum_{1\le h<H}\sum_{1\le q\le Q-h}(H-h)\big\{\alpha_q\bar{\alpha}_{q+h}+\bar{\alpha}_q\alpha_{q+h}\big\}\Big\}$$

$$=(H+Q-1)H\sum_{1\le q\le Q}|\alpha_q|^2+2(H+Q-1)\sum_{0<h<H}(H-h)\Big|\sum_{1\le q\le Q-h}\bar{\alpha}_q\alpha_{q+h}\Big|.$$

10\. i) Put $\alpha_q = e^{2\pi i s_q}$, then, for $0 < H < Q$, using #9,

$$\frac{1}{Q^2}\left|\sum_{1\le q\le Q} e^{2\pi i s_q}\right|^2 \le$$

$$\frac{H+Q-1}{HQ} + 2\sum_{0<h<H}\frac{(H+Q-1)(H-h)}{H^2Q^2}\left|\sum_{1\le q\le Q-h} e^{2\pi i (s_{q+h}-s_q)}\right|;$$

for fixed $H$ the right side tends to $\frac{1}{H}$ as $Q \to \infty$; since this is true for any $H$ the left side tends to $0$ and we are done;

ii) by our earlier work we know $e^{2\pi i k (s_{n+h}-s_n)} \rightsquigarrow 0$ for all positive integers $k$ and $h$; thus, by (i), $e^{2\pi i k s_n} \rightsquigarrow 0$ for all positive integers $k$; the conclusion now follows from Weyl's theorem, #5(vii).

11\. (A) Suppose $a_r$ is irrational; when $r=1$ the result follows from #6(ii); thus suppose $r>1$ and that the result has been proved for $r-1$; for each fixed positive integer $h$ the quantity $f(n+h)-f(n)$ is a polynomial of degree $r-1$ with irrational leading coefficient $ha_r$; thus the result follows from that for $r-1$ and #10(ii);

(B) suppose $a_r, \cdots, a_{s+1}$ are rational and $a_s$ is irrational, $0 < s < r$; let $M$ be such that $Ma_r, \cdots, Ma_{s+1}$ are integers; if we can show $\{(f(Mn+m))\}$ is uniformly distributed for each $m = 0, 1, \cdots, M-1$ then $\{(f(n))\}$ is uniformly distributed; but, modulo 1,

$$f(Mn+m) = a_0 + a_1(Mn+m) + \cdots + a_r(Mn+m)^r$$

$$\equiv a_0 + a_1(Mn+m) + \cdots + a_s(Mn+m)^s + a_{s+1}m^{s+1} + \cdots + a_r m^r$$

$$\equiv \beta_0 + \beta_1 n + \cdots + \beta_s n^s,$$

where the $\beta_j$ are independent of $n$; in particular, $\beta_s = M^s a_s$ is irrational; this is the first case of the result (see (A)) so $\{(f(Mn+m))\}$ is uniformly distributed for each $m$ and, as we have already observed, this implies $\{(f(n))\}$ is uniformly distributed.

1. i-a) If $d_1, \cdots, d_k$ are the divisors of $m$ then the coefficient of $x^m$ on the right side, after multiplying out and collecting terms, is $a_{d_1} + \cdots + a_{d_k}$; thus the right side is $\sum_{m=1}^{\infty} (\sum_{d|m} a_d) x^m$; identifying coefficients on left and right yields the desired conclusion;

b) this is shown by induction; for $s = 1$ and any $t$ this follows from (a); suppose true for $s < n$ and all $t$; then $(s = n, t > 1)$

$$0 = \sum_{d|nt} a_d = \sum_{d|n, \delta|t} a_{d\delta} = \sum_{\substack{d|n, \delta|t \\ d\delta \neq nt}} a_d a_\delta + a_{nt}$$

$$= (\sum_{d|n} a_d)(\sum_{\delta|t} a_\delta) - a_n a_t + a_{nt} = a_{nt} - a_n a_t ;$$

c) $a_{p^0} = a_1 = 1$; by (a), $a_1 + a_p = 0$ so $a_p = -1$; also by (a), $a_1 + a_p + a_{p^2} = 0$ so $a_{p^2} = 0$; if $a_{p^k} = 0$ for $2 \le k \le n$, then, by (a), $a_1 + \cdots + a_{p^n} + a_{p^{n+1}} = 0$

so $a_{p^{n+1}} = 0$;

d) this is immediate from (b) and (c);

ii-a) this follows from the fact that $f(x^m) = \frac{x^m}{1-x^m}$ and the definition of the $a_j$;

b) for $x = \frac{1}{10}$ the result in (a) is true not only formally but also in the sense of convergence; putting $x = \frac{1}{10}$ in the expression in (a) yields

$$\frac{1}{10} = \sum_{m=1}^{\infty} \frac{a_m}{10^m - 1} = \frac{1}{9} - \frac{1}{99} - \frac{1}{999} - \frac{1}{99999} + \frac{1}{999999}$$

$$- \frac{1}{9999999} + \frac{1}{9999999999} - \cdots ;$$

c) multiplying the expression in (b) by 9 and then subtracting from 1 yields

$$\frac{1}{10} = 1 - \sum_{m=1}^{\infty} \frac{9a_m}{10^m - 1} = \frac{1}{11} + \frac{1}{111} + \frac{1}{11111} - \frac{1}{111111}$$

$$+ \frac{1}{1111111} - \frac{1}{1111111111} + \cdots .$$

2.i) This follows from the definition and from # 1(i-b, d) since these show $\mu(n) = a_n$ and that $a_n$ is a multiplicative function.

ii) if $n = p_1^{\alpha_1} \cdots p_k^{\alpha_k}$ then

$$\sum_{d|n} \mu(d) f(d) = \sum_{d|p_1 \cdots p_k} \mu(d) f(d) = (1 - f(p_1)) \cdots (1 - f(p_k));$$

(a) - (d) put $f(d)$ respectively equal to $1, d, \frac{1}{d}, \mu(d)$ in (ii); in (c) also recall the expression for $\varphi(n)$; compare this with XIV#18 (viii).

3. i). From left to right we have

$$\sum_{d|n} \mu(d) f\left(\frac{n}{d}\right) = \sum_{d|n} \mu(d) \sum_{m|\frac{n}{d}} g(m) = \sum_{m|n} g(m) \sum_{d|\frac{n}{m}} \mu(d)$$
$$= g(n);$$

in the other direction

$$\sum_{d|n} g(d) = \sum_{d|n} \sum_{m|d} \mu(m) f\left(\frac{d}{m}\right) = \sum_{d|n} \sum_{m|d} \mu\left(\frac{d}{m}\right) f(m)$$
$$= \sum_{d|n} \sum_{m|\frac{n}{d}} \mu\left(\frac{n}{dm}\right) f(m) = \sum_{m|n} f(m) \sum_{d|\frac{n}{m}} \mu\left(\frac{n}{dm}\right) = \sum_{m|n} f(m) \sum_{d|\frac{n}{m}} \mu(d)$$
$$= f(n);$$

a) if $n = p_1^{\alpha_1} \cdots p_k^{\alpha_k}$ then

$$\sum_{d|n} \Lambda(d) = \sum_{j=1}^{k} \sum_{d|p_j^{\alpha_j}} \Lambda(d) = \sum_{j=1}^{k} \alpha_j \ln p_j = \ln n;$$

now apply (i) and use #2 (ii-a);

b) apply (i) to the expression $\varphi(n) = \sum_{d|n} \mu(d) \frac{n}{d}$ obtained in #2 (ii-c); (compare with VIII#19);

c) imitate the proof of (i) exactly; note that if all quantities were positive we could take logs and obtain (i) from this result;

d) apply (c) to the formula $x^n - 1 = \prod_{d|n} F_d(x)$ of XIV# 17 (i); compare with XIV# 18 (viii);

e) from (i) we see that $\Psi(n) = \sum_{d|n} \mu(d)\mu(\frac{n}{d})$; if $n$ contains $p^3$ for some $p$ then either $d$ or $\frac{n}{d}$ contains $p^2$ so each summand is 0 and $\Psi(n) = 0$; otherwise, if $n = s^2 v$, $(s, v) = 1$, we have

$$\Psi(n) = \sum_{d|v} \mu(sd)\mu(\tfrac{n}{sd}) = \sum_{d|v} \mu(sd)\mu(s\tfrac{v}{d})$$

$$= \sum_{d|v} \mu(d)\mu(\tfrac{v}{d}) = \sum_{d|v} (-1)^t = \begin{cases} (-2)^t & \text{if } v > 1 \\ 1 & \text{if } v = 1 \end{cases};$$

ii) $$g(n) = \sum_{d|n} \sum_{\substack{r \le n \\ (r,n)=d}} \psi(\tfrac{r}{n}) = \sum_{d|n} \sum_{\substack{r/d \le n/d \\ (r/d, n/d)=1}} \psi(\tfrac{r/d}{n/d})$$

$$= \sum_{d|n} f(\tfrac{n}{d}) = \sum_{d|n} f(d);$$

now apply (i) to obtain the desired result;

a) put $\psi(x) = e^{2\pi i x}$ in (ii) and observe that

$$g(1) = 1,\ g(n) = 0 \text{ for } n > 1;$$

b) with $S_k(n)/n^k = \sum_{\substack{(r,n)=1 \\ r \le n}} (\frac{r}{n})^k$,

$g(n) = \sum_{r \le n} (\frac{r}{n})^k = \frac{1^k + \cdots + n^k}{n^k}$ use (ii) to obtain $S_k(n)/n^k = \sum_{d|n} \mu(d) g(\frac{n}{d})$, from which the desired result follows;

1) put $k=1$ in (b) and use #2 (ii - a, c) plus $1+2+\cdots+d = \frac{d(d+1)}{2}$;

2) put $k=2$ in (b) and use #2 (ii - a, b, c) plus $1^2+2^2+\cdots+d^2 = \frac{d(d+1)(2d+1)}{6}$;

3) put $k=3$ in (b) and use #2 (ii - a, b, c) plus $1^3+2^3+\cdots+d^3 = \frac{d^2(d+1)^2}{4}$;

4) apply (i) to the result in (b);

c) use (ii) with $f(n) = \sum_{\substack{(r,n)=1 \\ r\le n}} \ln \frac{r}{n}$, $g(n) = \ln \frac{n!}{n^n}$;

iii) $\sum_d \mu(d) \sum_{d|k_i} f(k_i) = \sum_{i=1}^{N} f(k_i) \sum_{d|k_i} \mu(d) = \alpha f(1)$;

a) in (iii), put $N=[x]$, $k_i = (i, P_y)$, and $f(k_i)=1$, $1 \le i \le N$; then $S_d = [\frac{x}{d}]$ and

$$\varphi(x,y) = \sum_d \mu(d) S_d;$$

1) $1 = \varphi(n,n) = \sum_{m|P_n} \mu(m)[\frac{n}{m}] = \sum_{m\le n} \mu(m)[\frac{n}{m}]$;

2) $\left|\sum_{m\le n} \frac{\mu(m)}{m}\right| = \left|\frac{1}{n}\sum_{m\le n-1} \mu(m)(\frac{n}{m} - [\frac{n}{m}]) + \frac{1}{n}\right| \le \frac{n-1}{n} + \frac{1}{n} = 1$, where we have used (1) at the 1st equality;

3) $\pi(x) - \pi(\sqrt{x}) + 1 = \varphi(x, \sqrt{x})$; compare VIII #24.

4. i) $\sum_{mn\le x} \mu(n) h(x, mn) = \sum_{k=1}^{[x]} \sum_{mn=k} \mu(n) h(x, mn)$

$$= \sum_{k=1}^{[x]} \sum_{m|k} \mu\left(\frac{k}{m}\right) h(x,k) = \sum_{k=1}^{[x]} h(x,k) \sum_{m|k} \mu\left(\frac{k}{m}\right)$$

$$= \sum_{k=1}^{[x]} h(x,k) \sum_{m|k} \mu(m) = h(x,1) \; ;$$

ii) from left to right we have, using (i) at the last equality,

$$\sum_{n\le x} \mu(n) P(n) g\left(\frac{x}{n}\right) = \sum_{n\le x} \mu(n) P(n) \sum_{m\le \frac{x}{n}} P(m) f\left(\frac{x}{mn}\right)$$

$$= \sum_{mn\le x} \mu(n) P(nm) f\left(\frac{x}{nm}\right) = f(x) \; ;$$

in the other direction, using (i) at the last equality,

$$\sum_{n\le x} P(n) f\left(\frac{x}{n}\right) = \sum_{n\le x} P(n) \sum_{m\le \frac{x}{n}} \mu(m) P(m) g\left(\frac{x}{nm}\right)$$

$$= \sum_{nm\le x} \mu(m) P(nm) g\left(\frac{x}{nm}\right) = g(x) \; ;$$

iii) from left to right we have

$$\sum_{n=1}^{\infty} \mu(n) g(nx) = \sum_{n=1}^{\infty} \mu(n) \sum_{m=1}^{\infty} f(mnx)$$

$$= \sum_{\nu=1}^{\infty} f(\nu x) \sum_{n|\nu} \mu(n) = f(x) \; ;$$

in the other direction,

$$\sum_{m=1}^{\infty} f(mx) = \sum_{m=1}^{\infty} \sum_{n=1}^{\infty} \mu(n) g(mnx) = g(x) \; ;$$

iv) from left to right we have

$$\sum_{\substack{t \\ td\in\mathcal{D}}} N(t)F(td) = \sum_{\substack{t \\ td\in\mathcal{D}}} N(t) \sum_{\substack{s \\ std\in\mathcal{D}}} G(std)$$

$$= \sum_{\substack{v \\ vd\in\mathcal{D}}} G(vd) \sum_{t|v} N(t) = G(d) \;;$$

in the other direction,

$$\sum_{\substack{\delta \\ d|\delta,\,\delta\in\mathcal{D}}} G(\delta) = \sum_{\substack{\delta \\ d|\delta,\,\delta\in\mathcal{D}}} \sum_{\substack{t \\ t\delta\in\mathcal{D}}} N(t)F(t\delta)$$

$$= \sum_{\substack{v \\ vd\in\mathcal{D}}} F(vd) \sum_{t|v} N(t) = F(d).$$

1. i) $\sum_{m=M+1}^{N} F(m)(g(m+1)-g(m)) =$

$\sum_{m=M+1}^{N} F(m)\,g(m+1) - \sum_{m=M+1}^{N} F(m)\,g(m) =$

$F(N)g(N+1) + \sum_{m=M+1}^{N}(F(m-1)-F(m))g(m) - F(M)g(M+1) =$

$F(N)\,g(N+1) - \sum_{m=M+1}^{N} f(m)g(m)$ , since $F(M) = 0$ ;

ii) $\left|\sum_{m=M+1}^{N} f(m)\,g(m)\right| \le$

$|F(N)|\,g(N) + \sum_{m=M+1}^{N-1} |F(m)||\,g(m+1)-g(m)|$

$\le \max_{M<m\le N} |F(m)| \left\{g(N) + \sum_{m=M+1}^{N-1} |\,g(m+1)-g(m)|\right\}$

$= \begin{cases} \max_{M<m\le N} |F(m)| \{g(N)+g(M+1)-g(N)\} & \text{if } g \text{ is decreasing;} \\ \max_{M<m\le N} |F(m)| \{g(N)-g(M+1)+g(N)\} & \text{if } g \text{ is increasing;} \end{cases}$

from which the desired result follows ;

iii-a) the result in (ii), first part, shows the sequence of partial sums is a Cauchy sequence ;

b) $\sum_{n=1}^{\infty} f(n)g(n) = \sum_{n\le x} f(n)g(n) + \sum_{n=[x]+1}^{\infty} f(n)g(n)$

and, since $\left|-\sum_{n=[x]+1}^{N} f(n)g(n)\right| \le g([x]) \max_{[x]<m\le N} |F(m)|$,

we know $-\sum_{n=[x]+1}^{\infty} f(n)g(n) = O(g([x]))$ ;

iv) let $r$ be the largest index $m$ for which $\lambda_m \le x$ ; then $\int_{\lambda_1}^{x} F(t)\, g'(t)\, dt =$

$$\sum_{m=1}^{r-1} \int_{\lambda_m}^{\lambda_{m+1}} F(t)\, g'(t)\, dt + \int_{\lambda_r}^{x} F(t)\, g'(t)\, dt$$

$$= \sum_{m=1}^{r-1} F(\lambda_m) \int_{\lambda_m}^{\lambda_{m+1}} g'(t)\, dt + F(\lambda_r) \int_{\lambda_r}^{x} g'(t)\, dt$$

$$= \sum_{m=1}^{r-1} F(\lambda_m)\,(g(\lambda_{m+1}) - g(\lambda_m)) + F(x)(g(x) - g(\lambda_r))$$

$$= \sum_{m=1}^{r} (F(\lambda_{m-1}) - F(\lambda_m))\, g(\lambda_m) + F(x)\, g(x)$$

$$= F(x)\, g(x) - \sum_{m=1}^{r} f(m)\, g(\lambda_m) \; ;$$

v) in (iv) we put $\lambda_1 = a$, $\lambda_j = a + j - 1$, $f(m) = 1$ for all $m$ and obtain

$$\sum_{a \le m \le x} g(m) = ([x] - a + 1)\, g(x) - \int_a^x ([t] - a + 1)\, g'(t)\, dt,$$

and, after noting that

$$\int_a^x g(t)\, dt = x g(x) - a g(a) - \int_a^x t\, g'(t)\, dt,$$

we see that this agrees with the given expression;

vi) by (v),

$$\left| \sum_{a \le m \le x} g(m) - \int_a^x g(t)\, dt \right| \le \int_a^x |g'(t)|\, dt + |g(a)| + |g(x)|$$

$$= \pm (g(x) - g(a)) + |g(a)| + |g(x)| = O(|g(a)| + |g(x)|) \; ;$$

vii) $\sum_{0<m\le x} g(m) - \int_a^x g(t)\,dt - c =$
$$-\int_x^\infty (t-[t])\,g'(t)\,dt - (x-[x])\,g(x)$$
and, since $|t-[t]|<1$, $\int_x^\infty g'(t)dt = g(x)$ we know the right side is, in absolute value, smaller than $2|g(x)|$; we are using $g(x)\to 0$ as $x\to\infty$ in guaranteeing the infinite integrals exist.

2. i) Put $a=1$, $g(x)=x^{-s}$ in #1(vi);

ii) put $a=1$, $g(x)=\frac{1}{x}$ in #1(vii) to obtain the expression for $\sum_{n\le x}\frac{1}{n}$ with $\gamma = 1-\int_1^\infty \frac{t-[t]}{t^2}\,dt$; the other expression for $\gamma$ follows from the obtained expression;

iii) put $a=1$, $g(x)=\ln x$ in #1(vi);

iv) $\sum_{n\le x}\ln\frac{x}{n} = \sum_{n\le x}\ln x - \sum_{n\le x}\ln n$
$= [x]\ln x - (x\ln x - x + O(\ln x)) = ([x]-x)\ln x + x + O(\ln x)$, and this last is clearly $O(x)$;

v) first of all note

$$\frac{1}{x}\sum_{p\le x}\ln p\left\{\left[\frac{x}{p}\right]+\left[\frac{x}{p^2}\right]+\cdots\right\}\le\sum_{p\le x}\frac{\ln p}{p}+\sum_{p\le x}\ln p\left\{\frac{1}{p^2}+\frac{1}{p^3}+\cdots\right\}$$
$$=\sum_{p\le x}\frac{\ln p}{p}+\sum_{p\le x}\frac{\ln p}{p(p-1)}\quad;$$

using XIV#11(v) the last sum tends to a limit as $x\to\infty$ and, therefore, our 1st equality is proved; the 2nd equality is immediate from the definition of $\Lambda$; using XXII#3(i-a) we see that $\ln n=\sum_{d|n}\Lambda(d)$ so $\frac{1}{x}\sum_{n\le x}\ln n=\frac{1}{x}\sum_{n\le x}\sum_{d|n}\Lambda(d)$; for each $d\le x$, $\Lambda(d)$ will occur in the last sum precisely as many times as there are $n\le x$ with $d|n$; i.e. $\frac{1}{x}\sum_{n\le x}\sum_{d|n}\Lambda(d)=\frac{1}{x}\sum_{d\le x}\left[\frac{x}{d}\right]\Lambda(d)$; but this, with different notation, yields our 3rd equality; the last equality is a consequence of (iii);

vi) the 1st equality is contained in (v); for the 2nd we have, where we use (v) and the Chebyshev inequality, see XIV#10(ix), the inequality, $\sum_{n\le x}\frac{\Lambda(n)}{n}=\frac{1}{x}\sum_{n\le x}\left(\frac{x}{n}-\left[\frac{x}{n}\right]\right)\Lambda(n)+\frac{1}{x}\sum_{n\le x}\left[\frac{x}{n}\right]\Lambda(n)$

$$\le\frac{1}{x}\cdot\frac{Bx}{\ln x}\cdot\ln x+\ln x+O(1)=\ln x+O(1)\quad;$$

vii) put $\lambda_j = p_j$, $f(m) = 1$ for all $m$,
$g(x) = \frac{\ln x}{x}$ in #1 (iv) ; then

$$\sum_{p \le x} \frac{\ln p}{p} = \pi(x)\frac{\ln x}{x} - \int_2^x \pi(t)\frac{1-\ln t}{t^2}\,dt$$
$$= O(1) + \int_2^x \pi(t)\frac{\ln t - 1}{t^2}\,dt\ ;$$

the result follows immediately by taking a difference .

3. i) The 1st equality follows immediately from #2 (vi) by taking a difference ; the 2nd equality is #2 (vii) ; the 1st inequality follows (for N sufficiently large) from our assumption that the $\lim_{x\to\infty} \frac{\pi(x)}{x/\ln x}$ exists and is less than 1 ; the 2nd inequality follows from suppressing the term $\frac{1}{t \ln t}$ and integrating ; the result is false because it says that for sufficiently large N,

$$\frac{1-\beta}{2}\ln N = O(1)\ ;$$

ii) parallel to (i) ;

iii) immediate from (i) and (ii) .

4. i) $\sum_{\substack{n \le N \\ d | n}} n = d + 2d + \cdots + \left[\frac{N}{d}\right] d$

$= \frac{d}{2}\left[\frac{N}{d}\right]\left(\left[\frac{N}{d}\right]+1\right) = \frac{d}{2}\left(\frac{N}{d} + O(1)\right)^2$ ;

ii) clearly $N^* = \sum_{n=1}^{N} \varphi(n)$ ; now, using XXII # 2 (ii-c), we find this latter expression equals

$$\sum_{n=1}^{N} \sum_{d|n} N(d) \frac{n}{d} = \sum_{d=1}^{N} \frac{N(d)}{d} \sum_{\substack{n \le N \\ d|n}} n \quad ;$$

iii) substituting the result of (i) into (ii) yields $N^* = \sum_{d=1}^{N} \frac{N(d)}{2}\left(\frac{N}{d} + O(1)\right)^2 =$

$$\frac{N^2}{2} \sum_{d=1}^{N} \frac{N(d)}{d^2} + \left\{ N \sum_{d=1}^{N} \frac{N(d)}{d} + \frac{1}{2} \sum_{d=1}^{N} N(d) \right\} O(1)$$

$$= \frac{N^2}{2} \sum_{d=1}^{N} \frac{N(d)}{d^2} + O(N),$$

where we have used XXII # 3 (iii-2) ; this last equals $\frac{N^2}{2} \sum_{d=1}^{\infty} \frac{N(d)}{d^2} - \frac{N^2}{2} \sum_{d=N+1}^{\infty} \frac{N(d)}{d^2} + O(N)$

$$= \frac{N^2}{2} \sum_{d=1}^{\infty} \frac{N(d)}{d^2} + N^2 g(N),$$

where $g(N) = -\frac{1}{2} \sum_{d=N+1}^{\infty} \frac{N(d)}{d^2} + \frac{1}{N^2} O(N) \to 0$ as $N \to \infty$ ; finally, note that

$$\left(\sum_{n=1}^{\infty} \frac{1}{n^2}\right)\left(\sum_{n=1}^{\infty} \frac{N(n)}{n^2}\right) = \sum_{n=1}^{\infty} \sum_{j|n} N(j) \frac{1}{n^2} = 1$$

so the desired result is correct ;

iv) $N' = \frac{N(N+1)}{2}$ and $\frac{N^2}{2N'} \to 1$ as $N \to \infty$ ; therefore, using (iii), the result is immediate ;

v) this is just a restatement of (iv) after replacing $\sum_{n=1}^{\infty} \frac{1}{n^2}$ by its value $\frac{\pi^2}{6}$ .

5. i) The series is merely the Taylor expansion of $\ln \frac{1}{1-x}$ ; since

$$x \le x + \frac{x^2}{2} + \frac{x^3}{2} + \cdots \le x(1 + x + x^2 + \cdots) = \frac{x}{1-x} \le 2x$$

the inequalities are correct ;

ii) first we note that if either series converges it must be true that all $x_j$ with $j$ sufficiently large satisfy the conditions on $x$ in (i) ; hence we may, without loss of generality, assume $0 \le x_j \le \frac{1}{2}$ for all $j$ ; hence, by (i),

$$\sum_{j=1}^{n} x_j \le \sum_{j=1}^{n} \ln \frac{1}{1-x_j} \le 2 \sum_{j=1}^{n} x_j$$

and the desired conclusion follows ;

iii) this follows from (ii) and the continuity of the logarithm function since

$$\sum_{j=1}^{n} \ln \frac{1}{1-x_j} = \ln \prod_{j=1}^{n} \frac{1}{1-x_j} .$$

6. i) If for some integer s, $|f(s)| > 1$ then by the complete multiplicativity of f we would have f unbounded on the sequence $s, s^2, s^3, \cdots$ and this would contradict the convergence of

$$\sum_{j=1}^{\infty} f(j) ;$$

ii) this follows from

$$\prod_{p \le m} (1-f(p))^{-1} = \prod_{p \le m} (1+f(p)+f(p^2)+\cdots) = \sum_{j=1}^{m} f(j) + \Sigma f(j),$$

where the 2nd sum is over all those j exceeding m and having no prime factors exceeding m ;

iii) this follows from (ii) by allowing m to tend to infinity.

7. i-a, b) These follow immediately from the existence of the integral $\int_1^\infty x^{-s}\,dx$ and the inequality $\int_1^\infty x^{-s}\,dx < \sum_{n=1}^\infty \frac{1}{n^s} < 1 + \int_1^\infty x^{-s}\,dx$;

c) from (b), $1 < (s-1)\zeta(s) < s$, and the result follows;

d) put $f(x) = x^{-s}$ in #6 (iii);

e) take logarithms in (d) and then use the equality of #5 (i);

f) from (e) we find

$$0 \le \ln \zeta(s) - \sum_p \frac{1}{p^s} = \sum_p \sum_{n=2}^\infty \frac{1}{np^{ns}};$$

but $\sum_{n=2}^\infty \frac{1}{np^{ns}} < \frac{1}{p^{2s}}\left(1 + \frac{1}{p^s} + \cdots\right) = \frac{1}{p^s(p^s-1)}$; hence $\sum_p \sum_{n=2}^\infty \frac{1}{np^{ns}} < \sum_p \frac{1}{p(p-1)} < \sum_{n=2}^\infty \frac{1}{n(n-1)} = 1$ and the conclusion follows;

ii-a) the result in (i-c) tells us $\zeta(s)$ must tend to $\infty$ as $s \to 1^+$; the resut in (d) says this could not happen if there were only finitely many primes;

b) since $\ln \zeta(s)$ exists for $s>1$, (i-f) shows us $\sum_p \frac{1}{p^s}$ exists; by (i-f) we see that

$$\sum_p \frac{1}{p^s} \to \infty \text{ as } s \to 1^+ ;$$

since for all $s>1$, $\sum_{p\leq n} \frac{1}{p^s} < \sum_{p\leq n} \frac{1}{p}$ we see that

$$\sum_p \frac{1}{p} \text{ diverges .}$$

8. i-a) $\sum_{n=1}^{\infty} \frac{\chi(n)}{n^s} = 1 - \frac{1}{3^s} + \frac{1}{5^s} - \frac{1}{7^s} + \cdots$ is an alternating series with terms in absolute value tending strictly monotonically to 0; hence the convergence follows by the

Leibniz' test ;

b) these follow immediately from the fact that in such a series the error made in a partial sum approximation never exceeds the magnitude of the 1st omitted term and

has the same algebraic sign ;

c) this follows from the uniformity of the convergence of the series in a small

neighborhood of 1 ;

d) by #7(i-c), $\zeta(s) \to \infty$ as $s \to 1^+$ and by (b) and (c), $L(s)$ tends to a finite positive limit ;

e) $\prod_{p \equiv 3 \pmod 4} (1 - p^{-2s})^{-1} = \Sigma \frac{1}{n^{2s}}$, where the sum is over all integers $n$ all of whose prime factors are of the form $4k+3$ ; thus the product, which is monotonically increasing as $s \to 1^+$, is always bounded above by $\zeta(2)$ ; thus the assertion is correct ;

f) by #5(iii) and #7(i-a) all products converge ; the result for $\zeta(s)$ now follows from the truth of the identity for all integers $n$ when all primes are restricted to be $\leq n$ ; the result for $L(s)$ follows from the complete multiplicativity of $\chi$, an argument like that leading to #7(i-d) (except that one now takes $f(x) = \frac{\chi(x)}{x^s}$), and the above argument for the $\zeta(s)$ identity ;

g) from (f) we see that

$$\zeta(s) L(s) = \frac{1}{1-2^s} \prod_{p \equiv 1 \pmod 4} (1 - p^{-s})^{-2} \prod_{p \equiv 3 \pmod 4} (1 - p^{-2s})^{-1} ;$$

now the left side, by (d), tends to $\infty$ as $s \to 1^+$ while, by (e), the product on the far right of the right side tends to a finite limit; hence since $\frac{1}{1-2^s}$ also tends to a finite limit it must be the case that $\prod_{p \equiv 1 \pmod 4} (1-p^{-s})^{-2} \to \infty$ as $s \to 1^+$; but this implies there must be infinitely many $4k+1$ primes; for the $4k+3$ primes we consider $\zeta(s)L(s)^{-1} = \frac{1}{1-2^s} \prod_{p \equiv 3 \pmod 4} \left(\frac{1+p^{-s}}{1-p^{-s}}\right)$; the left side tends to infinity as $s \to 1^+$ and, therefore, so must the right, and that implies the existence of infinitely many primes of the form $4k+3$;

ii-a, b) follow from the same argument used to prove #7(e, f);

c, d) $\ln \zeta(s) + \chi^{-1}(a) \ln L(s) = \sum_p \frac{1}{p^s} + \sum_p \frac{\chi^{-1}(a)\chi(p)}{p^s} + O(1)$

$$= \begin{cases} 2 \sum_{p \equiv 1 \pmod 4} \frac{1}{p^s} + O(1) & \text{for } a = 4k+1; \\ 2 \sum_{p \equiv 3 \pmod 4} \frac{1}{p^s} + O(1) & \text{for } a = 4k+3; \end{cases}$$

since, using (i-c) and #7(i-c), the left side tends to infinity as $s \to 1^+$ independent of a the conclusion follows.

9. i-a) Direct checking ;

b) when $a \equiv n \pmod 5$ this is clear since in this case each summand is 1 and there are 4 summands ; when $a \not\equiv n \pmod 5$ the sum is just a sum of column entrees in our table after each entry is divided by the corresponding entry in another column ; direct inspection yields the result ;

c) for each $\chi$ under consideration each series breaks into a real and a complex part ; the series of the two parts are each alternating and converge by the same argument used in the proof of #8 (i-a) ; that they are not zero follows as in the proof of #8 (i-b) ;

d) these follow from (c) and the alternating character of the real and complex parts of the series ;

e) from the 1st formula of (d) we have

$$\sum_{n \le x} \chi(n) \frac{x}{n} = x L_0(\chi) + O(1) ;$$

now put $P(n) = \chi(n)$, $f(x) = x$ in the Shapiro form of the Möbius inversion formula, see XXII #4 (ii) , to obtain

$$x = \sum_{n \le x} \mu(n)\chi(n)\left\{\frac{x}{n} L_0(\chi) + O(1)\right\};$$

this implies $\sum_{n \le x} \frac{\mu(n)\chi(n)}{n} = L_0(\chi)^{-1}\left\{1 - \frac{O(1)}{x}\sum_{n \le x} \mu(n)\chi(n)\right\};$
the conclusion follows since $|\mu(n)\chi(n)| \le 1$ for all $n$ so the right hand side of this last equation is $O(1)$;

f) from XXII, #3(i-a-2) and #2(ii-a), we have $\Lambda(n) = \sum_{d|n} \mu(d) \ln \frac{n}{d}$; this along with the 2nd formula of (d) yields

$$\sum_{n \le x} \frac{\chi(n)\Lambda(n)}{n} = \sum_{n \le x} \frac{\chi(n)}{n} \sum_{d|n} \mu(d) \ln \frac{n}{d} = \sum_{d \le x} \mu(d) \sum_{j \le \frac{x}{d}} \frac{\chi(jd)}{jd} \ln j$$

$$= \sum_{d \le x} \frac{\mu(d)\chi(d)}{d} \sum_{j \le \frac{x}{d}} \frac{\chi(j) \ln j}{j} = \sum_{d \le x} \frac{\mu(d)\chi(d)}{d}\left\{L_1(\chi) + O\left(\frac{\ln x/d}{x/d}\right)\right\};$$

the result now follows from (e) since $L_1(\chi)$ is a constant and $\sum_{d \le x} \frac{\mu(d)\chi(d)}{d} O\left(\frac{\ln x/d}{x/d}\right)$ is dominated by an expression $\frac{1}{x}\sum_{d \le x} O\left(\ln \frac{x}{d}\right) = O(1)$, by #2(iv);

ii-a) by the definition of $\Lambda(n)$, the 1st sum on the right is clearly equal to the left hand sum plus the 2nd sum on the right; the rest follows from (i-f) and

$$\sum_{j=2}^{\infty}\sum_{p}\frac{\ln p}{p^{j}}=\sum_{p}\frac{\ln p}{p^{2}}+\sum_{j=2}^{\infty}\sum_{p}\frac{\ln p}{p^{j+1}}<\sum_{p}\frac{\ln p}{p^{2}}+\sum_{j=2}^{\infty}\sum_{p}\frac{1}{p^{j}}$$

$$<\sum_{p}\frac{\ln p}{p^{2}}+\sum_{j=2}^{\infty}\left\{\frac{1}{2^{j}}+\sum_{n=3}^{\infty}\frac{1}{n^{j}}\right\}<\sum_{p}\frac{\ln p}{p^{2}}+\frac{1}{2}+\sum_{j=2}^{\infty}\int_{2}^{\infty}\frac{dx}{x^{j}}$$

$$=\sum_{p}\frac{\ln p}{p^{2}}+\frac{1}{2}+\sum_{j=2}^{\infty}\frac{1}{(j-1)2^{j-1}}=O(1);$$

b) multiplying the expression in (ii-a) by $\chi(a)^{-1}$ and summing over all characters, including $\chi_0$, we find, after using (i-b),

$$\sum_{\chi}\chi(a)^{-1}\sum_{p\le x}\frac{\chi(p)\ln p}{p}=4\sum_{\substack{p\le x\\ p\equiv a\,(\mathrm{mod}\,5)}}\frac{\ln p}{p};$$

on the other hand the left side is equal to, using (a),

$$\sum_{p\le x}\frac{\ln p}{p}+\sum_{\substack{\chi\\ \chi\ne\chi_0}}\chi(a)^{-1}\sum_{p\le x}\frac{\chi(p)\ln p}{p}=\sum_{p\le x}\frac{\ln p}{p}+O(1);$$

finally, putting these together with #2 (ii-v) we have

$$\sum_{\substack{p\le x\\ p\equiv a(\mathrm{mod}\,5)}}\frac{\ln p}{p}=\frac{1}{4}\sum_{p\le x}\frac{\ln p}{p}+O(1)=\frac{1}{4}\ln x+O(1);$$

c) this follows immediately from (b) by taking $a=1,2,3,4$.

# XXIV Numerical Characters
# and the Dirichlet Theorem - Solutions

1\. *i*) This is clear;

*ii*) by complete multiplicativity

$$\chi(1) = \chi(1 \cdot 1) = \chi(1)\chi(1);$$

thus either $\chi(1) = 0$ or $\chi(1) = 1$; but $\chi(1) = 0$ is prohibited from the definition since $(1, k) = 1$;

*iii*) by Euler's generalization of Fermat's theorem (see IX #7(*iii*)) we know

$$a^{\varphi(k)} \equiv 1 \pmod{k};$$

by periodicity and multiplicativity of $\chi$ we have $(\chi(a))^{\varphi(k)} = \chi(a^{\varphi(k)}) = \chi(1)$; but by (*ii*) $\chi(1) = 1$ so we are done;

*iv*, *v*) by direct checking;

vi) by (iii) for $(a,k)=1$, $\chi(a)$ is one of the $\varphi(k)$, $\varphi(k)^{th}$ roots of unity; since $(a,k)>1$ implies $\chi(a)=0$ we clearly have no more than $\varphi(k)^{\varphi(k)}$ possible specifications for $\chi$ on the set of $a$ prime to $k$;

vii) $(a,k)=1$ implies $(a,d)=1$ so $\chi^*(a)=\chi(a)\neq 0$; if $(a,k)>1$ then $\chi^*(a)=0$; if $(ab,k)>1$ then $\chi^*(ab)=0=\chi^*(a)\chi^*(b)$; if $(ab,k)=1$ then $\chi^*(ab)=\chi(ab)=\chi(a)\chi(b)=\chi^*(a)\chi^*(b)$; finally if $a\equiv b \pmod{k}$ then $(a,k)=(b,k)$ so if this is $>1$, $\chi^*(a)=\chi^*(b)=0$, while if this is equal to 1, $\chi^*(a)=\chi(a)$, $\chi^*(b)=\chi(b)$ and $a\equiv b \pmod{d}$, so $\chi^*(a)=\chi(a)=\chi(b)=\chi^*(b)$;

viii) since $\chi$ is not principal there is an $a$ with $\chi(a)\neq 1$, $(a,k)=1$; now $a, 2a, \cdots, ka$ run over a complete system of residues modulo $k$ so $\chi(a)\sum_{n=1}^{k}\chi(n)=\sum_{n=1}^{k}\chi(an)=\sum_{n=1}^{k}\chi(n)$ and, therefore,

$$(\chi(a)-1)\sum_{n=1}^{k}\chi(n)=0\ ;$$

since $\chi(a)-1\neq 0$ the conclusion follows ;

ix) direct checking ;

x) if not then there are two mod $k$ characters, say $\chi, \chi'$, such that $\chi(a)\chi_1(a)=\chi'(a)\chi_1(a)$ ; but for $(a,k)=1$ this means $\chi(a)=\chi'(a)$; since $\chi$ and $\chi'$ are zero on all $a$, $(a,k)>1$, this implies

$$\chi=\chi'.$$

2. i-a) This is clear by direct checking ;

b) for $\chi(d)=1$ with $(d, p^{\beta})=1$ we would have to have $\lambda=0$, which implies $d\equiv 1 \pmod{p^{\beta}}$;

c) $(d,k)=1$ and $p^{\beta}\,|\,k$ implies $(d,p^{\beta})=1$; thus by (b), $\chi(d)\neq 1$ ; if we let $\chi^*$ be the mod $k$ extension of $\chi$, see #1 (vii), then

$$\chi^*(d)=\chi(d)\neq 1\quad ;$$

ii) since $d \equiv -1 \pmod 4$ we know $(d, k) = 1$ and $\chi^*(d) = \chi(d) = -1$ ;

iii-a) first of all we note that, by the proof of XVIII# 15 (iii), $\chi$ is in fact defined for all odd $n$ ; since the product of two odd numbers is of the form $4k+3$ precisely when exactly one of them is of this form we see that $\chi$, as defined, is completely multiplicative (this is clear when a factor is even); the rest is clear ;

b) since $(d, k) = 1$ we know $(d, 2^\alpha) = 1$ and, therefore if $\chi^*$ is the mod $k$ extension of $\chi$ we have $\chi^*(d) = \chi(d)$ ; if $\chi(d) = 1$ then $t = 0$ and $d \equiv \pm 1 \pmod{2^\alpha}$, contrary to assumption; thus $\chi^*(d) \neq 1$ ;

iv) the hypotheses imply $k$ is an integer greater than 2 ; thus either some odd prime divides $k$, or $k = 4$, or $k$ is divisible by 8 ;

in any event, by (i), (ii), (iii) there is a mod $k$ character such that $\chi(d) \neq 1$.

3. i) If $(a,k) \neq 1$ the result is clear; otherwise, if $a \equiv 1 \pmod{k}$ then the sum equals $\sum_{\chi} \chi(1) = c$, while, if $a \not\equiv 1 \pmod{k}$ then, by #2(iv), there is a $\chi_1$ with $\chi_1(a) \neq 1$; multiplying by $\chi_1(a)$ and recalling #1(x) we see

$$\chi_1(a) \sum_{\chi} \chi(a) = \sum_{\chi} (\chi_1 \chi)(a) = \sum_{\chi} \chi(a)$$

and, therefore, $(\chi_1(a) - 1) \sum_{\chi} \chi(a) = 0$; since $\chi_1(a) - 1 \neq 0$ the conclusion follows;

ii) using (i) at the 1st equality and #1(viii) at the 3rd equality we have

$$c = \sum_{a=1}^{k} \sum_{\chi} \chi(a) = \sum_{\chi} \sum_{a=1}^{k} \chi(a) = \sum_{a=1}^{k} \chi_0(a) = \varphi(k);$$

iii) if $(n,k) \neq 1$ then the sum is zero; if $(n,k) = 1$ we may select $m$ so that $am \equiv n \pmod{k}$; note then, using (i) and (ii), that

$$\sum_{\chi} \chi(a)^{-1}\chi(n) = \sum_{\chi} \chi(a)^{-1}\chi(a)\chi(m) = \sum_{\chi} \chi(m)$$

$$= \begin{cases} \varphi(k) & \text{if } m \equiv 1 \pmod{k}; \\ 0 & \text{if } m \not\equiv 1 \pmod{k}; \end{cases}$$

but $m \equiv 1 \pmod{k}$ is equivalent to $a \equiv n \pmod{k}$ and we are done;

iv) since $\chi(a)$ is a root of unity, see #1 (iii), $\chi(a)\overline{\chi(a)} = 1$; thus $\bar{\chi}(a) = \overline{\chi(a)} = \chi(a)^{-1}$ and the result is the same as (iii).

4. i) $\frac{\chi(a)}{m}$ is an $m^{th}$ root of unity and with the exception of 1 all of them satisfy the equation $x \cdot \frac{x^m - 1}{x-1} = x^m + \cdots + x = 0$;

ii) the 1st equality is immediate from (i); on the other hand

$$\sum_{\chi} \sum_{j=1}^{m} \left(\frac{\chi(a)}{\omega}\right)^j = \sum_{j=1}^{m} \frac{1}{\omega^j} \sum_{\chi} \chi(a)^j = \sum_{j=1}^{m} \frac{1}{\omega^j} \sum_{\chi} \chi(a^j);$$

but, by #3 (i), (ii), the inside sum on the right is 0 unless $a^j \equiv 1 \pmod{k}$ when it is $\varphi(k)$; since only for $j=m$ is the latter true we see that the right hand expression is merely

$$\frac{1}{\omega^m}\varphi(k)=\varphi(k).$$

5. Taking $f$ to be $\chi$ and $g(n)$, respectively, to be $\frac{1}{n}$, $\frac{\ln n}{n}$, $\frac{1}{\sqrt{n}}$ in XXIII #1 (iii) we obtain all of these results.

6. i) Since $\chi$ is multiplicative the multiplicativity of F follows from VIII #27 (i); the expression given for $F(p_1^{\alpha_1}\cdots p_s^{\alpha_s})$ is immediate from the multiplicativity;

ii) this follows from the expression in (i) and the observation

$$\sum_{j=0}^{\alpha_i}\chi(p_i{}^j)=\chi(1)+\chi(p_i)+\chi(p_i)^2+\cdots+\chi(p_i)^{\alpha_i}$$

$$=\begin{cases}\alpha_i+1 \text{ if } \chi(p_i)=1\\ 1 \text{ if } \chi(p_i)=-1 \text{ and } \alpha_i \text{ is even or if } \chi(p_i)=0\,;\\ 0 \text{ if } \chi(p_i)=-1 \text{ and } \alpha_i \text{ is odd,}\end{cases}$$

plus the fact that for $n^2$ all $\alpha_i$ are even;

iii) this is an immediate consequence of (ii) since $\sum_{n=1}^{\infty} \frac{F(n)}{\sqrt{n}}$ has a subseries which dominates the harmonic series $\sum_{n=1}^{\infty} \frac{1}{n}$ ;

iv) the 1st equality is clear from the definitions of G and F ; the 2nd follows from a consideration of the lattice points under the graph of $d\delta = x$ in the graph to the right ;

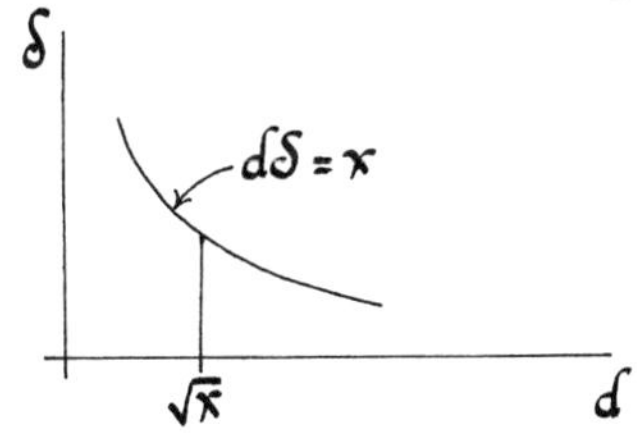


v) by comparing $\sum_{\delta \le \frac{x}{d}} \frac{1}{\sqrt{\delta}}$ and $\sum_{\delta < \sqrt{x}} \frac{1}{\sqrt{\delta}}$ with $\int_0^{\frac{x}{d}} x^{-\frac{1}{2}} dx$ and $\int_0^{\sqrt{x}} x^{-\frac{1}{2}} dx$ we have

$$\sum_{\delta \le \frac{x}{d}} \frac{1}{\sqrt{\delta}} = 2\sqrt{\frac{x}{d}} + O(1) + O\left(\sqrt{\frac{d}{x}}\right), \quad \sum_{\delta < \sqrt{x}} \frac{1}{\sqrt{\delta}} = 2\sqrt[4]{x} + O(1) + O\left(\frac{1}{\sqrt[4]{x}}\right);$$

from the last part of #5 we have

$$\sum_{\sqrt{x} < d \le \frac{x}{\delta}} \frac{\chi(d)}{\sqrt{d}} = O\left(\sqrt{\frac{\delta}{x}}\right) + O\left(\frac{1}{\sqrt[4]{x}}\right) ;$$

substituting these into (iv) and simplifying yields

$$G(x) = 2\sqrt{x} \sum_{d \le \sqrt{x}} \frac{\chi(d)}{d} + O(1) \sum_{d \le \sqrt{x}} \frac{\chi(d)}{\sqrt{d}} +$$

$$O\left(\frac{1}{\sqrt{x}}\right)\left(\sum_{\delta \le \sqrt{x}} \chi(d) + \sum_{\delta < \sqrt{x}} 1\right) + O\left(\frac{1}{\sqrt[4]{x}}\right) \sum_{\delta < \sqrt{x}} \frac{1}{\sqrt{\delta}}$$

$$= 2\sqrt{x}\, L_0(\chi) + O(1) + O\left(\frac{1}{\sqrt[4]{x}}\right)\left(2\sqrt[4]{x} + O(1) + O\left(\frac{1}{\sqrt[4]{x}}\right)\right)$$

$$= 2\sqrt{x}\, L_0(\chi) + O(1) ;$$

vi) if $L_0(X) = 0$ then, from (v), $G(x) = O(1)$ and this contradicts (iii).

7. i) Put $P(n) = X(n)$, $f(x) = x$, $g(x) = \sum_{n \le x} \frac{X(n)x}{n}$ into XXII#4 (ii), to obtain $x = \sum_{m \le x} N(m) X(m) g(\frac{x}{m})$; now noting that, by #5, $g(x) = x L_0(X) + O(1)$ and substituting we have

$$x = \sum_{m \le x} N(m) X(m) \left(\frac{x}{m} L_0(X) + O(1)\right)$$

$$= x \sum_{m \le x} \frac{N(m)X(m)}{m} L_0(X) + O(1) \sum_{m \le x} N(m) X(m) \;;$$

since the last sum is $o(x)$ the desired result follows;

ii-a) $$g(x) = x \ln x \sum_{n \le x} \frac{X(n)}{n} - x \sum_{n \le x} \frac{X(n) \ln n}{n}$$

$$= x \ln x \left(L_0(X) + O\left(\tfrac{1}{x}\right)\right) - x \left(L_1(X) + O\left(\tfrac{\ln x}{x}\right)\right)$$

$$= -x L_1(X) + O(\ln x) \;;$$

b) using XXII#4 (ii) on the expression for $g(x)$, taking $f(x) = x \ln x$, we find, using (a),

$$x \ln x = \sum_{n \le x} N(n) X(n) g\left(\tfrac{x}{n}\right) = \sum_{n \le x} N(n) X(n) \left(-\tfrac{x}{n} L_1(X) + O\left(\ln \tfrac{x}{n}\right)\right)$$

$$= -x L_1(X) \sum_{n \le x} \frac{N(n)X(n)}{n} + O\left(\sum_{n \le x} \ln \frac{x}{n}\right);$$

the desired conclusion now follows from XXIII#2 (iv);

iii) for $L_0(\chi) = 0$ this follows from (ii) and for $L_0(\chi) \neq 0$ this follows from (i).

8. This proceeds along lines similar to XXIII #9 (ii);

$$\sum_{p \le x} \frac{\chi(p)\ln p}{p} + O(1) = \sum_{p \le x} \frac{\chi(p)\ln p}{p} + \sum_{j=2}^{\infty} \sum_{p^j \le x} \frac{\chi(p^j)\ln p}{p^j}$$

$$= \sum_{n \le x} \frac{\chi(n)\Lambda(n)}{n} = \sum_{d \le x} \frac{N(d)\chi(d)}{d} \sum_{\delta \le \frac{x}{d}} \frac{\chi(\delta)\ln\delta}{\delta}$$

$$= \sum_{d \le x} \frac{N(d)\chi(d)}{d} \left\{ L_1(\chi) + O\left(\frac{\ln x/d}{x/d}\right) \right\} = L_1(\chi) \sum_{d \le x} \frac{N(d)\chi(d)}{d} + \frac{1}{x} O\left(\sum_{d \le x} \ln \frac{x}{d}\right)$$

$$= L_1(\chi) \sum_{d \le x} \frac{N(d)\chi(d)}{d} + O(1) \ ;$$

using #7 (iii) completes the proof.

9. i) By #6 (vi) no such $\chi$ may be real; thus if $L_0(\chi) \neq 0$ then $L_0(\bar{\chi}) \neq 0$ and $\chi \neq \bar{\chi}$ ; thus N is at least 2 ;

ii) the left inequality is clear; the 1st equality follows from #3 (i), (ii) since

$$\sum_{\chi} \sum_{p \le x} \frac{\chi(p)\ln p}{p} = \sum_{p \le x} \frac{\ln p}{p} \sum_{\chi} \chi(p) \ ;$$

finally, making use of #8 and XXIII #2(v) we see

$$\sum_{\chi}\sum_{p\le x}\frac{\chi(p)\ln p}{p}=\sum_{p\le x}\frac{\ln p}{p}+\sum_{\substack{\chi\\ L_0(\chi)=0\\ \chi\ne\chi_0}}\frac{\chi(p)\ln p}{p}+\sum_{\substack{\chi\\ L_0(\chi)\ne 0\\ \chi\ne\chi_0}}\frac{\chi(p)\ln p}{p}$$

$$=\ln x+O(1)+N(-\ln x+O(1))+O(1)=(1-N)\ln x+O(1);$$

iii-a) if $N>1$ then the right side in (ii) would tend to $-\infty$, contrary to the inequality there stated;

b) by #8 every contributor to the sum $Q(x)$ is either $O(1)$ or of the form $-\ln x+O(1)$; this means that if some contributor were of the form $-\ln x+O(1)$ then $Q(x)$ would tend to $-\infty$ as $x\to\infty$; this contradicts (a); but then, by #8, every $\chi\ne\chi_0$ satisfies

$$L_0(\chi)\ne 0\;;$$

c) this follows from (b) and #8.

10. The 1st equality follows from #3 and the 2nd equality from #9(iii) and XXIII #2(v); if there were only finitely many primes $p$, $p\equiv a \pmod{k}$ then the left side would be finite in contradiction to its being equal to $\ln x+O(1)$.

# References


R. C. Archibald, The golden section, AMM 25 (1918) 232-8 ; II.

P. Bachman, Niedere Zahlentheorie 2 vols. 1902, 1910, Chelsea 1968.

T. Bang, On the sequence [nα], n = 1, 2, ···, Math. Scand. 5 (1957) 69-76 ; IV, VI 14, 21.

P. T. Bateman, (1) Remark on a recent note on linear forms, AMM 65 (1958) 517-8 ; IV 32,
(2) Elementary Problem E 2051, AMM 76 (1969) 190-1 ; XV 9 (iii).

S. Beatty, Problem 3173, AMM 34 (1927) 159 ; VI 6.

N. G. W. H. Beeger, (1) Report on some calculations of prime numbers, Niew Archief voor Wiskundig genootshap te Amsterdam 20 (1939) 48-50 ; XVII 17,
(2) On even numbers m dividing $2^m - 2$, AMM 58 (1951) 553-5 ; IX 19.

M. Beiter, Magnitude of the coefficients of the cyclotomic polynomials $F_{pqr}$ II, Duke Math. J. 38 (1971) 591-4 ; XIV R.

C. L. Bouton, Nim, a game with a complete mathematical theory, Annals of Math. 3 (1902) 35-9 ; VII R.

Z. I. Borevich, I. R. Schafarevich, Number Theory, Acad. Press 1966 ; IX 20.

A. Brauer, J. Shockley, On a problem of Frobenius, J. für reine und augew. Mat. 211 (1962) 215-220 ; IV 32.

J. L. Brenner, Zolotarev's theorem and the Legendre symbol, Pac. J. Math. 45 (1973) 413-4 ; XVII 18.

J. Brillhart, J. Tomascia, P. Weinberger, On the Fermat quotient, pp 213-222 of Computors in Number Theory, Ed. by A. O. L. Atkin, B. J. Birch, Acad. Press 1971 ; IX 18.

J. L. Brown, On Lamé's theorem, Fib. Q. 5 (1967) 153-160; I 6, XIII 18.

Br. A. Brousseau, Continued fractions of quadratic Fibonacci ratios, Fib. Q. 9 (1971) 427-435; XIII 6 (X).

V. Brun, La serie $\frac{1}{5}+\frac{1}{7}+\frac{1}{11}+\frac{1}{13}+\cdots$ est convergente ou finie, Bull. des Sci. Math. 43 (1919) 104-104, 124-8; XVI.

A. A. Бухштаб, Теория Чисел, Moscow 1966; XIII 9, XVIII.

R. D. Carmichael, Diophantine Analysis, Wiley 1915; XX 6.

P. Cartier, Sur une généralisation des symboles de Legendre-Jacobi, L'Enseignement mat. 16 (1970) 31-48; XVII 18.

J. W. S. Cassels, An Introduction to Diophantine Approximation, Cambridge 1957; XIII 16, XXI.

E. W. Cheney, Introduction to Approximation Theory, McGraw-Hill 1966; XIII 25.

G. Chrystal, Textbook of Algebra, 2 vols. (1904) Chelsea 1952; XIII.

P. L. Cijsouw, R. Tijdeman, Distinct prime factors of consecutive integers, Diophantine Approx. Appli., Proc. Conf. Washington 1972, pp. 59-76; III 12.

P. A. Clement, Congruences for sets of primes, AMM 56 (1949) 23-5; IX 15.

A. J. Cole, A. J. T. Davie, A game based on the Euclidean algorithm and a winning strategy for it, Math. Gaz. LIII (1969) 354-7; I 1, 2, 3.

I. G. Connell, (1) A generalization of Wythoff's game, Can. Math. Bull. 2 (1959) 181-190; VII,
(2) Some properties of Beatty sequences I, Can. Math. Bull. 2 (1959) 190-197; VI,
(3) " " " " " II, " " " 3 (1960) 17-22; VI.

H.S.M. Coxeter, (1) Integral Cayley numbers, Duke J. 13 (1946) 561-578; XI R, XV R,
(2) The golden section, phyllotaxis, and Wythoff's game, Scripta Math. 19 (1953) 135-143; II, VII.

C. Curtis, The four and eight square problem and division algebras, pp 100-125 of MAA *Studies in Modern Algebra* vol. 2, 1963; XI R, XV R.

T.W. Cusick, The largest gaps in the lower Markoff spectrum, Duke J. 41 (1974) 453-463; XIII 16.

H. Davenport, *The Higher Arithmetic*, Hutchinson's 1952.

M. Davis, Hilbert's tenth problem is unsolvable, AMM 80 (1973) 233-269; III R.

L.E. Dickson, (1) On quaternions and their generalization and the history of the eight square theorem, Annals of Math. 20 (1919) 155-171; XI R, XV R,
(2) *Algebren und ihre Zahlentheorie*, Füssli 1927; XV R,
(3) *Algebras and their Arithmetics*, U. of Chi. 1923; XV R,
(4) *History of the Theory of Numbers* 3 vols., Chelsea 1952.

J.D. Dixon, A simple estimate for the number of steps in the Euclidean algorithm, AMM 78 (1971) 374-6; I 6, XIII 18.

R.E. Dressler, (1) A stronger Bertrand's postulate with an application to partitions, PAMS 33 (1972) 226-228, 38 (1973) 667; XI R, XIV 19,
(2) Sums of distinct primes, Nordisk mat. Tidskr. 21 (1973) 31-2; XIV 19.

R.E. Dressler, T. Parker, 12758, Math. Comp. 28 (1974) 313-4; XI R.

S. Drobot, A note on continued fractions, PAMS 14 (1963) 197-8; XIII 13 (iv).

R. Dubisch, Elementary Problem E 852, AMM 56 (1949) 554-5; I 6, XIII 18.

V. Dudley, History of a formula for primes, AMM 76 (1969) 23-8; XIV.

P. Erdös, (1) Beweis eines Satzes von Tschebyshef, Acta Sci. Math. 5 (1930-32) 194-8; XIV 12, 13,

(2) Über die Reihe $\Sigma\frac{1}{p}$, Mathematica B7 (1938) 1-2 ; XIV 4,
(3) On the coefficients of the cyclotomic polynomial, BAMS 52 (1946) 179-184 ; XIV R,
(4) Advanced Problem 4319, AMM 57 (1950) 346 ; IX 9.

P. Erdös, R.L. Graham, On a linear diophantine problem of Frobenius, Acta Arith. XXI (1972) 399-408 ; IV 32.

P. Finsler, Über die Primzahlen zwischen n und 2n, pp 118-122 of *Festschrift zum 60. Geburtstag von Prof. Andreas Speiser*, Füssli 1945 ; XIV 12, 13.

L.R. Ford, A geometric proof of a theorem of Hurwitz, Proc. Edin. Acad. Sci. 35 (1917) 59-65 ; XIII 19, 20.

H.G. Forder, A simple proof of a result on diophantine approximation, Math. Gaz. 47 (1963) 237-8 ; XIII 16(iii).

A.S. Fraenkel, The bracket function and complementary sets of integers, Con. J. Math. 21 (1969) 6-27 ; VI.

A.S. Fraenkel, J. Levitt, M. Shimshoni, Characterizations of the set of values $f(n) = [n\alpha]$, $n = 1, 2, \cdots$, Discrete Math. 2 (1972) 335-45 ; IV.

F. Frobenius, Über das quadratishe Reziprozitätgesetz (1914), Gesammelte Abhandlungen III (1968) 628-649 ; XVII 18.

H. Furstenberg, On the infinitude of primes, AMM 62 (1955) 353 ; XIV 20.

J.M. Gandhi, The number of representations of a number as a sum of ten squares, Research Problem 5, BAMS 72 (1966) 220-1 ; XV 9.

M. Gardner, (1) Mathematical Games, Sci. Amer. Aug. 1959, pp 128-134 ; II.
(2) *The 2nd Scientific American Book of Mathematical Puzzles and Diversions*, Simon and Schuster 1960.

C.F. Gauss, *Disquisitiones Arithmeticae*, Yale 1965 ; XVII.

A. O. Gelfond, The Solution of Equations in Integers, Freeman 1961; XX.

A. O. Gelfond, Yu. V. Linnik, Elementary Methods in the Analytic Theory of Numbers, Rand McNally 1965; XVI.

I. Gessel, Advanced Problem H 187, Fib. Q. 10 (1972) 417-9; XIII 15 (v).

A. Gloden, Mehrgradige Gleichungen, Noordhoff 1944; XII.

S. W. Golomb, Combinatorial proof of Fermat's "little" theorem, AMM 63 (1956) 718; IX 4.

R. L. Graham, (1) On a theorem of Uspensky, AMM 70 (1963) 407-9; VI 15,
(2) Covering the positive integers by disjoint sets of the form $\{[n\alpha+\beta] \mid n=1,2,\cdots\}$, J. Comb. Thy. A 15 (1973) 354-8; VI R.

R. L. Graham, H. O. Pollak, Note on a nonlinear recurrence related to $\sqrt{2}$, Math. Mag. 43 (1970) 143-5; IV R.

C. A. Grimm, (1) A note on consecutive composite numbers, AMM 68 (1961) 781; III 12,
(2) A conjecture on consecutive composite numbers, AMM 76 (1969) 1126-8; III 12.

R. K. Guy, The primes 1093 and 3511, The Math. Student XXXV (1967) 204-6; IX 18.

P. Hagis, A lower bound for the set of odd perfect numbers, Math. Comp. 27 (1973) 951-3; VIII R.

H. Halberstam, K. F. Roth, Sequences I, Oxford 1966; XXII 4 (iv), XVI R.

A. Hall, Geneology of Pythagorean triples, Math. Gaz. LIV (1970) 377-9; XI 15.

G. H. Hardy, (1) A formula for the prime factors of any number (1906), Collected Papers II 4-5; XIV 6.
(2) Divergent Series, Oxford 1949; XXI.

G. H. Hardy, E. M. Wright, An Introduction to the Theory of Numbers, Oxford 1962.

E. Härtter, Über die Verallgemeinerung eines Sätzes von Sierpinski, El. der Math. 16 (1961) 123-7 ; XIV 9.

H. Hasse, Vorlesungen über Zahlentheorie, Springer 1950 ; XXIV.

G. Hofmeister, Zu einem problem von Frobenius, Norske Vide. Sels. Skr. Nr 5 (1966) 37 pp; IV 32.

J. C. Holladay, Some generalizations of Wythoff's game and related games, Math. Mag. 41 (1968) 7-13 ; VII R.

H. E. Huntley, The Divine Proportion, Dover 1970 ; II.

A. Hurwitz, (1) Über der Zahlentheorie der Quaternionem (1896) Math. Werke II 303-330 ; XV R.
(2) Vorlesungen Über die Zahlentheorie, J. Springer 1919 ; XV R.

A. E. Ingham, The Distribution of Prime Numbers (1932), Stechert 1964.

E. Just, (1) A note on the $n^{th}$ term of the Fibonacci sequence, Math. Mag. 44 (1971) 199 ; II 6,
(2) Elementary Problem E 2279, AMM 79 (1972) 92-3 ; III 12.

M. Kac, Statistical Independence in Probability Analysis and Number Theory, Carus Monograph 12, Wiley 1959 ; XIII R.

A. Ya. Khinchin, Continued Fractions, U. of Chi. 1964. XIII.

F. Klein, Elementary Mathematics from an Advanced Standpoint vol. I, Dover [1924 Dover 1945] ; XI 3, XIII 21.

J. F. Koksma, Diophantische Approximationen, J. Springer 1936; XXI.

P. Laborde, A note on the even perfect numbers, AMM 62 (1955) 348-9 ; VIII 26.

G. Lamé, Note sur la limite du nombre des divisions dans la recherche du plus grand commun diviseur entre deux nombres entiers, C.R. Acad. Sci. XIX (1844) 867-870 ; I 6, XIII 18.

E. Landau, (1) *Primzahlen* 2 vols. (1909), Chelsea 1953; XIV 2, 17, 18,
(2) *Elementary Number Theory*, Chelsea 1958; XVI.

L. J. Lander, T. R. Parkin, J. L. Selfridge, A survey of equal sums of like powers, Math. Comp. 21 (1967) 446-459; XII.

D. H. Lehmer, The Tarry-Escott problem, Scripta Math. 13 (1947) 37-41; XII.

E. Lehmer, On the magnitude of the coefficients of the cyclotomic polynomial, BAMS 42 (1936) 389-392; XIV 18.

W. LeVeque, (1) *Topics in Number Theory* 2 vols., Addison-Wesley 1956,
(2) *Reviews in Number Theory* vol. 1-6, Amer. Math. Soc. 1974.

N. Levinson, A motivated account of an elementary proof of the prime number theorem, AMM 76 (1969) 225-245; XXIII 3.

M. Lewin, (1) A bound for a solution of a linear Diophantine problem, JLMS 6 (1972) 61-9; IV 32.
(2) On a linear Diophantine problem, BLMS 5 (1973) 75-8; IV 32.

E. Lieuwens, Do there exist composite numbers M for which $k\varphi(M) = M-1$ holds?, Niew Arch. von Wiskunde 18 (1970) 165-9; IX 9R.

Ю. В. Линник, Кватернионы и числа кэли; некоторые приложения арифметики кватернионов, Успехи матем. наук IV (1949) 49-98; XV R.

Е. Л. Литвер, Г. Е. Юдина, Первообразные корни для простых чисел первого милиона и их степеней, Math. Anal. and Applic. (Russian) Rostov Univ. III (1971) 106-109; XVIII R.

R. S. Luthar, (1) Elementary Problem E 2164, AMM 76 (1969) 1151; III 11,
(2) Elementary Problem E 2316, AMM 79 (1972) 910-1; VIII 15.

C. C. MacDuffee, *Abstract Algebra*, Wiley 1940; XV R.

K. R. Matthews, R. F. C. Walters, Some properties of the continued fraction expansions of $(m/n)e^{1/q}$, Proc. Camb. Phil. Soc. 67 (1970) 67-74; XIII 23.

P. J. McCarthy, Odd perfect numbers, Scripta Math. 23 (1957) 43-7; VIII R.

W. H. Mills, A prime representing function, BAMS 53 (1947) 604; XIV R.

A. F. Möbius, Über eine besondere Art von Umkehrung der Reihen, J. reine augew. 9 (1832) 105-123; XXII 1.

M. A. Morrison, J. Brillhart, The factorization of $F_7$, BAMS 77 (1971) 264; III R.

L. Moser, A prime representing function, Math. Mag. XXIII (1950) 163-4; XIV 7.

T. Nagell, Introduction to Number Theory, Wiley 1951; XIV 17, 18.

K. S. S. Namboodiripad, A note on formulae for the $n^{th}$ prime, Monat. für Math. 75 (1971) 256-262; XIV R.

E. Netto, Lehrbuch der Combinatorik, [2nd ed. 1927, Chelsea]; IV R.

J. von Neumann, Gleichmässig dichte Zahlen folgen, Mat. fiz. Lap. 32 (1925) 32-40; XXI 7 (iii).

C. A. Nicol, Elementary Problem E2425, AMM 81 (1974) 778-9; XIV 21.

I. Niven, (1) Irrational Numbers, Carus Monograph 11, Wiley 1956, (2) Diophantine Approximations, Interscience 1963; V, VI, XXI.

I. Niven, B. Powell, Primes in certain arithmetic progressions, AMM 83 (1976) 467-9; XIV R.

I. Niven, H. S. Zuckerman, An Introduction to the Theory of Numbers, Wiley 1966; XIII 17 (vii)(g), XX.

L. Pacioli, De Divina Proportione, Venice 1509, Buenos Aires 1946, Milan 1956; II.

G. Pall, On sums of squares, AMM 40 (1933) 10-18; XI R.

O. Perron, Die Lehre von den Kettenbrüchen I, II, 3rd ed., Teubner 1954; XIII.

B. Plankensteiner, Untersuchung über die Schrittanzahl des euklidischen Algorithmus bei Anwendung auf echte Brüche, Monat. für Math. 74 (1970) 244-257; I6, XIII 18.

G. Pólya, G. Szegö, Aufgaben und Lehrsätze aus der Analysis, 2 vol., Springer 1925.

K. Prachar, Primzahlverteilung, Springer 1957; XIV, XVI, XXII-XXIV.

M. E. Prouhet, Mémoire sur quelques relations entre les puissances des nombres, C. R. Acad. Sci. 33 (1851) 225; IV 32.

H. Rademacher, Lectures on Elementary Number Theory, Blaisdell 1964; XIII 19, 20, XVI.

L. Redei, Algebra vol. I, Pergamon 1967; XV R.

I. Richards, On the incompatibility of two conjectures concerning primes; a discussion of computors in attacking a theoretical problem, BAMS 80 (1974) 419-438; XVI R.

H. E. Richert, Über Zerfällungen in ungleiche Primzahlen, Math. Zeit. 52 (1941) 342-3; XIV 19.

H. Riesel, Note on the congruence $a^{p-1} \equiv 1 \pmod{p^2}$, Math. Comp. 18 (1964) 149-150; XVIII R.

M. Riesz, Sur le lemme de Zolotareff et sur la loi de réciprocité des Restes quadratiques, Math. Scand. 11 (1953) 159-169; XVII 18.

J. B. Roberts, (1) Note on linear forms, PAMS 7 (1956) 465-9; IV 32, (2) Relations between the digits of numbers and equal sums of like powers, Can. J. Math. 16 (1964) 626-636; XII,

(3) Integral power residues as permutations, AMM 76 (1969) 379-385; XVII 18.
(4) Advanced problem H-196, Fib. Q. 11 (1973) 506-7.

G-C. Rota, On the foundations of combinatorial theory I, Zeit. für Wahrsh. und Verwandte Gebiete 2 (1963-4) 340-368; XXII.

A. Rotkiewicz, *Pseudoprime Numbers and their Generalizations*, U. of Novi Sad 1972; IX R.

D. Sato, E. G. Strauss, P-adic proof of non-existence of proper prime representing algebraic functions and related problems, JLMS ser. 2, 2 (1970) 45-8; XIV R.

V. Schlegel, Ein geometrisches Paradoxen, Zeit. Math. Phys. Leipzig 13 (1868) 162; II 11.

A. Scholz, B. Schoenberg, *Einführung in die Zahlentheorie*, Gruyter 1966; XI 6.

G. Szekeres, On the number of divisors of $x^2+x+A$, J.N.T. 6 (1974) 432-442; XVII 17.

H. N. Shapiro, On primes in arithmetic progression II, Annals Math. 52 (1950) 231-243; XXII 3, 4, XXIV.

W. Sierpinski, (1) Sur une formule donnant tous les nombres premiers, C. R. Acad. Sci. 235 (1952) 1078-1079; XIV 8,
(2) *Pythagorean triangles*, Yeshiva 1962; XI R, XX R,
(3) *Elementary Theory of Numbers*, Paúst. Wyd. Nauk. 1964 (a),
(4) *What we know and what we do not know About Prime Numbers*, in *A Selection of Problems in the Theory of Numbers*, Pergamon 1964 (b); III, IX 5.

Th. Skolem, On certain distributions of integers in pairs with given differences, Math. Scand. 5 (1957) 57-68; IV 22, VI 9, 19, 20.

W. Specht, Elementare Beweise der Primzahlsätze, Deutscher V. 1956; XXIV R.

E. L. Spitznagel, Properties of a game based on Euclid's algorithm, Math. Mag. 46 (1973) 87-92; I 1, 2, 3.

R. Sprague, (1) Über Zerlegungen in ungleiche Quadratzahlen, Math. Zeit. 51 (1947-9) 289-290 (a); XI 2,
(2) Über Zerlegungen in n-te Potenzen mit lauter verscheidenem Grundzahlen, Math. Zeit. 51 (1947-9) 466-8 (b); XII.

H. N. Stark, An Introduction to Number Theory, Markham 1970; XIII R.

O. Taussky, (1) A determinantal identity for quaternions and a new eight square identity, J. of Math. Anal. and Applic. 15 (1966) 162-4; XV,
(2) Sums of squares, AMM 77 (1970) 805-830; XI R,
(3) (1,2,4,8)-sums of squares and Hadamard matrices, pp. 229-233 of Combinatorics vol. XIX A.M.S. Proc. Symp. on Pure Math. 1971; XI R.

D'Arcy W. Thompson, On Growth and Form 2 vols., Cambridge 1952; II.

E. Trost, Primzahlen, Birkhäuser 1968; XIV.

H. S. Uhler, A brief history of the investigations on Mersenne numbers and the latest immense primes, Scripta Math. 18 (1952) 122-131; XIX R.

J. V. Uspensky, On a problem arising out of the theory of a certain game, AMM 34 (1927) 516-521; VI 15.

J. V. Uspensky, M. A. Heaslet, Elementary Number Theory, McGraw-Hill 1939.

I. M. Vinogradov, (1) On a general theorem concerning the distribution of the residues and non-residues of powers, TAMS 29 (1927) 209-217; XXI R,
(2) Elements of Number Theory, Dover 1954.

B. L. van der Waerden, *Science Awakening*, Noordhoff 1954; II R.

R. F. C. Walters, Alternate derivation of some regular continued fractions, J. Austral. Math. Soc. 8 (1968) 205-212; XIII 23.

G. L. Watson, On integers $n$ relatively prime to $[\alpha n]$, Can. J. Math. 5 (1953) 451-5; IV R.

A. Weil, Sur les sommes de trois et quatre carrés, Ens. Math. 20 (1974) 215-222; XV R.

H. Weyl, Über die Gleichverteilung von Zahlen modulo Eins, Math. Annalen 77 (1916) 313-352; XXI.

C. P. Willans, On formulae for the $n^{th}$ prime number, Math. Gaz. 48 (1964) 413-5; XIV 22.

E. M. Wright, (1) Prouhet's 1851 solution of the Tarry-Escott problem of 1910, AMM 66 (1959) 199-201; XII, (2) Approximation of irrationals by rationals, Math. Gaz. 48 (1964) 288-9; XIII 16 (iii-v).

W. A. Wythoff, A modification of the game of Nim, Nieuw Arch. voor Wisk. 7 (1907) 199-202; VII.

H. Zassenhaus, W. Eichhorn, Herleitung von Acht- und Sechzehn-Quadrate-Identitätem mit Hilfe von Eigenschaften der verallgemeinerten Quaternionen und der Cayley-Dicksonschen Zahlen, Arch. Math. 17 (1966) 492-6; XI R.

D. Zeitlin, On coefficient identities for cyclotomic polynomials $F_{pq}(x)$, AMM 75 (1968) 976-980; XIV R.

M. Zolotareff, Nouvelle démonstation de la loi de réciprocité de Legendre, Nouvelles Annales de Math. 11 (1872) 354-362; XVII 18.

J. Züllig, Geometrische Deutung Unendlicher Kettenbrüche, Füssli 1928; XIII.

# Index

(Numbers followed by s refer to solutions.)

This book was scripted by
Gregory Maskarinec
& completed on
Midsummer, 1975